Ball Lightning:
Paradox of Physics

Ball Lightning: Paradox of Physics

Paul Sagan

iUniverse, Inc.
New York Lincoln Shanghai

Ball Lightning: Paradox of Physics

iUniverse, Inc.

For information address:
iUniverse, Inc.
2021 Pine Lake Road, Suite 100
Lincoln, NE 68512
www.iuniverse.com

Includes bibliographical references.

ISBN: 0-595-31394-9

Printed in the United States of America

To Carl Sagan

May he forever in seas of uncharted thought,
boldly voyage the stars that he so loved.

And dedicated to all
who seek the final Theory of Everything in physics.

"A paradox
is not
a conflict within reality.
It is a conflict between reality
and
your feeling
of what reality should be like."

—Richard Feynman

Contents

Preface…Fireballs Endanger Hill Air Force Base

During the dark days of the Cold War, when the world stood balanced on the edge of the black abyss of total annihilation and an ensuing darkness of nuclear winter, the nuclear-tipped Minuteman ICBM intercontinental ballistic missiles were a critical link in the defense of the United States against the Soviet Union.

Unknown to most Americans, Hill Air Force Base in northern Utah played an important part in the Cold War. Inside Hill's Top Secret Liniac test facility, the solid-fuel rocket motors were periodically x-rayed for dangerous cracks because such fissures cause uneven burning that renders a Minuteman useless. Since these propellants age, they must be periodically re-inspected.

In late 1975, many frightening and mysterious appearances inside the accelerator chamber forced a dangerous decision that was unknown to the Soviets. The Air Force shut down the Varian Liniac x-ray linear accelerator, a huge vacuum-tube super x-ray machine, for nearly one year. Secretly, for the past decade, USAF officers and scientists had wrestled with a dangerous situation that threatened to blow up building 985 and half of Hill Air Force Base—a mysterious ball of glowing blue fire (usually on clear-weather days) kept dropping out of a space adjacent to the Liniac's high voltage power supply, usually when the Liniac had arcing problems. Each time, the fireball glided silently to the floor, floated around the room, and then rose again to the power supply. Each time it silently vanished. Although the rocket engineers worried, testing continued.

Then, during a storm, lightning hit the building—a large concrete structure with a 60-foot ceiling—and a three-inch blue ball of fire popped out of the supply and followed the wall for thirty feet. A terrified technician was in the room but could not escape. As the fireball floated toward him, he put his back to the wall. The ball floated out and around his shoulders; and, continued along until it reached a duplex outlet on the dry concrete wall, where the ball instantly exploded. Electric motors, breaker circuits, lamps and equipment burned out. Fortunately, no rocket motor was in the bay.

The Minuteman rocket motor is a solid fuel propellant more powerful than a quarter-ton of C-4 or Sematek plastic explosives. Since the solid-fuel rocket motor lies in the x-ray bay with igniter port open and the x-ray film inserted under the handling ring, the propellant is often exposed to ball lightning—a very dangerous situation. In fact, so dangerous that Issie Cantor, then Chief of Explosives Safety Function of the Ogden AF Logistics Command (AFLC) Safety Office documented (copies in author's files) this in an AFLC document on January 28, 1975.

The Air Force called in Varian Associates, the company that designed the Liniac, to stop it, and they installed several safety devices in an attempt to stop the fireballs. It failed. Next, a fireball destroyed a large 220-volt transformer. Next time, a fireball wiped out a voltage regulator. After several more disastrous appearances, as one Air Force scientist put it: "We knew something was wrong here—'radically wrong conditions' was the phrase we used."

Time was running out. Sooner or later a Minuteman rocket motor would blow up. The USAF had seen enough, and just before Christmas of 1975, they closed the Liniac down. They tore down the machine and rebuilt it, using solid-state controls. That ended the visitations, and as far as we know, the troubles ceased.

Later in July of 1977, when the Hill AFB commander and his assistant flew by helicopter to the decommissioned Wendover AFB and witnessed a lightning demonstration at the Project Tesla hangar, director Robert Golka asked him about the status of the Liniac. In the conversation, the commander said he personally had witnessed ball lightning. Because of that frightening incident, he was reluctant to fly anywhere near storms.

In his conversation, he stated that someone should collect and document the different qualities of fireballs, and that this would be of interest to the Air Force and also to the airlines. This book fulfills that request.

Introduction...Fireballs Defy Gravity

Ball lightning is a paradox of physics. By its mysterious propulsion, navigation, confinement and quasi (as if) intelligent behavior, ball lightning violates known physics and "defies" gravity—and provides clues to a unified theory, the Theory of Everything—the sacred holy grail of physics.

Just a decade ago, most scientists doubted that ball lightning even existed. But this is changing. Even the famous skeptics' association, the *Committee for the Scientific Investigation of Claims of the Paranormal* states on its Web site that ball lightning does indeed exist—but is unexplainable.

All theories have one thing in common—none work. Even Nobel Prize winners such as Oppenheimer knew of its existence and were puzzled. Some attempted to explain the phenomena. Nobelist Pyotor Kapitza developed a reflective microwave theory of interacting nodes and anti-nodes. But when a Brazilian experimentalist could not detect the predicted frequencies, Kapitza rejected his own theory. But, undeterred by Kapitza's rejection, lesser physicists modify the old theory—and it resurfaces constantly disguised as a retread. All existing theories are fatally flawed. Many theories begin by making restrictions and assumptions on boundary conditions that are untrue. But all theories ignore certain qualities, such as propulsion.

Ball lightning qualities defy conventional physical interpretation, which leaves physicists shaking their heads in amazement and frustration. As a result of these contradictions, physicists respond to ball lightning in one of several ways. Some avoid it; others deny it. Those who accept it attempt to explain it. This is impossible. The qualities are too contradictory and seem to violate physics. So they ignore or deny unexplainable qualities and selectively accept only those properties that physics might explain. These theories are respectable. Some get published. None work.

Eleven Paradoxes

The great physicist Richard Feynman once remarked, "A paradox is not a conflict within reality. It is a conflict between reality and your feeling of what reality should be like." Although the following brief summary of eleven conflicting and unexplainable qualities are genuine paradoxes that might not give legitimate physicists nervous breakdowns, it is a paradoxical reality that will puzzle them greatly. Physicists who examined ball lighting cases puzzle over the following eleven paradoxical qualities.

1. Lightning storms and ball lightning are often (but not always) connected—somehow. However, no one knows how. Despite incorrect claims of close correlations, they are not frequently caused by nearby (up to 300 feet away) lightning strokes. Otherwise, any electromagnetic emissions from the lightning would be enormous, would damage electronics and induce enormous voltages and currents into surrounding conductors, and also create great damage. Such monstrous electromagnetic emissions have never been detected, nor does any reputable physicist dare claim they exist. Emissions from lightning are not the usual cause.

2. Shapes vary. Although some fireballs resemble oblate spheroids, sausages and rods, most are spheres. A sphere is symmetrical: it has no handedness—no left and right, no up and down, nor front and back. And if a fireball is concentrically homogenous—that is, it is uniform within—then there is no possible explanation for their abilities to sense their environments and navigate within them. The sensing, the linear and nonlinear feedback servomechanisms, the flight control, the mechanism of propulsion—these all are not satisfactorily explained by any of the proposed theories, and cannot be explained by existing physics.

3. Fireballs' awareness of their surroundings is legendary. If your author did not believe otherwise, he would suspect that some fireballs somehow "see" their surroundings and possess quasi-"intelligence." If so, is this quasi (as if) "intelligence" governed by rules of cellular automata?

4. Fireballs appear instantly and often vanish instantly, although some explode either slowly or suddenly and violently. A few shoot out sparks or glowing pieces as they come apart. Most just silently vanish. They can be dim or dazzling, big or small, move fast or slow, float or fly against a rapid wind, fade out slowly or quickly vanish.

5. Sometimes they explode, occasionally doing great damage. A few electrocute and burn animals and humans, sometimes horribly. When they explode, a

huge surge of electrical current may flow in nearby conductors. Some balls dive faster than a cannonball, then level off and float against a gale wind. Some fly effortlessly close beside a subsonic jet. Some materialize inside a closed metal or wire-mesh enclosure, such as inside an airliner and can last a split second or five minutes, although most last ten seconds. A few escaped by melting a hole in the fuselage and bouncing along the wing.

6. Sizes vary. A few are a quarter-inch across; most, six inches; others, three feet; and some, fifteen feet. Some balls burst in two or break into over eight tiny glowing pieces. Some balls leave an ozone odor and a few leave a mist that may be blue in reflected light and brown in transmitted light, but may be white in damp air. Most balls float, many bounce, and others roll along surfaces. Some are drawn to metal conductors; others avoid them. Some seek and enter closed spaces, chimneys or cracks.

7. Some deform and pour slowly through tiny openings, such as old skeleton-type keyholes, and then expand into a floating fire-sausage that usually transforms into a ball. Others pass through plastic and metal window screens—occasionally damaging the screen and leaving a hole with melted edges, but more likely leave no evidence. Some pass through glass effortlessly, usually leaving no visible remnant, although some melt holes in glass. Some balls roll atop—some go under or envelop—a metal line atop a transmission tower, phone line, or clothesline for up to three minutes.

8. Although most roll in one direction, hopping over anything in their way, a few roll or slide back and forth. Sometimes a parade of balls follows a power line. Some balls fall into and submerge under water, boiling for up to several minutes. However, most avoid diving into water. Some balls may be supersonic but make no sound.

9. Although many balls act dumb, others exhibit puzzling behavior that seems almost supernatural and quasi-intelligent. Some balls chase people across fields and up or down stairs. A few circle people, spiraling up from the knee level to the head and then dart out the nearest window or up a chimney, then explode. Some grow larger, sometimes quickly but often slowly, but most keep the same size.

10. Some twin balls travel in unison, one atop the other, sometimes with a glowing filament or "striae" between them. On rare sightings, several tiny balls or beads are in the filament, like beads on a necklace. The bottom twin usually floats down, stretching the filament thinner until it disappears and the bottom ball flies or floats independently. Although most balls float as if immune to wind and gravity, a few rapidly fall and then level out.

11. Fireballs that briefly touch people usually do no permanent harm, but some shock, burn or do worse. If the ball clings to the person, it may just "sting" them, but in rare cases has electrocuted them and charred their flesh. If a person reaches out to touch a fireball, it usually explodes, sometimes injuring the person. Some balls exhibit a Jekyl and Hyde, kitten and lion, dual nature and may change in an instant from harmless to violent. Most are silent; others hiss, sizzle, and crackle. A few roar. Some give off intense heat felt ten feet away, but others come within inches and witnesses felt no heat.

Extensive Evidence

The mathematics of physics is conceptual. Some of the views that I take are not the only ones and I do not claim them to be the only correct ones, just that they are comfortable for me. I am not dogmatic. As physicists Richard Feynman and Isidor Rabi advised students: accept physics as it is—crazy and making no sense. Modern physics is a collection of algorithms and equations that accurately describe experiments. It works. Use it. Do not worry.

Wrapped inside the hard-core nucleus of advanced mathematics, we love to create models and analogs to explain things. Einstein and Hawking thought in terms of pictures, then did the math. Feynman warned that those who come first as mathematicians into physics and attempt to treat physics as mathematics would understand little and accomplish less—until they become conceptual.

Finally, in the following anthology of cases, why did I purposely include so many cases? First, most of them—the McNally Oak Ridge cases, in particular—are valuable and of historical and scientific importance. Second, cases are like fingerprints—all slightly different that exhibit new clues from different perspectives. Second, extraordinary claims require extraordinary evidence. Third, examining so many cases provides extensive evidence of extraordinary qualities that enable fireballs to "defy" gravity.

PART I
Defying Gravity

1

Clues Puzzle Physicists

At first, no one noticed, but in the late nineteenth century, clues were rampant that classical physics was in trouble. Physicists condemned the supposedly sloppy measurements made by planetary astronomers of the perihelion shift in mercury's orbit—a mysterious extra 0.10-arc-second shift for each orbit—exceeding the tug of Jupiter and other planets. But little did the critics realize that it was another strange hint, another clue of the approaching birth of a new physics.

Already, by then the cracks in the edifice of Newtonian physics forecast major change. The Michelson-Morley interferometer experiments already dispensed with the ether. Although Einstein was unaware of the experiments, other physicists were puzzled. Maxwell's treatise also hinted at problems. And physicists such as Max Planck puzzled over blackbody cavity radiators' odd emissions and the strange meaning of the arbitrary curve-fitting equation created by Wien and Stefan.

Likewise, in the same way, today there are huge clues—such as cosmological dark matter and energy, perhaps an alleged fifth force—probably operating by the exchange of messenger particles—that accelerates cosmic expansion, the inadequacies of String-Membrane M-Theory, the successes and inadequacies of the Standard Model, the four choices of possible Multiverse levels, and other difficulties. Now, add another puzzling clue—fireballs—which because they "defy" gravity, some physicists ignored them or explained them with existing physics. But fireballs have embarrassing properties. Not only do they defy gravity and fly easily against even gale winds, but they can possess clear edges, sense their surroundings, navigate, and sometimes seem to exhibit quasi (as if) intelligence.

Faced with the puzzling qualities of fireballs, physicists pretend that this phenomena or its qualities do not exist—simply because the available physics cannot explain these qualities. You might disagree with my conclusions, and that is fine, but the facts of ball lightning qualities are undisputable. Physics moves forward by challenging the inexplicable with new explanations if it cannot incorporate the new challenge into the old framework of existing theory.

Existing Physics Sacrosanct

First, let me be perfectly clear. I do not advocate "alternative physics"—it is very mysterious to me. I accept the Standard Model, and Quantum Theory, Quantum Field Theory, Special Relativity and General Relativity. These four great theories are not independent of each other because GR was built upon SR, and QFT was constructed upon the inputs of QT and SR. These pillars of physics are not debatable. QFT is accurate to one part in one trillion and GR is tested accurate to one part in one hundred trillion. And, the limitations are only from the inherent inaccuracies of our best clocks and instruments.

I am a positivist, not a Platonist. I believe that in its deepest roots, physical theory is only a mathematical model, and that to expect it to correspond to reality is an expectation without meaning. String theory is elegant and looks promising, but makes few testable predictions, and must go through several more iterations that will explain mysteries like ball lightning, dark matter and dark energy, and other things.

Let me make this perfectly clear—ball lightning will not and cannot over-turn established physics. Certainly, established theories in physics often become subsets of larger more inclusive theories. Once verified by observations or experiments, these theories are never proven wrong. Newtonian physics became a special case of Special and General Relativity or SR and GR.

Special Relativity and quantum theory or QT were incorporated into quantum field theory. Today, as just described above, although QFT and General Relativity are tested and proven accurate, QT and QFT do not correspond to our reality. QT has no reality in our perception. We must never demand a theory to reflect our "reality." If so, we misuse our perception as a comparison by which to judge theories. A theory has one goal—to predict.

QT predicts results of measurements very successfully. And, despite its problems, the success of the Standard Model cannot be denied. When one day the dreams of a final theory of physics are realized, it will incorporate but not invalidate SR, GR, QT and QFT. All proposed final theories—including Penrosian and Bohmian Mechanics—are inadequate. They are just some of the proposed bridges across the great divide—the Grand Canyon of physics. On one side is quantum physics; on the other, relativity—isolated.

Critics of string theory point out that it cannot yet adequately describe black holes, the sun's structure, or make any distinctive observationally testable predictions that are not already predicted by General Relativity. So, string theory is probably not the end of the road, and there will be further iterations, some of them major, and today's physicists hypothetically time-transported into the early Twenty-Second Century may not at all recognize today's great progenitor's final successor.

Some iconoclastic ideas can be assimilated, certainly, but those that require physicists to restructure their picture of the universe only provoke hostility. For example, Einstein's theory of Special Relativity was not cataclysmic to Classical physics—physicists could live with it—but not with General Relativity, which was far more important. It altered physics forever. But Einstein did not receive a Nobel Prize for his greatest achievement. It was too radical at that time. Critics are not converted—they just die off, and the younger physicists are of the new physics.

I cannot propose any new Theory of Everything in physics—it is beyond my skills. And, I certainly cannot determine which of the competing Theories of Everything will best succeed. Nor do I know if loop quantum gravity with its discontinuous spacetime networks and spin foams is testable. And, as promising as it looks, I certainly do not know how Multiverse Theory will evolve, or if present string theory will survive extant, or if in its future iterations—perhaps in its fourth generation—if it will be recognizable to today's physicists. I am a rational skeptic. I am cautious. I borrow my physics from leading physicists.

Puzzling Enigmas

Many fireball enigmas puzzle physicists. Earlier we briefly listed just eleven paradoxes. But, now we will further examine these and many other puzzling enigmas—but in much greater detail. There is a good reason.

Make no mistake about this: ball lightning asks important questions—and it challenges the very foundation of our physics. Yes, there are exceptions to the following properties, which only make things for any proposed theory even more difficult. A true theory of ball lightning must obey existing physics and yet still satisfactorily explain these following enigmatic properties.

Propulsion...

- Fireballs exhibit mysterious flight that "defies" gravity. They fly against storm winds and fly rapidly without deformation beside aircraft. Rarely do they randomly drift; rarely does wind drive them.

Origination...

- Most fireballs suddenly appear out of thin air, out of nowhere—usually near or during a storm, but usually when lighting is not nearby.

Flight...

- Can drop fast like a cannonball and then either fly slowly or bounce back up into the sky.
- Flies against winds and beside planes at 300-plus mph and may possibly be silently supersonic.
- Navigates usually with deliberateness in complex environments. Some balls lose their deliberateness if they bump into and bounce off a tree, which is rare, after which they spout a burst of sparks and then fly carelessly.

Energy...

- Energy manifestation is sometimes massive, but its source is probably not completely internal (stored and generated) or external. Simply to fly at such speeds against air resistance, not to mention their explosive energy, requires the energy of multiply ionized air far in excess of calculated estimates.

Confinement...

- Confinement (stability): many last 20 seconds, but a few last minutes.

Servomechanisms...

- Controlled flight. Interacts and communicates internally between propulsion, sensing, feedback-servomechanism control and navigation quasi "systems" within the fireballs to maintain control and steering, symmetry and orientation during hovering and flying. Everything else with controlled flight—living or man-made, bird or aircraft—controls its pitch, yaw and roll.

Appearances…

- Shapes are of sphere, and sometimes of sausage, rod or pear, but can change shape. Broken or disintegrating fireballs sometimes briefly become other shapes, such as half moon with smaller pieces that are unaffected by winds.

- Dual vertical balls rarely have a glowing filament or strand between them, sometimes with several glowing beads in the strand, which will stretch thinner as the lower ball floats down, until the strand disappears.

- Colors. Gases in air cannot explain the wide range of colors. Some are white or multi-colored.

- Corona point discharges are sometimes reported suppressed in the neighborhood of a fireball.

- Surfaces. Solid or fuzzy, sometimes enclosed in multiple shells.

- Holes or jets—usually one, sometimes two—open in fireballs and continue spouting fire, often hissing. Most do not immediately explode. Such fireballs seem stable and in equilibrium. There seems to be no action-reaction. Although these jets hiss—some roar—the balls do not zig-zag about from action-reaction forces like an unsealed toy balloon let go. Some holes slowly open until the ball fragments apart or explodes.

- Some fireballs have protuberances; send out lumps of fiery material, rays, streamers of fire, jets, and horns. If fireballs are purely electrical phenomena, they could not throw off lumps of fiery material, and plasma theory, as we know it, is incomplete.

- Rarely deformed when bouncing off floor and never deformed in flight, even at high speeds.

- Some are translucent; most are opaque. Some are enclosed in a translucent shell; some have two or three concentric nested shells.

- Most (about 90 percent of fireballs) form half an hour before or after, or during lightning storms. Some occur on sunny days.

- Many fireballs form inside closed rooms or enclosures that resemble Faraday cages. Some fall into barrels full of water and glow under the surface and boil off much of the water.

- They ooze through crevices and skeleton keyholes, morphing their shape. When they exit, they often are rod or sausage shaped, but sometimes resume their spherical shape. Although on occasion the fireball later causes damage—in one case it set hay afire—after it passes through a crevice, it usually oozes through crevices without setting anything afire or doing damage—although it is in immediate contact.

Movement...

- Sealed toy balloons set free will roll and bounce erratically across a field in a brisk wind, differently than fireballs. Your author attempted to simulate realistic fireball flight and bouncing, but all attempts failed, both indoors before a fan and outdoors in a wind. Bouncing fireballs do not make any bouncing sounds.

Intelligence...

- Quasi-Intelligent, that is, sometimes they behave "as if" they are intelligent.

Longevity...

- Typical observed lifespans are in the 2 to 10 second range, but one minute is not uncommon, and a few lasting for several minutes do occur. Reported lifespans may be in error, because although most witnesses see when a fireball departs, many do not see its origination, thus skewing statistics and increasing actual lifespans.

Departure...

- Termination. Most maintain constant brightness to the end, but a few do pulsate or change intensity.

- In a few cases, mildly exploding balls hurled witnesses to the ground; but at other times, even violently exploding ones may not affect anyone. One witness touched by a fireball may feel a minor shock; another, pain; a rare few, horrifying burns or death.

- Undamped bouncing—although it occasionally is damped—where each successive bounce is usually of the same height, as if bouncing is a totally elastic collision or is impervious to gravity, or kinetic gravitational energy is added to precisely balance out frictional losses. It is almost unheard of for the height of successive bounces to be higher or erratic. In several cases, the fireball bounced for a hundred bounces of equal height.

- Lightning strokes sometimes hit an existing fireball, perhaps forty feet above the ground, and terminate—no lightning continues to the ground!

Such rare fireballs stop lightning and absorb the charge of a lightning stroke. Other reports exist of lightning stopping in mid-air to form a fireball. This is puzzling because of the massive charge terminating and disappearing into an existing fireball. Where does the charge go? Five feet of solid rubber cannot stop a cloud-to-ground discharge.

- Fireballs do bounce along the ground and then rapidly fly, or instead fly and then bounce several times (for example, down some stairs) and then fly again.

- Several originated and flew in a vacuum. One entered the outside atmosphere after melting a small hole in the containers' glass walls.

Electronics…

- Fireballs are attracted to electronics—both operating and inactive—and such fireballs damage silicon chips, triacs, cables, resistors, connectors, printed circuit boards, fiber optics, magneto-optical storage media, and resident software. One destroyed a supercomputer.

- Most fireballs act electrically neutral and ignore grounded conductors, but some change their mind and suddenly dart to a conductor and disappear into it, generating a large transient surge of intense current.

- Electrostatic effects like skin crawling and hair standing up sometimes occurs, but usually does not.

- After a ball disintegrates, sometimes tiny ballettes or glowing remnants remain that are unaffected by the wind.

- Most fireballs leave no residue, no charcoal, no lamp black, no soot.

- Sometimes, after the departure or passage of a fireball, there is lingering smoke, dust or colloidal suspension. Some witnesses report light scattering as if from the Tyndall Effect.

- Some balls pass through solid walls, doors, and one even descended down into a marble tile floor and rose back up.

- Why do most fireballs not penetrate through walls and doors? But those that do penetrate once may do it again, and a few have passed through three walls.

Spooky Action…

- Some fireballs instantly detect electrical changes near them, such as when a witness turns on a switch, and may react instantly.

- Fireballs occasionally cause nearby—within fifty feet—wires and electronics to burn out, even without a storm.

- A few fireballs exhibit the uncanny ability to detect and follow buried pipes.

- Other concurrent, unexplainable effects occur that hint at a deeper cause—at a possible, underlying Hidden Progenitor.

Mechanics…

- Some fireballs seem to spin about a vertical axis—why? Many do spin but have random or chaotic lights moving within.

- Why don't spinning fireballs slow down from internal viscous friction and external friction? Spinning fireballs possess angular momentum. Assuming that those seeming to spin do indeed spin inside, then there is viscosity or gaseous friction. And if not, then viscosity still exists. Try to rotate a bowl with dye dropped into this bowl of water mounted on a caster, and you will see what happens. Or spin a cooked egg and watch it spin, but try it with a raw egg and see how its spin is damped. The viscosity of spinning objects possess angular momentum—that is, "I" times omega, which differs for different shapes and for non-homogenous densities.

Chimerical Clues

Your author questions (personal prejudice) if fireballs pass through solid walls—in spite of rare reports to the contrary—and is prejudiced against such witnesses and cases (including two reported at Oak Ridge by physicists). But when he kept coming across witnesses who saw a fireball pass through a wall (sometimes seen on both sides), this made him wonder if occasionally fireballs do indeed pass through solid walls. Other chimerical clues are easier to accept, although puzzling.

Fireballs often instantly and silently appear and disappear. This suggests that there is no significant displacement of air. In contrast, in lighting, the rapid heating and supersonic expansion of air create thunder. Even a mild slower displacement of air creates a sound.

Damage is puzzling. Although reports of fireballs that scorch or burn grass occur, but are uncommon, there are no detailed reports of balls that float near or touch tree leaves and scorch or set them afire. Even though some balls do fly into tree trunks, usually emitting sparks with each collision, they are not reported to go through the leaves. If fireballs floated through leaves and touched against them, would the leaves ignite or simply harmlessly move aside? Some witnesses report fireballs set fires. Could some fireballs set forest fires?

Why do fireballs roll? Why do some roll slowly but others roll fast? Roll on the ground, floor, trees, wires, and cables? Why do some maintain a steady state size, but others grow or shrink? Shrink or expand slowly or fast? Why do a few shrink on each bounce, while most keep the same size? Why do most bouncers not lose height with each bounce?

Despite close contact with metal, they may not electrically short out or give off sparks. For example, many reports describe how a fireball slipped through the skeleton metal keyhole in an old-style door and emerged as a rod on the other side that gathered itself into a fireball, although others retained their rod shape. Others slipped through small crevices in wood. One such case was in a barn, yet the unbaled hay did not ignite. Yet, oddly, in other cases, fireballs have set such loose hay on fire. Whether or not such loose stored hay was sprinkled with rock salt, to discourage spontaneous combustion, was not stated, but almost certainly was done and had no effect in either case.

The flight of lightning balls is frustrating. They can fly like a thrown ball and even silently bounce off a floor or lawn, and many fly straight, methodical, and deliberate. Some chase people and may circle them. Some exhibit a quasi-intelligent flight pattern, and some witnesses report a feeling that the fireballs watched them and possessed some form of intelligence.

Many fireballs exhibit quasi-deliberateness. Some pursue fleeing witnesses, but others circle them, rise to eye level, and then dart off through a window or ascend a chimney and instantly explode upon reaching the outdoors. Others behave equally bizarre. They do not "dumbly" drift with air currents nor flow with electromagnetic fields.

Fireballs are electrically neutral for they usually are not attracted to metal conductors. When they are attracted to wires, they frequently only follow them and do not merge into them. But if they should merge into a conductor such as a cable, they convert into a huge surge of current.

Although most fireballs are silent, some hum, hiss, sizzle—even roar. Loud humming noises are associated with an oscillatory process.

Although most fireballs appear to vanish instantly, it is not instantaneous, only so quickly that the human eye cannot see the extinction. A few fireballs explode, often suddenly, and others disintegrate by shooting out tongues of fire, or go through fitful behavior. They do not shrink in size. This infers that the internal process cannot sustain its existence because something, perhaps the available energy rate, drops below a critical threshold.

Fireballs bounce. Bouncing is mysterious and physics cannot explain it. They commonly bounce for six or more bounces—some bounce dozens of times—on soft and hard surfaces, on both lawns and asphalt, yet they often keep bouncing to the same height. But a few exhibit damped parabolic bouncing with each bounce lower, thus apparently demonstrating loss of gravitational kinetic energy. A few bounce up and down, but in slow motion, remaining in one spot, bouncing for dozens of times. Most do not change shape when they hit the ground or floor, but a rare few do deform and briefly flatten upon contact with the floor.

Fireballs do not exhibit net electrical charge. If they had the slightest unequal charge, as in the case of static on a toy balloon that would be attracted to, touch and cling to objects. If the gaseous ball contained particles with a net positive or negative charge, the excess charge would move to the sphere's exterior or envelope. If charges were uniformly distributed internally (uniform charge density), Gauss' Law gives an internal electrical field inside the ball at a distance from the center times the charge density divided three times the permitivity constant.

The field would increase linearly with distance until outside the envelope, and then obey Coulomb's Law and rapidly weaken as the inverse of the distance squared. Outside the ball, the field is radial and spherically symmetric, which it would be, so then Gauss' Law shows that outside the shell, it can be treated as a point charge. If the ball were a shell of charge but with no charge within, then the field anywhere inside would be zero.

Many fireballs (perhaps one in twenty) appear without a storm—during sunny days, and sometimes inside a Faraday Cage, that is, a metal enclosure or equivalent. For many obvious reasons, this really puzzles and vexes physicists.

These enigmas challenge science. There are many eyewitness cases that support each property. Those from Top Secret Oak Ridge National Laboratories are among the most authoritative.

Oak Ridge Cases Resurface

Your author possesses the only known copies of the "lost" McNally Oak Ridge National Laboratory cases. Cases of this caliber may never again be collected. These witnesses—Top Secret scientists, technicians and engineers—were repeatedly screened by the FBI for integrity and came from a time when integrity was a matter of life and death. The deadly Cold War was under way. People were honest. America as it was then is now what is forgotten. To lie was unthinkable. It would cost them their job, security clearance, and permanently blacklist them.

Dr. James McNally's 630-case study of Tennessee's Oak Ridge National Laboratory (ORNL) employees—many of them physicists, chemists, mathematicians, engineers and technicians—went mostly unpublished and was nearly lost, and is partly published here for the first time. These Oak Ridge cases are from the landmark study in the early 1960s that never received the recognition that it deserved.

In studying the Oak Ridge cases, there were surprises, and these cases differ from other studies. A large number of fireballs followed a metal conductor, usually a line. The fireballs lasted longer and were larger than those that other studies report. It was surprising that a few grew brighter and larger in half a minute or less. But most did not change size. Although many balls made noises, hissing, sizzling or crackling, most remained silent. There were more color changers than anticipated.

The typical ball measured five to twelve inches in diameter, but two inches was not uncommon. A few measured five to ten feet in diameter, and even thirty feet across. Few witnesses were close enough to feel heat or electrical charge. Some balls contained internal motion. Many had glowing fuzzy surfaces, but many were smooth. Some rapidly followed buried pipes—a feat unequaled by the best dowsers.

All Oak Ridge reports were unclassified. Still, I took precautions. I deleted most building numbers and identification—for example, such as the fireballs that materialized inside or traveled through the nuclear-bomb plutonium or uranium-enrichment apparatus modules.

There are two categories of witnesses—first, those from the Oak Ridge McNally study and second, other cases that often came from my own files. In the Oak Ridge cases, I listed case numbers and witnesses' last names. For example, the case "*Fireball rides line to nuclear weapons power house.* [25 Derleaux]" identifies this as report number 25 reported officially to Dr. McNally at Oak Ridge by a man named Derleaux. Most of these witnesses came from the 1920-through-1960 decades, and three of them even saw fireballs when Woodrow Wilson was still president. Avoid contacting them for more details—most are dead.

Most cases came from Oak Ridge National Laboratories. However, to totally protect the privacy of witnesses not from Oak Ridge, I omitted all identifying information and names, and totally rewrote these non-ORNL accounts—from whatever sources, especially the Russian and Italian cases and a few from some Internet sites—to totally render them unrecognizable. Besides, you will require no additional information.

The Oak Ridge study was conducted in 1963 when truthfulness meant more. The FBI bestowed a Top Secret clearance upon those with a need-to-know and who continued to pass the strictest background investigations. (Four years later, your author received a Secret clearance for military electronic research. The FBI maintained on-going checks and interviews with neighbors.) The Oak Ridge witnesses mostly had Top Secret clearances and possessed integrity.

The Oak Ridge questionnaires were filled out during the presidency of John F. Kennedy. Most never saw the light of day and are printed in this book for the first time. However, in a brief paper, Dr. McNally published a mere handful of these cases and this generated great interest. But this was forgotten with time. Today, when your author examined them, this box of Oak Ridge questionnaires were faded with time and held together by rusty paper clips.

After McNally gave the original study of 630 cases to Golka, it lay dormant in his lab except for a brief time he tried to convert them to text and categorize them. In the 1980s, Golka asked an MIT student (for graduate credit) to extract the data and create a database. The attempt was a total disaster and was discarded because of great difficulty in categorizing incomplete data. it was like nailing jello to a tree. In 1999, with Golka's permission (in a written contract) to photocopy, edit and publish the best 200 accounts, I undertook this extensive task. Not the least obstacle facing me was the witnesses' mysterious handwriting.

How did I group the cases and categorize them? I examined how everyone else did it, then with that in mind, I instead let the cases fall naturally into categories, that is to say, they organized themselves. This is arbitrary, because in most cases, several qualities exist in each case. When two or more qualities existed in each case, I determined which one was most significant and listed that in its title and category.

Substantial Evidence For Claims

To convince skeptics, I list many cases. As Carl Sagan pointed out, extraordinary claims require extraordinary evidence. Second, most of these Oak Ridge cases are original, of historical importance, and were never published. Third, each fireball has a unique and different nature. Most cases, if they have any detail, are like fingerprints—no two are alike.

Unfortunately, since the best 230 cases (out of the 630 Oak Ridge cases) were incomplete, the gaps were filled with selections from interviews and questionnaires. Some came from other sources, which were summarized, neutered of identification, and totally rewritten. The difference is this—Oak Ridge cases list report numbers; others do not.

Details resemble fingerprints—no two are alike. Today, fireballs emerge and enter televisions, computers and monitors, often doing extensive damage. And, one fireball destroyed a costly supercomputer. They are videotaped and photographed with a quality that exceeds earlier photographs. Other related phenomena, such as glowing fogs that travel, are still widely reported. And witnesses still report that most fireballs avoid walls—but some pass right through plastics and solid walls, even descending into a marble floor, but often without leaving any visible trace. So, to focus on the details of each case, I removed the nonessentials.

Let me say it once again. Never before published, except for two-dozen small cases, the Oak Ridge reports eventually were nearly discarded, and barely escaped destruction. Finally, James McNally gave the questionnaires with permission to publish, if desired, to Robert Golka, who later provided them to your author with permission to edit and publish them here for the very first time. Unfortunately, I found much of these witnesses' grammar and handwriting so tortured, even after using a magnifying glass and much patience, that it took hundreds of hours of copying and editing to make them presentable.

First, I edited cases by cleaning up grammar, pruned redundant phrases, and deleted similar cases, then in some cases fleshed out laconic descriptions, but in others, removed fleshy words. For example, most witnesses redundantly describe thunderstorms as "severe, particularly severe, intense and very intense, it had tremendous flashes with loud crashes" and so on—not particularly helpful descriptions since everyone else writes this.

Few are the storms described as mild, and these your author noted. In deleting most of these tedious superlatives and descriptions, he purged qualifying phrases, like "it seemed to me" or "to the best of my recollection" or "I am not a hundred percent sure of its exact size, but I think it was…" Witnesses used exclamatory language, like "I was thoroughly frightened…terrified…shaking in fear…never forgot…never want to see anything like this again." Saying this a few times is fine. Beyond that, this litany is boring. Occasionally, I left several of their phrases in to keep the flavor of witnesses' descriptions. Rest assured, if they do not describe the storm as severe, it was so.

As for terminology, "Ball lightning" is an inadequate phrase. "Fireball" is more accurate. Most are not connected with lightning flashes. In fact, five percent occur during clear weather. Some even appear in sunny weather. And even those that materialize or vanish simultaneous with lightning are often at a distance from the flash—so electromagnetic radiation (sferics) cannot make most fireballs. Many appear inside closed spaces, even inside Faraday cages.

The phrase "Artificial Electric Ball" refers only to those created by Nikola Tesla and the Corums, using three-coil Magnifiers. These artificial balls are not true fireballs. This book has cases and theory. In both halves, I use the three phrases "fireball, ball lightning, lightning ball" interchangeably.

Finally, Carl Sagan was a member of CSICOP, whose acronym stands for the excellent skeptics organization, *Committee for the Scientific Investigation of Claims of the Paranormal*, which publishes a bimonthly journal, the *Skeptical Inquirer*. On its website, CSICOP states that although ball lightning exists, it is unexplainable.

Mostly, witnesses notice fireball movement. This includes rolling along lines, along the floor and ground, atop or inside gutters, and along or attached to pipes, such as downspouts. A few that chase witnesses sometimes circle them, spiraling upward, while others float up or down stairs. Some bounce along streets, but others bounce off trees. Some ascend or descend trees and towers. Anything that moves attracts immediate attention. It is no surprise that, aside from appearances, in most cases witnesses vividly describe movement. Therefore, of all the following categories of Oak Ridge sightings, more cases fall into the category of movement—especially rolling along lines.

PART II

Oak Ridge Provides Evidence

2

Line Rollers

Movement is the most noticed property, which is not surprising, because change of location involves motion. Fireballs are stationary or move. Some follow lines, usually metal, phone and electrical power lines, and also high-tension lines and cow fences—both electrified and barbed wire—and also clotheslines. Some move by rolling on floors, lawns, ground and roofs, and inside and atop gutters—in short, along anything. A few chase people. Some circle inside perimeters, like rooms, while others shoot up like rockets, or fall like stones. A few fly beside aircraft, sometimes at great speed. Many bounce. Most ignore winds, even fierce gales. And some enter and leave chimneys.

Oak Ridge National Laboratories

Oak Ridge National Laboratories is the home of the atomic bomb. In three years, ORNL became an "instant city" built from cow pastures up into the fifth largest city in Tennessee, with its peak population of 75,000. It ravenously devoured electricity beyond what the large TVA Norris Dam could generate.

With a total area of 58 square miles, Oak Ridge had 10,000 family units, 13,000 dormitory spaces, and over 18,000 hutment and barrack spaces. ORNL was established in 1943 to produce and separate uranium and plutonium for the World War II Manhattan Project and for the Cold War. Even after the war, the separate buildings and functions were kept compartmentalized, so that few could grasp its overall magnitude. Most of the many scientists and engineers came from outside, but many of the production workers were locals.

Many of the following cases came from a famous ball lightning study conducted by Dr. James McNally of ORNL employees, and are published here for the first time. Such cases contain [brackets], which enclose the ORNL case number and witness surname. Of the cases, more dealt with movement; and, of these, those that followed lines (either above or below) seemed to roll along the lines.

Rolls Along Electric, Phone Lines

Why do fireballs frequently roll on or slide on or along metal lines? They appear on metal clotheslines, power lines, telephone wires, hanging chains, and guy wires that support antennas, but rarely near cord clotheslines or suspended rope. It makes no difference if a metal cable or wire is bare or insulated, grounded or not, or whether it is electrically active. Rarely do fireballs float through the air and attach themselves to a line, although such cases do exist.

They usually ignore lightning arrestors, earthing wires, and ground lines. Usually fireballs glide, slide or roll atop or on lines; although some few do follow below a line or slide and envelope it like a bead. They do not slavishly follow the lines, but occasionally float or jump off, sometimes before reaching an insulator—or may do so for no apparent reason. Although they then float, fly or fall, they usually float or fall to the ground, but some follow the pole or tower to the ground. What they do varies. There are no iron rules.

Do they sense and follow electrical fields? Some seem to. But why do they follow wires not carrying currents? If they were attracted to electrical or electromagnetic fields around wires, then they should prefer the higher voltage lines of 400,000 volts and virtually ignore telephone lines. They show no such preference. They seem to sense wire conductors and follow them, but usually do not merge with them. However, if they do merge into the wire, they shortly thereafter disappear and a large sudden or transient surge of current flows in that wire. But a few fireballs can slide along the wire like a bead on a string and not create a current in that wire, unlike most of the following cases.

Slow-slider melts phone-wire. [353 Capp, Oak Ridge National Laboratories or ORNL] Lightning struck telephone wires on ground and a 15-to-20-inch ball of fire slowly moved a hundred feet on the wire, lasting one minute [1.7 feet per second]. It burned and melted them.

Melts phone wire. [269 McCarty and McWagon, ORNL] This fall, lightning hit a telephone line about 150 yards from the entrance to our house. The eight- to ten-inch ball was orange-white. The wire was disconnected from the phone and the wire was on the ground and melted to the end.

Night ball leisurely rolls along phone line. [559 Huntington] At night I was sitting before the window, glanced up and saw this bright light ball of fire roll leisurely along the telephone cable outside, 25 feet away. The volleyball-sized fireball did not change size and remained on the cable. It vanished without a sound after two seconds and did no damage.

Slow phone line ball. [317 Overton] During a thunderstorm, a ball appeared on the telephone line extending from a house to a pole in the street. Larger than a foot but smaller than two feet in diameter, the ball appeared at one end of the conductor and slowly moved across it to the other end and disappeared.

Purple slider jumps off phone line near house. [32 Farnham about 1930] A group of children in Tennessee were playing in a neighbor's dining room when a six-inch bright, pale purple lightning ball appeared on a telephone line leading to the house. It rolled or slid along the line toward the house for about twelve feet, in three seconds. Just at the point contacting the house wall, the ball fell and disappeared instantly. This was all visible through an open door.

Jumps over phone line insulators. [356 Kennedy] I have seen ball lighting run on old rural telephone lines. They would follow the conductor but appear to bounce. The 12-inch ball would jump over the old insulators and roll down the ground near the house. It usually disappeared with a crack and would loosen the earth around the copper ground rod.

Author—These fireballs sensed each glass insulator before reaching it, and despite the balls' momentum, jumped over the insulators. How did these fireballs sense objects, almost like seeing them? And what about the balls' guidance and propulsion mechanisms?

Phone-line slider destroys transformer. [Burnhouse] During a lightning storm on June 6, 1962 in Boys, Maryland, while swimming when a storm came up, I looked north and saw a three-foot ball of orange fire slid along on top of a telephone line. After passing five or more poles, it hit a transformer and exploded. The ruined transformer smelled burned out.

Hydroelectric phone-line fireball. [263 Zotmann] During a lightning storm, I was in a hydroelectric generating station control room some 30 feet above the ground on a parallel plane with the Hi-line Telephone lines, some 12 to 15 feet from the windows. A 30-inch dull red fireball moved along and above a line for a short time. Sometimes it changed in size.

Telephone line fireball splits pole, breaks insulator, and flies to lawn, sparkle. [184 English] Lightning struck a country telephone line. A four-to-six-inch fireball followed a quarter-mile line to where it was disconnected at the house, rolled off the end of the line, hit 12 feet out into the yard and burst like a giant sparkler. It broke the insulator and split the pole, located at the end of the line. The pole where the line was disconnected was split and the insulator broken. It lasted six seconds.

Author—Did the lightning surge break the insulator and split the pole? The witness blames the fireball. Detailed damage description is needed. From this it would be possible to calculate the energy needed. The ball fell to the ground. Did the impact cause it to burst?

Monster jumps up and down along power line. [366 Schimmel] The noise of the thunderbolts was deafening and stunning. During a lightning flash, a giant five-meter [over 16-foot] fireball suddenly appeared. It jumped up and down along the power line and after five seconds suddenly vanished.

Schimmel wrote that the giant jumping fireball moved along the line and disappeared without explosion. The volume was about 4,400 times greater than a one-foot ball, which can do extensive damage. If the explosive power is directly related to volume, then could terrorists use explosive devices based on such discoveries to topple skyscrapers? Probably not.

Strikes wire, hugs building. [127 Thomas] A twelve-inch blue ball of fire struck some electrical overhead wires, ran to the corner of a building, emitting a loud popping, and ran down the side of this building to the ground. It flew along the ground for a short distance, then after three seconds, it disappeared.

Disintegrates Metal, Scars Tree. [152 Whitmore] A blue eight-inch fireball traveled horizontally in a curving path to hit—with a loud crack and blinding flash at contact—a hammock ring and plate on a hardwood tree, disintegrating the iron and bark beneath it. The underlying wood was slightly charred. I also saw a yellow ball travel vertically and disappear at the water surface above a submarine rock ledge.

Tailed contour-hugger scorches earth. [212 Nicley] Just after a storm, lightning struck a huge tree and from this came a 20-inch orange fireball that flew parallel to the earth. It followed the surface contours. Trailing a long tail, it came down off a high ridge by the way of a curving hollow or ravine, then crossed a valley and after ten seconds hit the ground on the hillside across from the hollow. The spot where this ball hit the ground was burnt for five feet in diameter.

Dazzler plays around oak treetop. [305 Wakefield] During a thunderstorm in 1937, from 300 yards away, I saw a basketball-sized fireball extremely bright playing around at the top of an Oak tree for three seconds.

Light cord to sink. [310 Michaels] A reddish cannon-ball-sized fireball flew from the light cord to the sink, about twenty feet, and went down the sink, which was hot afterward. It did no damage, but there was a loud crash and it blew fuses in the switch box.

Ball originates six feet from lightning. [336 Henry] At Lake Killarney at Winter Park in Florida, during a June 1943 thunderstorm, lightning struck the beach, hitting a metal stake at 125 feet from two witnesses under a lifeguard stand. Near the water's edge, the stake had a boat tied to it. Six feet from the lightning stroke, a ten-inch light blue-white glowing ball materialized and drifted slowly at about five feet per second, at ten feet above the beach. After three seconds it faded. After the original lightning stroke, there was no noise except the wind, distant thunder, and lake surf. In then minutes, the storm left and the witnesses examined the half-inch rope, by which the boat was tied to the stake. It was charred around the stake and broke at that point when pulled.

Power Line Sliders, Gliders

This category is important—not just because fireballs occur frequently on towers and lines. Fireballs occur frequently on high-tension lines and metal towers and even sometimes in substations. They roll or slide along lines—usually above or on them (rarely under or enveloping lines like a bead)—and glide down towers.

Although exactly what causes high-tension-line fireball formation is unknown, whether impedance mismatches, sudden changes and faults—either man-made or from lightning—in power lines, or from generation and switching, all may create opportunities for fireball formation. Before we look at such cases, we must examine power distribution. By examining them, we will better understand power line fireballs.

Sixty- and fifty-hertz alternating electricity powers the world. Tall high-tension towers of metal, wood and concrete that cross our landscape carry three lines of three-phase electricity to power our cities and homes. When, as an electrical engineering student, your author took two courses on this subject of generators, motors and power lines, and studied how electrical current comes from inside the wire windings on giant armatures that spin through powerful-but-invisible magnetic fields. It is amazing, that to get a stationary machine like the huge spinning armature—of these giant generators—up to speed takes time; and, with older steam turbo-generators, this took nearly an hour. Such huge masses of metal have great rotational inertia.

Not surprisingly, this created problems. Sudden demand is not easy to meet, more so if it is greater than anticipated. To provide for such sudden increases in consumer demand or load, a "spinning reserve" of generating capacity is required. Generator stations for this purpose are kept running, and shared into the network, rather than let each station keep its own spinning reserve. More recently, we have gone over to power sharing over vast interconnected grids.

In the earlier times and southern areas where many of these fireball cases took place, the generator voltages were stepped up by transformers into the low range of 11 to 25 kilovolts or KV. Today, over 69 KV is high voltage, and the cost-effective trend is to replace all wood poles with steel posts—small posts weigh 3,600 pounds—which are hoisted in place by crane.

Geography greatly influences distribution topology and power system layout and voltage magnitudes. Very long transmission links transmit with greater voltages from 765 KV to 1.5 MV, because transmission at lower voltages cannot travel the longer distances without great resistive losses. In geographically smaller areas, the upper voltages are about 480 KV. There are substations along the way. Voltages are reduced for secondary distribution locally—residences get two-wire single-phase and industries, three-wire three-phase. House voltage is often single-phase at 110-and-220 volts rms.

How long can overhead lines span between two towers? It depends on line sag. Some towers are 140 feet tall. Most steel towers are made of virgin steel (it rarely rusts) and are 1200 to 1500 feet apart. For lower voltages, wooden and reinforced concrete poles are used in a single-or double-pole tower, with cross members.

The lines are insulated from the towers by pin or suspension-type insulators. Pin types have two or three cap-like small porcelain "sheds" or "petticoats" stacked and mounted over a steel pin. For higher voltages, suspension insulators have a string of interlocking porcelain or glass disks. For a 400-KV line, 19 disks with the conductor held from the bottom of the string, which is suspended from a tower cross-arm.

On drizzling foggy nights, leakage currents flow over the porcelain, and each disk glows with a white star-like corona and hisses loudly, and you can hear them 200 feet away. On such nights, the corona emissions destroy nearby AM radio reception. Each ceramic disk acts like a resistor in parallel with a capacitor, and even on the best-weather days, some current flows from the towers to earth, and is lost. This leakage resistance depends on the weather, salt and soot.

The nonlinear magnetizing behavior of transformers distorts the three-phase sine waves and pollutes them with higher-frequency harmonics. The even and odd multiples of a fundamental frequency are harmonics of the fundamental frequency. A current surge in a line from an exploding fireball can create harmonics or higher frequency currents. Under certain conditions not yet understood, harmonics are associated with some fireballs.

Some might suggest that the increasing harmonic emissions in industry and homes are associated with increased ball lightning sightings near control devices and modern motors. At the very least, they provide a possible clue. Until 1985, harmonics were uncommon, and most electrical waveforms from the power lines were clean 60-hertz sine waves. These clean sine waves did not cause too much trouble, that is, any appreciable spurious emissions. However, starting in 1980, a certain category of new energy-conserving control semiconductors—including power transistors, triacs, SCRs and power control integrated circuits—improved efficiency of equipment and reduced energy consumption.

Instead of over consuming electricity and using rheostats that heat up and waste energy, these new adjustable speed drives, control circuits, and other equipment draw current in short pulses. They chop up the sine wave. This is great for efficiency, but it has a downside—annoying harmonic emissions at higher frequencies that we engineers must minimize with various techniques, such as filters. But the techniques are only partially successful.

While at a major firm, your author once co-designed an industrial three-phase ten-horsepower solid-state motor controller that dramatically improved efficiency, but chopped up and distorted the voltage and current sine waves, thus generating many harmonics. We did what we could to reduce emissions. This type of controller is now in universal usage.

Semiconductor light dimmers and motor controllers are far more efficient and energy conserving than older potentiometers ("pots") and rheostats, but they chop sinusoidal waves and thus generate harmonics—higher frequency sinusoids—that interfere with AM radio reception. The static drowns out reception. Most FM is not affected, except by some electrically "noisy" electric heater pads.

Although these spurious radio emissions are an annoyance, there is a far more serious problem. Mangled sine waves create load currents rich in harmonics that overheat transformers, damage neutral lines, and if you are lucky, only trip circuit breakers. If unlucky, they could start a widespread blackout. These non-sinusoidal waves are highly nonlinear and composed of harmonics.

Harmonic frequencies are integer multiples of the fundamental power frequency. At 60 hertz, the second harmonic is 120 hertz; the third, at 180 hertz; and so on. Drawing current in abrupt pulses rather than in smooth sinusoids, nonlinear loads create harmonics.

Medical instruments, printers and computers—in fact, almost all modern electronics and even motor controls—now all possess diode-capacitor-type power supplies that charge capacitors and draw down the current and then recharge in the next half cycle. When the capacitor is below the capacitor residual, it draws no current. Voltage harmonics can damage induction motors and power factor correction capacitors. These capacitors for a resonant (oscillating) circuit with the inductive elements to the distribution network, and its resonant frequency is near that of the harmonic voltage, then the resulting harmonic current will greatly increase and blow the capacitor fuses or exceed their ratings. Once the fuses blow or the capacitors are destroyed, this detunes the circuits and resonance stops.

There are ways by which this can create conditions favorable to formation of fireballs.

In three-phase, four-wire polyphase systems or with single-phase two-wire nonlinear loads, some odd harmonics or triplens—odd multiples of the third harmonic—do not cancel—as occurs with balanced linear loads—but add in the neutral wire. If there are many single-phase nonlinear loads, then the neutral current will exceed the phase current. Since the neutral line lacks a circuit breaker, excessive heating occurs and excessive voltage drops occur between the neutral line and ground.

These higher frequency fields mechanically resonate the panels, which whistle and buzz. The telecommunications (telephone, fax, computer) cable is run close to the neutral line of power cables. This normally reduces inductive interference from phase current because it carries only small imbalance currents from the three phase lines. But when triplens on the neutral wire cause inductive interference on the phone lines, it is a clue that worse trouble is coming.

Industrial and commercial buildings have a 210/120-volt transformer in a delta-wye configuration. Within the building, single-phase nonlinear loads produce triplen harmonics, which algebraically add up inside the neutral line. In the past, this would be reflected back into the delta primary inside the transformer and not be sizeable—unlike today, where this neutral line current is sizeable. And, although loads are balanced, office building and manufacturing plant transformers are mysteriously overheating and failing.

Most power transformers are rated for a 60-hertz current load, not for higher-frequency harmonics. These harmonic currents increase core loss because of eddy currents and core hysteresis. Either the transformers must be derated or replaced. These imbalances cause standby generators to fail. Overheating is not their only problem. Some harmonics produce distortion at the zero-crossover and thus may create instability in the generators' control circuits.

Costly and massive circuit breakers throughout the network protect it and automatically trip open to isolate faulty equipment and thus limit damage and provide continuous service, and also permit maintenance. Because most transmission and distribution faults are transient or brief, auto-reclose circuit breakers are widely used. They open on a fault and re-close later. Abnormal high over-current surges on a power line caused by an exploding fireball would be detected and trigger a circuit breaker.

Is it possible that some fireballs have triggered imbalances on power grids and tripped circuit breakers (they have) on a grid that also led to massive blackouts? It is probably possible, but no one knows if this every occurred, nor is it likely that there is any way to detect this without a lucky eyewitness account.

Fireballs have exploded or merged into lines, causing a sudden surge (transient) of enormous current. In some cases, a second fireball can re-form at another location, perhaps at the end of the wire, or from equipment that the wire feeds into. Secondary electrical effects of some power line fireballs damage switches and circuit breakers so that they cannot reset.

Because flowing current has "momentum," whenever the contacts of a switch carrying a current are opened or separated, an arc or flaming spark begins in 10 microseconds (millionths of a second). This arc contains vaporized metal and ionized gas or air and conducts electricity. Most circuits possess inductance—a quality that causes inertia of the current flow, sometimes provided by coils of wire—that keeps current flowing though the arc, but does not affect arc voltages. Since arcing damages equipment, much research has gone into arcing and switching, and much is known about it. Arcing, gaseous conductors and plasmas are also important to ball lightning researchers because arcs—somehow in ways yet not fully understood—occasionally generate fireballs in electrical machinery.

Once an arc starts, a mere ten volts will maintain it. As a switch opens, the last point to break contact will exceed 3,000 degrees C or 5,432 degrees F and vaporize metal. There are four categories of circuit breakers: bulk oil, air-blast, low oil-volume, and vacuum. Arc suppression is important to the power industry.

But arcs also are associated with fireball formation. In World War II submarines, green fireballs floated from switches that arced. Early fireball researchers such as Nielsen used huge arcs to generate fireballs. Eminent fusion researcher at Los Alamos, James Tuck used arcs in an attempt to create ball lightning.

Arcs that are not suppressed can create resonances in the generators and lines to create fireballs in a manner slightly different, but similar to that created by Tesla. The following cases are not uncommon. Cases from ORNL list a case number and last name of its witness, all within [square brackets], followed by the witnesses written account.

Red ball follows ground-wire of high-tension line. [16 J. Long, about 1939] During a local electrical storm, I observed a ball of fire following the upper ground wire of a high-tension line a little less than a half-mile away. This ball was immediately preceded by a stroke of lightning that I presume hit the high-tension line. The ball lasted for two or three seconds. I don't remember its disappearance, but I think it was sudden.

The reddish ball, two to three feet diameter, did not seem to change shape and remained spherical. It followed the conductor as though it were rolling on top of it for about 100 yards. I do not remember any unusual sounds or visual effects. The electrical transmission line was knocked out but I do not know for sure of any other disturbances or after effects. Some of our neighbors also saw this phenomenon.

The top line on the traditional steel towers is well grounded at each tower to the earth. If they are double-poles of wood with a cross-member, then there are two top lines, each with a braided cable that goes down each respective pole to a solid ground in the earth. Lightning strikes are directed to ground. Most steel towers have two pairs of three-wire three-phase lines, with three cables on each side of the tower. The wooden towers usually carry only one. This had no effect on the fireball.

"Bowling ball" hisses on TVA power line. [398 Purkey] Lightning struck a line in the TVA switchyard, knocking out a circuit breaker in an oil-type transformer. A brilliant bowling-ball sized fireball moved along the power line leading from the circuit breaker, hissing for five seconds and suddenly vanishing.

Blue ball turns green, jumps conductor, knocks horse down. [399 Stewart] The two-to-three inch blue ball jumped from one conductor to another, a distance exceeding two feet. When I first saw it, it was the color of an arc welder. It then changed to a greenish yellow. It lasted ten seconds.

Tower ball descends to water. [185 S. Johnson] The last week of March 10 in '31 was preceded by typical storms in west Alabama, and a tornado had struck the edge of my hometown, Tuscaloosa, and the small town of Northport across the Warrior River. There had been casualties and heavy damage. The following Sunday afternoon (the storm was last Tuesday), I was caddying with four members of the University of Alabama faculty at the public golf course near Lock No. 10, one of the navigation locks on that stretch of the river. The weather was still very unsettled with considerable overcast. We had started the back nine, and were on the twelfth hole, when a sudden windstorm arose, accompanied by an impressive electrical display.

At that point we were fifty yards from where some Alabama Power Company Transmission lines crossed the river. The wind blew up-river, roughly northeast, into our faces. The transmission lines were swaying vigorously, especially the span across the river, and all play stopped, while we anxiously watched them and the very dark clouds overhead.

When I first noticed the fireball, it was ten feet from the tower on our side of the river, and so about fifty yards from where we were standing. It traveled rapidly along one of the wires to the low point of the span and dropped off, disappearing before it reached the water. The river is about one hundred yards wide at this point, so it traveled about fifty yards after I first noticed it. Later, it turned out that three others of the party had also seen it. To all of us it appeared to be about six or eight inches in diameter at fifty yards, but this cannot be right, so it was considerably larger. No other noteworthy effects were noticed.

Blue fireball forms on 5.5-KV power line, accelerates, shatters insulators. [12 McDuffee 1956] I twice observed ball lightning during early summer afternoons in Pensacola, Florida, prior to local thunderstorms (quite frequent here). I was 150 feet from where the ball appeared on the eastern top string of (I believe) a 5.5-KV power line. There was a flash of lightning accompanied by thunder. The bolt struck one string of the six lines, three courses of two separated by three feet.

When the bolt struck the line, a brilliant blue-white fireball appeared at 15 feet (north) from a creosote-impregnated, wooden power pole. It remained there, momentarily fizzing and popping, and then slowly moved toward the pole, accelerating as it moved uphill. It did not damage the power line or its insulation. When the ball approached near to or touched the porcelain insulator supporting the power line, there was a loud, sharp report, like a high-powered rifle shot, and the ball disappeared. Shattered, the insulator needed replacement. Although the initial bolt maybe damaged it, this is unlikely because I saw pieces flying when the ball disappeared.

Author—Taking into account that thunder clap echoes had died before the ball began to move from its initial position, and knowing that the echoing lasts three seconds, the fireball life span calculates from five to seven seconds. The brilliance resembled an electric-welding torch, although much larger in diameter, and the sound resembled that of an electric "burning" torch. The power line was not the highest object in the vicinity. Within a thousand-foot radius there were numerous 75-foot metal smoke stacks, steel industrial buildings up to six floors high, and a water tower about 150 feet high

Transformer Pours Out Big Red Balls. [ORNL] Lightning struck a power line transformer. One after another, reddish orange balls of fire poured from the transformer and fell onto the ground.

Follows High Tension Lines. [377] While atop a tower of a high-tension power line—which was not connected to the generating station—a line worker saw lightning strike a tower 12 miles away and felt his tower vibrate. A red glowing fireball moved from the distant tower towards him along a cable. After he quickly descended the tower, a three-foot red ball—moving on the middle outside cable—passed by, three minutes after the lightning flash, and then reached the last tower, and exploded in a large roar.

Slowly Follows Cable. [12] A fireball suddenly appeared on a 5500-volt transmission cable, at 20 feet from a pole and moved slowly along the line cable to the insulator and then exploded.

Local Power Line Rollers

Unlike the majestic towering high-tension towers threading the land with their long spans of sagging cables, the diminutive wooden poles of local power lines on most streets are at considerably lower voltages—providing 110 and 220 volts for homes and small businesses. Fireballs also follow these lines, usually on top, but sometimes below, sliding or rolling along. Usually such rollers or sliders travel at a slow or moderate speed—rarely fast. They can jump over any obstructions. Some reach a wire and follow it to a house. A few float free or hop off and fly away. Some simply vanish. Others explode. Most that encounter transformers or other equipment do so without damage—but not always.

Hot power line ball damages substation. [2 Robertson] Lightning hit a 154-KV transmission line and a fireball followed it to a power substation 40 yards away. Choke coils and lightning arrestors with horn gaps at the substation failed to attenuate the ball, although ground cables from arrestors were burned slightly, insulators broken, copper conductors on low-tension bus melted, and 3-2000 KVA transformer completely burned out.

The substation suffered extensive damage. At 80 yards, the witnesses (mostly technical workers) heard a deafening noise like a large high-frequency spark gap. Damage occurred quickly. Insulators on a low-tension bus shattered. Numerous discharges jumped from the low-tension bus and transformer cases to the structure. It is impossible to tell if the fireball or its electrical aspect did the damage.

How the fireball passed through or over a 154 KVA oil-switch, which was not damaged, on the line entering the substation is a mystery and was debated end-lessly by the engineers. Just as mysteriously, the switch did not open as occurs with a surge. The transmission line surface was deeply pitted on top. But the materials (steel, copper, porcelain) of the low-tension bus and its insulators with supporting steel structure were in part completely melted as if from a crucible. The same melting occurred on top of the transformer cases and with portions of both LT and HT bushings.

Fireball rides line to nuclear weapons power house. [25 Derleux] I was at the [deleted] powerhouse Building [deleted] on 24 June 1960, at 3:50 p.m. on June 24th, when a three-foot green-blue-yellow fireball suddenly appeared on a power line that it followed to the powerhouse.

Four-foot red fireball visits atomic bomb factory. [29 Ages, at X-10] I was at building [deleted] during an electric storm some months back on Friday at 3:40 P.M.. Lightning lit the sky. About 10 or 15 inches above a power pole appeared a bushel basket sized large red ball of fire at the end of the lightning bolt. It went away just like turning out a light bulb.

Line roller vanishes at nuclear power house. [268 Huffaker X-10 power-house, Building 3022, August 1960] Occurred at [deleted] powerhouse, viewed from Building [deleted] during August. A four-foot, yellow fireball rolled along a transmission line at the powerhouse. It started a hundred feet from the power-house and disappeared when it reached several feet from it and lasted three or four seconds. I viewed it from the second floor of Building [deleted]. It traveled between 22.7 and 17 mph, or the speed of a bicyclist.

Silver line slider disappears at guy wire. [69 Parrish 1935] An eight-inch sil-ver-yellow-red fireball appeared suddenly after a flash and lasted several seconds. There was no change, but it was noisy, with an odor. The ball rolled on an elec-tric line in air for 200 feet on a three-phase line and exploded at a corner.

Slides along line, stops at cross arm, fades. [199 Earley] After a lightning flash, a red and blue ball, a foot to a foot-and-half in diameter, appeared on a transmission line conductor at about the middle of a span and ran along the con-ductor to a cross arm assembly on a pole, a distance of 50 feet, and remained at this point for 30 to 45 seconds before it slowly died out. Another ball appeared on a conductor and moved along the conductor and disappeared without remain-ing at any fixed place for any length of time.

Dazzlers hug conductors. [203 Sternberg ORGDP at Oak Ridge National Lab] A baseball-size fireball bright like an electric arc lasted several seconds, with no color change. The several instances in which I observed these balls (some the size of a dazzling softball) were during severe electrical storms. They never were associated with any particular lightning stroke. These fireballs ran along the ground conductors along the roofs and ventilators of the [nuclear bomb] process buildings.

Power line six-foot fireball lasts 45 seconds. [349 Drake] Appearing simul-taneously with a nearby lightning flash, a six-foot orange to yellow fireball trav-eled along in contact with a power line and lasted 45 seconds. It caused scorching, melting metal and disturbed nonmetallic. It left an odor.

Back and forth on wires. [11B] Occurring after a lightning flash, one-foot diaphremous [sic] fireballs that gradually changed size traveled back and forth along horizontal electrical wires.

Power line ball quickly "burns up," leaves green smoke. [456 Busdges] During lightning this basketball-sized fireball appeared to be either on a power line or just above it. Airborne part of the time, it burned up quickly and left a yellowish green smoke, which disappeared quickly.

Blue fireball slides along power line. [Devan 57] Twice, I saw a fireball with a distinct blue cast that maintained a stable configuration. Neither left the power line. There were other lines nearby. Both moved along the line and out of sight very rapidly. There was no damage to the line. On a third occasion, while working underneath a car with an extension light hung nearby, I heard a lightning crack—there was a storm raging outside—and after two seconds saw a blue glow from the extension light. It flew from one point to another of the underside of the car. This blue light lasted for several seconds. I was hit with a major shock.

Rolls under cable and passes through glass. When lightning struck a transformer, a spinning three-inch, light green fireball appeared atop it, rolled down on the underside of a power line toward the house. After four bounces downward, it stabilized on the underside of the cable and continued to roll at 30 feet per second. When it hit the bottom of the drip loops (before it entered the service entrance head), the ball floated off directly and rapidly into the window, slowing only slightly as it harmlessly passed inside through the glass, barely missing an electronics technician. It flew 16 feet into a back wall, and with a small crackle it sent bright light-blue trails of static electricity over the wall for four feet in every direction around the impact site. It left no evidence or burns.

Power-line whips and forms blue fireball. Lightning struck a power-line and the wire began whipping around and a basketball-sized blue glowing fireball formed on the insulator of the first pole. After another second, the line on the other side of the pole began whipping around and another blue glow appeared on the insulators of the other pole. This whipping quickly repeated at the third span. The lines led to a major supercomputer firm a mile away.

Author—When a sudden and intense electrical current surge or short-circuit current or transient flows in a line, it is surrounded by a circular or concentric magnetic field. The field will expand and then collapse. This causes a hanging line to briefly leap up like a python and then drop, after which it may swing a bit.

While an electrical engineer in industry, your author experimented with sudden short circuits to see what they did to equipment and power lines. In experiments, he set up hanging power lines in the lab with a 15-foot span. Each time the transient spike of current passed through, the cable jumped up nine inches and dropped, writhing like a python. It was dramatic.

But why did it take one-second delays for the second and third line to whip around? And did it just leap up and fall, and then swung—or did it do more, as the wording implies? Nor did the report indicate what transpired at the computer firm, at which the witness worked. Though unnamed, the firm was either Cray or Cray Technology, both today defunct. As a journal editor, your author interviewed the founder of these two firms, the late Seymour Cray, for an engineering article, so was not unfamiliar with this unnamed firm.

Barbed and Electrified Lines

Fireballs are attracted to fences. For the past century, wooden posts that confined livestock carried one or two tiers of barbed wire. Earlier, in the American West, it was the invention of barbed wire—a little appreciated but major invention—that transformed the vast new land to the west and that made civilized livestock farming possible.

Today, it is very common to see miles of barbed wire fences throughout Oklahoma and much of the Midwest, with uncut shrubs growing up close against these fences. Barbed wire ended the wild days of the open range. Combined with the cattle railroad cars and slaughterhouses of Chicago, these inventions made possible the rapid growth of Eastern cities.

But, in colonial times, cows and sheep were confined by the stone walls of New England or the zig-zagged split rail fences to the south of here. With the advent of Tesla's polyphase ac system, lower voltage electrical cow lines—much easier to install than barbed wire fences—replaced some barbed wire, but not that much out west in the vast acreages of wide-open farmlands of places like Oklahoma and Texas. It is said that barbed wire won the west.

Recently, metal clips supplanted the white porcelain insulators—that your author knew on the farm—which were nailed to each post. Once wooden, these posts are now metal. But all of these livestock confinement systems have unique fireball cases told about them.

Skyball rolls along barbed wire fence, shrinks, scorches posts. [20 Massey About 1931} A fiery violet fireball (making a strange rustling sound) descended from the sky, struck a barbed wire fence attached to wooden posts, then traveled along the top of the wire for several minutes before vanishing. The fireball shrunk from 18 to 4 inches. When I examined the fence several minutes later, the post tops were warm and slightly scorched, but only the loose portion of the surface was scorched.

Electric fence shoots out three balls, dehydrates grass. [169 Melton] I was feeding cattle in middle Tennessee. Lightning struck an electric fence or close to it at 600 feet from its end, near where I stood. Three orange fireballs shot in rapid succession, each making a loud bang as it shot from the end of the fence, flew straight for 200 feet through the air and hit the ground. Where the balls hit, the grass was dehydrated. There was no evidence where the lightning had struck.

Author—Farmers left the end of the electrical cow line unconnected—there was no return circuit. The thick bare galvanized wires were wound once around the groove of white porcelain insulators, each about an inch in diameter nailed to wood posts, and about thirty inches above the ground. Today, newer attachment methods make installation and maintenance easier and faster than what your author once did.

Violet ball extracts fence staples. [416 Guice] I observed ball lightning twice, both times working as a lineman. The first time we were returning from a rural area. It appeared in a pasture, with slight slopes and very few trees. About fifty feet from the road stood a barbwire fence with four strands of wire, and a rural power line of 7260 volts six feet from the road. It was a warm day with a few scattered clouds.

A quick storm cloud appeared, very heavy and black. Immediately flashed six bolts of lightning, then a loud sound without the visibility of light, but at this moment appeared a three-inch ball. It was a dense color of ultra-violet with a haze-like corona, varying from an inch to an inch-and-a-half, all around the ball. It rolled along the top strand of wire on the fence with enough speed to keep it balanced on the fence, roughly 100 feet per minute.

Every staple on the top strand came out or burnt from the post. What phenomena can do this? The fireball lasted 30 seconds, traveling close to forty feet, erupting in the same manner as it started, but leaving the top strand hanging from the posts.

Author—What other phenomena instantly pulls nails and staples from doorways and wooden posts? It requires hundreds of pounds of force to extract nails from wood, and to do it instantly requires even more force. This instant pulling of nails from wood is reported in other cases. Although the witness stated that the ball rolled with enough speed to balance it atop the wire, this is not necessary: fireballs are not high-wire acrobats, and perform no balancing acts on wires. If fireballs spin and possess mass, as some seem to, they should behave like gyroscopes.

Witnesses report fireballs that spin, swirl like cream poured into tea (nonlinear chaos?), swarm around, or have one or two transparent layers, with the inner one spinning, or like bees swarming and flying inside a sphere. Without an internal source of energy, such spinning or random internal motion should cause viscous losses, and internal fluid friction losses should slow the internal motion. However, this does not occur.

Dusk lightning hits fence, sputtering ball crosses road. [453 Coopers] At dusk after a light shower, lightning struck a wire fence alongside the road and a baseball-sized fireball fell in the roadbed. Sputtering like an electric arc, it moved rapidly along the surface. It lasted two seconds. I was 20 feet from the fence and blinded somewhat from the bolt or ball.

Orange ball appears at fence, vanishes. [140 Rice 1923] At 2:00 P.M. in weather hot and humid, during a thunderstorm, in a car three men drove three miles southwest of Russellville, Kentucky. Thunder cracked and an orange-white fireball ten inches across appeared at an adjacent wire fence line 40 feet distant. After two seconds it vanished. The witnesses investigated, finding no evidence of searing, burning or damage—only an odor of ozone. Witnesses suffered temporary blindness.

Fence roller knocks wire from fence. [255 Durant] One afternoon in late September my boss and I were loading hay. A storm struck. We stood near woods. Lightning hit a tree to which was fastened a fence. A baseball-sized fireball traveled down the fence and knocked the wire from the posts. When it came to where this fence joined another, running horizontal, there came a loud cracking and the ball exploded. We examined the explosion location, but saw nothing unusual.

Travels between barbed wires. [105 Matlock] In Cains County about 20 miles east of Trinidad, Colorado, I sat in my car waiting out a summer shower. At five P.M., one storm had passed and another was approaching. I saw the one-foot fireball playing along between the top wire and the next under it on a three-wire barbed-wire fence. The blue-white loud snapping and crackling fireball traveled slowly between these two wires from one fence post to the next and then was lost on the second post.

Hisser follows barbed wire. [155 Woodward] Appearing suddenly, a 10-to-12-inch greenish white-yellow hissing fireball traveled 50 to 60 feet two-to-three inches above the top wire of a barbed wire fence, taking two seconds. It moved at 25-to-30 feet-per-second or 17-to-21 miles-per-hour. Formation and extinction was not associated with any lightning.

Barbed-wire fence-roller spits sparks. [271 Watson 1929] Lightning flashed and a five-to-six-inch orange ball of fire rolled along barbed wire for two seconds. Sparks swirled from the ball.

Rolls on barbed wire, splits trees. [66 Edwards] A lightning flash struck two trees. The first, 36-inches in diameter and quite tall, splintered; the second, split. Both later died. A large fireball came from the tree, ran along a nearby barbed wire fence, and then after several seconds loudly vanished.

Orange 20-second exploder wrecks fence. [167 Farmer] A six-inch ball of yellowish orange-white fire struck a fence and exploded into pieces. It had started rolling along a wire fence, burning the grass and knocking out some of the staples. The explosion was loud and very bright. It lasted 15 to 30 seconds. For a short time after this there was a stale odor and breathing was difficult. It completely demolished the fence post.

Fireball dances atop fence, hissing loudly like Tesla coil. [151 Carter 1941] A passenger and I were driving in southern Missouri on a rural road fenced on both sides. With a tremendous crash and bright flash, lightning struck the left-side fence, even with the car. A 10-to-15-inch, bluish-white glowing fireball danced on top of the fence for two seconds before it collapsed. It sounded like a Tesla Coil discharge, but louder. The surrounding terrain was level and comparatively open.

Transformer fireball rolls atop fence. Lightning struck a transformer near the end of a chain link fence and immediately a six-inch, bright sparkly white (with a touch of blue and yellow) fireball that formed—its core was denser—traveled along the top of a chain link fence about five feet from the house.

Rolls atop chain link fence. A hovering whitish blue, one-foot fireball drifted westerly and slowly rose to seven feet as it neared a chain link fence encircling the power station. Hissing, it rolled for 30 feet atop the fence, then rose in the air as it moved away. At 30 feet altitude, it grew bright white, exploded, and flashed out.

Dazzling fence roller blasts hole in wall. When lightning struck an oak, a basketball-sized dazzling fireball rolled down the tree, and at six feet above the ground leaped from the trunk to the ground. Loudly buzzing, it floated slowly until it hit a big air conditioner in the wall, destroying it, and a six-foot blue flame shot out. The ball floated onto a chain link fence and moved faster, jumping from fence-to-fence for two blocks—knocking a man off a porch swing—until it came to a corner at a stucco house and blew a big hole in its wall.

Dances atop garden fence. For five seconds, baseball-sized fireballs danced along the fence near a garden, then exploded and sparked with white light. There was no sound. A night lightning storm flickered in the distance.

Clotheslines

Fireballs that materialize on and travel along clotheslines may exhibit less energy. They usually form atop and travel on top of clotheslines, but not always. Although most clothesline fireballs are reported on metal lines, some were reported on rope.

Metal clothesline ball hot but dim. [21 Burger] A 12-inch quiet fireball stayed on a metal clothesline. I felt heat from the ball at a distance of 30 feet [sic]. It appeared in the morning. Although the ball seemed very hot, it was not too bright.

Author—Is Burger's estimate of distance wrong? Maybe it was a typo and he meant 30 *inches*. Metal clotheslines are made of aluminum. Why didn't the ball melt it? How could a dim ball radiate heat? Perhaps the witness stood on wet ground and felt induction heating. I doubt such cases, but do not dismiss them totally outright, simply because they fail to fit my prejudices. There are rare cases where distant witnesses reported heat from fireballs.

Four balls roll on metal clothesline, drop and bounce. [458 Keith] A wire clothesline was tied at one end to a wooden post and at the opposite end to a walnut tree. Shortly before the rain started, which would harbinger a violent summer thunderstorm, lightning struck the walnut tree. Out along the clothesline went four balls of fire about the size of basketballs. With a frying sound, they followed the clothesline for about 30 feet and then dropped to the ground. When they hit, each bounced several times and then dissipated.

Ball bobs under clothesline. For three minutes, a grapefruit-sized blue-white fireball with a smaller and brighter inside core whose size fluctuated from two to four inches, repeatedly bobbed from below the rope clothesline up until the core contacted the rope, then down again. This repeated for ten times, and then it suddenly dropped to the ground and vanished. There was no damage.

Aerial Antennas

Fireballs prefer to follow metal lines. Antennas are no exception. In many cases fireballs also follow horizontal antennas. Although they do follow vertical antennas, they do so less frequently, perhaps only because they are shorter and less likely to be followed or be observed.

Floor roller follows antenna. [ORNL Stowe] An egg-sized ball of fire followed an antenna lead in through the window to a [ham] radio set. It then rolled across the floor, into the bathroom, hit the water tank and disappeared. The event occurred in Paris, Texas, 95 miles northeast of Dallas and 20 miles south of Oklahoma.

Ball follows coax feeding broadcast antenna. [54 Affel] A two-foot fireball appeared on and noisily followed a coaxial transmission line feeding a commercial broadcast antenna. It lasted five seconds and gradually faded. There was no lightning flash.

Reciprocating Ball Hugs Car Antenna. [160 Weber 1956] While my wife and I were at a drive-in movie near Oak Ridge, the rain let up, when almost directly ahead of us a basketball-sized white fireball suddenly appeared. It moved up and down on the right side of and in contact with a radio antenna on a car about 100 feet ahead of us. The antenna, insulated from the ground, was on the car's right side. The fireball resembled a 40-to-60-watt light bulb placed inside a translucent ball. Its brightness did not demand attention; and my wife noticed it only after I pointed it out to her, even though it was within 15 degrees of her line-of-sight to the picture screen. I saw no evidence that others saw it, and it disappeared rapidly after 12 seconds. Its brightness was constant, except for the gradual fade out, diminishing in size.

Sparking Fireball Rolls On Army's Rhombic Antenna. Throwing off sparks similar to flames, a two-foot bright rotating fireball dropped onto an antenna wire and rolled 100 feet along it until it reached the next antenna mast and slid down it, and then after a couple hops on the ground, it broke up. Total observation time was ten seconds, agreed the eight witnesses who stood outside the operations building while watching the developing storm. A U.S. Army security unit near Nurnberg inside West Germany operated the antenna in 1967.

Author—The 360 rhombic antenna array in a "long wire" arrangement is used mostly for 800-Hz to 30-kHz reception, which is rather low; and it is curious what they were listening for. A traveling-wave rhombic antenna can be used in both transmission and reception. If a parallel-wire transmission line is pulled apart at its center to form a rhombus or diamond-shape, then one of the properties of a line—equal and opposite magnetic field at every given point in the neighborhood of the wires—ceases to exist and the line becomes an antenna.

The length of the antenna's four legs and the tilt angle are varied for maximum operation. Since the opposite end is terminated in the value of the line's characteristic impedance—its resistance and inductive and capacitive reactance, the latter, imaginary variables in the s-domain (using Laplace functions)—the antenna is wideband, that is, somewhat independent of frequency—and thus will give excellent results over a wide range. Although it is optimized at its center frequency, this type of antenna is forgiving and accommodates frequency changes.

Green fireball dances atop truck cab. During an afternoon thunderstorm, a man and his dad drove down a highway. Driving toward them, a pickup truck had two, one-foot CB antennas on the roof over each window. When the truck was 25 feet before them, lightning struck the driver's side antenna. A greenish white fireball jumped and danced back and forth atop the cab for five seconds. Then, with an explosion it disappeared.

On Inside Wire

Fireballs follow inside wires that are visible, such as wires that are stapled to a wall. In a few cases they follow along the wall where there is wiring inside the walls, but this is uncommon. Although it is pure speculation that they may follow wires within the walls and squeeze through holes in the wooden studs, no one can see it, and so this is unknown. Nor does anyone know if fireballs can and do pass through holes in metal studs. They have slipped through metal keyholes and crevices—morphing their shapes as they squeeze through—so who knows.

Leisure "grapefruit" on phone inside wall-wire evaporates. [420 Michelson] In 1935 I visited M/M Krom, later deceased in 1961. Their daughter was there. I was visiting a residence on Finger Street in Saugerties, New York, on a July mid-afternoon. During a severe thunderstorm, lightning may have struck the telephone lines in the neighborhood. The parents and their teen-age daughter and I were walking through the living room towards the hall, where the phone was installed, when came a thunderous crash and lightning flash. On the phone wire between the wall connection at the bell box and the handset I saw a large grapefruit-size ball of fire suddenly appear. It rolled leisurely along the wire to the phone handset and disappeared as though it evaporated. My cousin picked up the phone and called the operator. Since she could reach the operator, we assumed the phone was working.

Expanding fireball follows indoor bell wire. [161 Curtis] I stood watching a lightning storm in an open doorway of a building, in the left corner facing the street. Also along the left side of this building was an alley that also continues across the street. Lightning struck a transformer atop a pole in the alley across the street, a half-block away, and destroyed it. I held the closed metal screen door and was stunned slightly.

I turned left to face the nearby wall where wire ran from a doorbell button inside and along the left wall, back to a buzzer on the left rearward wall. The 12-foot wire was #14 strand, exposed, running along the left wall just beneath the ceiling. As I faced this wire, a fireball traveled along the wire, from front to rear, as fast as a person would run. When it reached the end of this wire, where the buzzer was, it jumped off into midair, traveled about three feet downward at a 45-degree angle, and then suddenly enlarged from softball to basketball size. The fireball traveled a path length of 14 feet in three seconds.

Those fireballs that roll along electric and power lines are common—in fact, based on the ORNL study, even more common than previously reported. They also have a predilection to slide or roll along barbed wires and electrical lines. But in the categories of movement, there are other cases of fireballs that either fly or bounce.

3

Flyers and Bouncers

As we just saw in the cases of the previous chapter, although many fireballs do roll, float or slide along lines and fences, instead most fireballs bounce or fly. Many ORNL witnesses reported fireballs that shot up like rockets, bounced off trees, rolled atop automobile hoods, descended trees and telephone poles, moved along roofs, even chased and circled people, or rolled along gutters and pipes, and on the ground.

On Ground, Gutters, Pipes

Fireballs roll on streets and floors, on lawns and earth, and along pipes and gutters—so much so that the Scandinavians call them "rolling lights." Although they often bounce, fireballs also roll along floors, on streets, and across grass fields. Although rolling friction slows down ordinary rolling balls, such as baseballs and basketballs, it rarely slows down fireballs. Why not? Something unknown propels sliding and rolling fireballs.

Floor-rolling fireball invades nuclear-plant turbine room. [173 Romines] Occurring after an outdoor flash, this silent 14-inch fireball remained in contact with the surface of the nuclear plant turbine room floor and slowly traveled 150 feet in a straight line for five seconds.

Three floor-rolling "ballettes" vanish under chest. [1-125 Keek] This occurred three times in succession after a lightning flash: large marble-sized, white glowing "ballettes" would start and vanish at the same exact locations—start rolling out of a closet and disappear under a chest on the other side of the room, without doing damage.

Gasoline pump ball rolls, boy faints. [209B Cottrell, about 1930] Just after dark, I stood on the concrete floor of a store porch, waiting for a thunderstorm to end. At my left stood an old-style gasoline pump, its intake pipe to the underground tank protruding from the ground about four feet before me. Lightning flashed out front, and an orange (some yellow) ball rolled toward me so fast I could not escape. It disintegrated at my feet. I thought I was hit and blanked out for an instant, feeling like floating in air, but the strange sensation soon passed. The basketball-sized fireball glided along the ground.

Fireball flies off surveyor's level rod, spins, rolls away. [10 Dehart] One summer afternoon in 1927 while I worked with the state highway surveying team, an electrical storm approached but there was not yet any rain. One man held a 13-foot-long, wooden level rod (with brass caps on top and bottom) exactly in the center of an asphalt road. Lightning may have struck the top of this rod. A spinning ball flew from the rod and struck the road 50 feet away, then rolled 100 feet down the road, seeming to lose its energy, and vanished.

Ball rolls from tree. [129 Rush] Looking out a window, I saw lightning strike the center of a large oak (about 230 feet distant from me), split off a limb, and shoot down the trunk. A white to orange-and-red eight-inch fireball rolled away from the tree along the ground for 20 feet, growing larger. It left white smoke and lasted three seconds.

Orange ball rolls from under electric range. [534 Johnson] My mother and aunt saw a softball-size orange-tinted fireball rolled out from under the electric range during a thunderstorm, rolling four feet and then in one second suddenly vanished. It did not scorch the floor.

Fireball rolls across flat farmland. [526 White] From where chain lightning struck the ground, a basketball-size fireball rolled across flat farmland for 260 feet before suddenly vanishing. No damage was visible.

Author—However, seemingly flat farmland is not "flat" to a rolling basketball, beach ball or balloon, and yet the rolling of fireballs across fields cannot be accurately duplicated. The field offers considerable resistance. Your author conducted experiments to simulate fireballs rolling across fields and ground in winds and before fans. Sometimes fireballs roll against the wind and scorch grass, even when wet; but, at other times, leave dry grass unharmed and leave no trail. However, your author sees no connection with crop circles and remains a skeptic.

Rolls down storm sewer. [536 Boyd] An eight-inch fireball in general contact with the surface disappeared into a storm sewer. I am with the TVA Research Staff, Chatl., Tennessee.

Blue ball rolls on clay, fades. [503 Barta] During a heavy rain and an intense lightning storm, a four-foot blue fireball rolled for four seconds on the bare clay ground of a ball field for 24 feet. It gradually faded. There was no grass or weeds in the area, and there were no detectable after effects.

Rolls back and forth, then away. [497 Simpson] In South Carolina I saw two 2.5-inch fireballs. At first they rolled back and forth about twenty feet from the house, then rolled for about thirty feet towards the walk and disappeared. After this there was a loud crash of thunder. In the second case, in Knoxville, Tennessee, a 2.5-inch fireball rolled down the middle of the street for a half block and then vanished in a loud explosion.

Rolls in open field. [341 Shefield] In early afternoon of a spring day, thunder preceded rain to come, when a tremendous bolt made me look. A 12-to-15-inch fireball rolled randomly along the ground for 150 feet in an open field, then vanished.

Ball rolls around pool. [501 Fagg] While in the Aleod swimming pool in 1959, a cloud came up and everyone left the pool, except one boy and my wife and two children. While walking around the pool, I saw lightning come down and bust over the area. It burned the boy on his stomach, were he stood in the water, and his legs, and my son and me on our legs. The ball hit and burst and rolled along on the water and lasted two seconds.

Pulsating roller explodes. [560 name omitted] A violent storm was in progress along Bottle Creek, when a fireball appeared suddenly on the side of the road and rolled for six seconds along the surface at the same speed we were traveling and then suddenly exploded. The size varied continuously between 12 and 15 inches.

Well-pump spews orange floor-rollers. [205 Long] During an afternoon summer electrical storm, I sat with two others at one end of a 50-foot porch that extended across the entire front of a rural farmhouse. On the far end of the porch was a manual metal water pump, atop a box-like platform over a well under the porch. Lightning struck the pump or nearby. Numerous orange and red golf-ball-sized fireballs shot from the pump in all directions. Three rolled along the porch at a rapid rate, almost reaching us when they vanished after five seconds. We found no residue or marks on the floor, no odor, and no damage to the pump.

Fuse panel emits ball, rolls under door. [50 Wyatt] An orange-red fireball the size of a bowling ball dropped from a fuse panel on a lighting circuit, during an electrical storm, to the floor and traveled rapidly in a straight line 20 feet in two seconds, and disappeared under a wooden door that led to a meeting hall, a vacant large room with steel-column supports for the roof. An ear-splitting thunderclap hit at the instant I saw the fireball. Neither panel door nor wooden door was scorched, nor was there any sign on the wooden floor.

Down the gutter, down the drain. During a heavy summer rainstorm, a basketball-sized fuzzy-covered, orange fireball moved from a house gutter, down the rainspout, then to the driveway's edge to the street, silently and slowly traveled 45 feet and vanished at the storm drain cover.

Ditch ball dazzler rolls 38 mph. [Henson Jr.] I was awakened by a storm. The brilliant white three-foot fireball struck at the top of a hill 300 yards from me, rolled along the highway ditch line for 150 yards in eight seconds or about 40 miles per hour.

Author—Fireballs are not heavy bowling balls—they contain light gases—and although they should never roll like this, they do. And, a rolling fireball should leave a trail in dirt or grass, and if hot, should also burn weeds and grass, even setting fires—especially if rolling slowly. Occasionally they do, but usually they do not, or are selective. Dissipative forces like dynamic friction should slow down a rolling ball, even one as heavy as a bowling ball, by reducing linear and angular momentum—but it does not. Why do they seem to "violate" our current Newtonian and quantum-relativistic physics?

Office ball rolls down hallway. [267 Fischer] In 1946, I was in the old Roane-Anderson Office Building in Oak Ridge National Labs during a severe summer electrical storm. A ten-inch fireball came down a large hallway that extended from the front of the building to the rear. I was standing in a large open office area next to the hallway near the rear. The fireball came in a flash down the hallway rolling or bouncing close to the floor and disappeared out the open back door.

Enters window, rolls out door. [221 Sawyer 1925 or 26] While living in Tennessee near North Carolina, I was sitting one summer day in the house working with my mother, when a ball of fire came in our window and rolled right on out through the front door, from there into the yard, and disappeared into the gravel. The phone line in the front yard where the ball disappeared was burned.

Gutter roller jumps to post. [193 Olka] A fireball appeared the instant lightning struck nearby. The size of a medium beach ball, it came directly from above, followed a gutter and jumped to a metal light-post, and then after two seconds, it disappeared. An ozone smell lingered and I experienced a "green after-vision."

Water-pipe fireball and one-mile wire-slider splinters poles. [302 Foster] During an early evening thundershower, I was in my grandfather's house, in the kitchen end. I noticed a glowing pinkish-red fireball following the water pipe along the wall. It came out to the end of the pipe and faucet, and disappeared with a mild pop, leaving an acid odor.

Also, twice during intense thunderstorms I saw similar one-foot fireballs traveling along telephone and electric wires. One other fireball—which I did not see, but did see the results, and talked with an eye witness—traveled along a tele-

phone line for a mile and at nearly every pole sent a discharge to ground of sufficient strength to rip out a splinter from each pole. All but three of the poles in
this line had a splinter removed. To me ball lighting is not a rare occurrence in
this area; and I was brought up to distinguish between ball and chain lightning.

Roller avoids metal. During a lightning storm, just after the lights went out,
a bright fireball slowly rolled down a hospital kitchen hallway, ignoring all the
metal, and lasted over one minute.

Spark-covered fireball rolls across hearth. A basketball-sized fireball materialized on the elevated brick hearth and immediately moved from left to right on
the elevated hearth surface. Many tiny electrical sparks an inch long coated the
ball.

Rolling shaggy tangle fireball trails sparking trail. Amidst the strong winds
and rain of a daytime thunderstorm, there appeared rolling rapidly across a vegetable garden and across the yard, a 20-inch shaggy yellowish red fireball. A glowing halo and shaggy spiral strips of fire surrounded the dense core. Yellow-red
sparks trailed 30 inches behind the rolling shaggy ball as it rolled out the open
gate and loudly exploded. It lasted several seconds.

Ball skids down road. During a storm with intense downpour, lightning hit
the roadside at a 45-degree angle, "rolled on itself" to create a fiery beach-ball-
sized fireball that "skidded or slid" (but not "rolled") down the road at 15
mph—keeping pace with the car for 30 seconds, then was gone.

Lightning spawns porch floor rollers. Lightning struck a metal well pump
located on a house porch. Several fireballs traveled on the floor the entire porch
length, about thirty feet, and then vanished, leaving no damage or indication.

Rain gutter roller rolls three. A grapefruit-sized white rullende lyne (Scandi-
navian for "rolling fireball") surrounded by a blue glow quickly rolled along the
rain gutters of three houses in series, in a straight line, jumping between the
houses, and then silently vanished at the end of the last gutter.

Chases People

Fireballs that chase people were reported for centuries, but these anecdotal stories
were later dismissed as fanciful old wives' tales. Whether they actually chase people is debatable because it may be that these witnesses were "dragging" along the
fireballs. However, some fireballs have chased people up and down stairs, and in
some cases overtaking and passing them. Those that chase people rarely hit them.

White ball chases lady upstairs. [Mrs. R] On July Third, Mrs. R of Cheltenham in England (70 miles northwest of London and close to Worcester) was watching an afternoon thunderstorm from her kitchen window. She turned. Behind her floated a white sphere of fire. She ran through the dining room and up the stairs. The sphere followed. It floated past her and entered the open door of a room and shot out the open window. Seconds later a flash lit up the window and there was a crash like thunder.

Fireball chases farmer's wife. Mrs. ER described an encounter that occurred when she was a young girl in rural Bridgewater, Massachusetts, 30 miles south of Boston. Around 1930, during an approaching thunderstorm, a sphere of white fire chased after mother and her daughter, who ran for the farmhouse. The sphere overtook and passed them, hit the house and exploded.

Fireball paces auto at 40 mph. At ten at night, driving at 40 mph east on the town's main street, two witnesses saw a fireball on the passenger side. It followed their car through intersections and finally veered down another street.

Circles Inside Perimeters

Why do some fireballs circle inside buildings? A few witnesses felt that the fireball was trying to escape. And since some fireballs do pass through walls and doors, why do these not also pass through? Or are some merely following electrostatic fields?

Orange ball runs around ceiling's perimeter. [149 Plengey] During the night we were in bed when lightning struck our apartment house. Persons sleeping in different rooms were awakened by orange fireballs of various sizes—six inches to two feet across—running around the top perimeter of the room on the ceiling, possibly following the electrical wiring. The balls faded away. An ozone odor like an electrical short lingered. They did no damage to the walls, wiring or ceiling.

Author—Unless this house had an eccentric wiring layout—very unlikely—the fireball was not following wiring. Your author once worked as a rough framer (homebuilding carpenter) and never heard of electricians installing wiring through the studs near all the top plates that circled the rooms like this—and certainly not in all the rooms!

Enters, circles garage, exits. [109 Mann] Lightning struck a tree several hundred feet away and a six-inch ball flashed across and into the open end of a garage, followed the wall counterclockwise around the garage at the floor or ground level. It faded as it left the garage.

Orange ball circles radio transmitter room. [564 Shirley] Lightning struck a radio transmitter antenna. I was in the transmitter room. Transmitter equipment was destroyed and melted. The fireball was not a flash, but an orange flame sphere that traveled quickly about the room leaving an ozone odor.

Author—The question is, did lightning cause the damage? Radio and TV stations take great precautions against lightning. Did ball lightning cause it? Or did some phenomena yet unrecognized, but associated with certain ball lightning, cause the intense current surges and high voltages that did this damage.

Garage blue ball floats back and forth. After an earlier storm in early June, in weather hot and humid, just before sunset, a man walked into his garage. A basketball-sized blue fireball with a large electrical coating floated in the left side, floated right, and then back to the left like a caged animal, then vanished. The garage was dark and the doors were closed. Was the ball there before he opened the door? Was the ball trying to escape?

Rocket Fireballs

Fireballs occasionally drop, dive down or fall like stones from clouds and then fly level. Occasionally these "divers" bounce back up to the sky. It is even less common for a ground-born fireball to ascend like a skyrocket.

River bank ball shoots up like rocket. [592 Yocum 1926] At my home in Parsons, WV, an electrical storm was descending. I stood about 300 feet from the Shavers Fork River. The ball of white-hot fire rose up vertically for 250 feet and then exploded with a tremendous roar. It was large, the size of the bottom of a home laundry tub or about thirty inches across. It flashed with a blindingly brilliant incandescence, and rose quickly like a rocket from the riverbank, which was sandy and pebbly, and of fast-moving clear water. The odor left was that of ozone, of large electrical arcs. The fireball was perfectly round and resembled the filament of an electric bulb. There were no point-type discharges.

Up, Down Stairs

Fireballs that fly up or down the stairs are commonly reported. Although fireballs sometimes roll and bounce down the stairs, no known cases exist of fireballs that roll or bounce up the stairs. (They do float up stairways, though.) As far as we know, these stairway fireballs do not bounce or float up or down outdoor stairs.

Shrinking sizzling "egg" descends stairs. [405 Puryear] While sitting in my living room during an electrical storm, there was a thunderclap that seemed to hit something, and I heard a sizzling sound. An egg-sized ball of fire came down the stairway (15 steps), diminishing all the time. It fell on the floor and vanished. It lasted 5 seconds.

Ball rolls down steps, out door. [11A] When lightning struck the second floor of her house, my mother saw a fireball roll down the steps and then turn and float out the front door.

Ball pair climbs stairs, explodes. During an early evening rainstorm, two boys were upstairs. The older boy sensed something beyond the bedroom door and looked out. Halfway up the stairs, two bright orange and almost-fiery ten-inch fireballs floated silently in mid-air a yard above the stairs. One fireball was 16 inches ahead of the other, and both floated up the stairs at a slow rate. The leader exploded first, quickly followed by the other. There was neither odor, nor ashes or burns. The downstairs and front doors were both closed. The parents returned later. The dad inspected the house and found nothing amiss.

Bounces On Streets, Off Trees

Classical Newtonian physics predicts the speed and height of bouncing and impulse; and, if enough data could be obtained on each bouncing fireball, we could make calculations. Oddly, although a few balls lose height when they bounce, most do not. Some bounce down the road better than any basketball. Yet others bounce down an incline, losing height like any ball would, with each bounce, even when they bounce off grass. Few if any balls bounce like a beach ball does—slowly downward with weak bounces. Instead, the balls mysteriously bounce as if they are solid.

In some cases, fireballs remain in place and bounce vertically without losing height. Whether all bouncers follow a parabolic trajectory is important. Basketballs curve through the air in a simple arc—a parabolic trajectory—that is guided by gravity, modified slightly by air resistance, and predicted mathematically by Newton's laws of dynamics. However, most fireballs are definitely not freely falling masses subject solely to buoyancy and gravity—so something else is involved.

Witnesses do not report bouncers that make bouncing noises, as does a basketball or even a balloon. Fireballs seem to bounce silently. It is no surprise that since they bounce off the floor or ground, they occasionally bounce off trees. When this occurs, though, they may emit sparks and suddenly lose their typical careful navigation skills and randomly bump into nearby trees, emitting more sparks with each collision, as if they suddenly started flying blind.

Orange ball bounces 150 feet. [23 Tirpak] In Paterson, New Jersey, during a heavy downpour a 20-inch glowing orange fireball struck the center of the pavement on Ellison Street just 30 feet from me and bounced three or four times (about four-feet high after the first bounce) down the street, traveling 10 to 15 feet with each bounce. It then rolled down the pavement for 25 feet and vanished, traveling a total distance of 150 feet.

Author—The distances add up to 75 feet, not 150. The report did not describe its origin, and the ball probably was visible before it struck the pavement. Note that it bounced four large bounces, then rolled.

Highway blue ball bounces eight. [107 Potts 1940] Walking along a concrete highway at night, I saw a lightning flash 80 yards ahead of me, followed by a blue-white fireball bouncing along the edge of the lightning in the same direction I was walking, but away from me at a slow rate, about ten miles per hour. Gradually changing size, the basketball-sized fireball hit the ground at five-foot intervals, bouncing about eight times up into the air two feet each time, and lasting a couple seconds before instantly vanishing.

Bounces tree to tree. [389A Northcutt] Lasting 20 seconds, this two-foot light green-and-orange fireball bounced from tree to tree.

Bounces off hickory tree. [2340A Coffman 1948] In the spring I was sitting on my front porch. Suddenly a thunderclap and lightning struck simultaneously. The lightning struck a larger hickory tree and a two-foot fireball bounced off the tree and traveled about 125 feet and disintegrated.

Randomly bounces off objects, repelled by others. [572 Lyon] My grandmother, now deceased, once saw ball lighting. The volleyball-sized fireball entered her bedroom window and in a few seconds—after randomly bouncing off several solid objects, but repelled by others—it passed out through the window in the other wall of her corner bedroom. The fireball did no damage, but lightning, which severely damaged a tree outside the window, preceded it.

Bounces like balloon. In Virginia's Blue Ridge Mountains, lightning struck a rock outcropping and a yellow fireball bounced up in a motion reminiscent of a balloon whacked from below, and then slowly descended to the rocks, bounced up and then flew into the nearby trees. This comparison is good but inaccurate. How can a toy balloon bounce up that high? I have futilely attempted to duplicate the motions, to no avail. A balloon is too light, and air resistance too great for it to bounce far.

Dull balloon sisses, bounces, and pops. [532] After coming in through an open door, this volleyball-sized fireball sissed [sic] when it bounced for 15 seconds and then burst like a white, bluish green toy balloon—only louder. We saw no mark at all where it touched and sissed. Not bright, it would not have been visible in bright light. Ozone odor lingered after it burst. We saw it during July in 1921, in Western Nebraska.

Four-foot yellow ball jumps up and down in street. [507 Jonakin] When lightning destroyed a chestnut telephone pole in front of my home, a four-foot bright yellow ball of fire bounced up and down on the surface of an oil-coated street for several seconds and then disappeared.

Shrinks with each bounce. [462 Edmonds] On one occasion, apparently following air currents through a window in an adjacent room, a basketball-sized fireball flew to the ground outside and bounced several times, then disappeared. It diminished in size on every bounce on the ground. The purple ball disappeared in a red flash.

Car Hood Rollers

As is well witnessed, fireballs do bounce and roll on both floors and pavement. And so it is no surprise that they would also bounce over, and roll along other odd surfaces—including car roofs and hoods.

Car hood roller. [307 Barringer] While driving with a passenger near woods, lightning struck. I heard a slight click, which occurs when the stepped leader connects with a rising discharge that was followed instantly by a deafening crash from nearby lightning. Simultaneously a two-foot bright orange-to-yellow fireball appeared on the center of my hood and rolled off the front right fender and in two seconds disappeared towards the ground. There was no damage.

Dazzler touches car hood. [95 Kertesz, Building 4500] While driving at night after a heavy rainfall, a one-foot, bright bluish-white and slightly blinding fireball seemed to touch the hood of the car. After ten seconds, it slowly decayed. There was no size change nor any associated lightning.

Globe of fire hugs auto. During an electrical storm, a whitish blue basketball-sized globe of fire slowly floated in at four feet above the ground, into the rear bumper of a parked car and hugging it, rapidly went up over the hood and out the front bumper, and floated slowly away. It lasted one minute. Another time, a fireball came through the window screen and bounced off the refrigerator and other objects before it bounced out the back door.

Descends Utility Poles, Trees

Since fireballs float above horizontal surfaces, sometimes rolling or sliding along them, it is not that surprising some ascend and descend trees, poles and towers—sometimes attached to surfaces, but often only near the surface.

Follows line, descends pole. [72 Kirby] The fireball came from the center of an overhead power line span and followed the line to a pole, where it followed the pole down to the ground.

Flaming pole rolls white ball. [335A Westbrook 1940] My brother, sisters and I visited friends who lived a quarter mile from our house in Mississippi. Summers here are usually stinking hot, dry and dusty. As the summer afternoon progressed, clouds rolled, and skies darkened, accompanied by great wind, thunder and lightning. We ran for home. It began to sprinkle. We almost made it when thunder crashed and the light pole in front of my house burst into flame and smoke. Out of it rolled a whitish-orange fireball, which rapidly grew smaller until it disappeared. Ball lightning frequently appears in front of my house. The balls follow a telephone line. They also affect the entrance power line and activate the thorite electrostatic arrester connected to the power line.

Skips along wire, descends pole into ground. [451 William T. Martin, Jr. Building 7500] Occurring after a lightning flash, a six-inch fireball skipped along on an electric wire and was airborne part of the time. It followed the wire to an electric pole, quietly went down it into the ground. It did not change size.

Descends pole, hits pavement, rolls. [552] A softball-sized yellowish-orange fireball followed conductors, remaining in contact with the surfaces. The ball came down a telephone pole, hit the wet pavement, and rolled out onto the street for 15 feet, then suddenly disappeared. This occurred during a severe electrical storm in a heavy downpour.

Author—It is odd that even dazzling fireballs do not sizzle when they roll on wet pavement. Yet fireballs that fell into barrels of water and remained submerged and glowing do give off steam and boil off considerable water [C.V. Boyes].

Water tank balls descend and parade across stove. [553] I lived in a house that attracted lightning, frequently burning out lights, with crackling and other noises. One day lightning hit, and I looked into the kitchen and saw three five-inch yellow fireballs in a row roll from the top of the water tank, down across the range top, lasting ten seconds. During another storm, two girls saw a yellow-orange fireball roll along the foot of their wooden bed. A light bulb near the foot of their bed burned out.

Rolls down tree, leaps to ground. [209A Cottrell] One stormy afternoon in 1930 we baseball players sat on the porch of Grant-Hall in Harrogate, Tennessee (about thirty miles north of Oak Ridge), waiting for the intense lightning storm to pass. I was staring at a large, tall oak about a hundred yards away that bore lightning scars. Suddenly lightning hit a top branch and streaked downward. A yellow basketball-sized fireball rolled down the tree trunk. At ten feet above ground, it leapt out and down at a 45-degree angle. It disappeared as though it had gone into the ground. A blue haze drifted from the tree, and a fresh scar marred an upper branch.

Falls from tree, slowly bounces, fades. [190 JW Hurd] I stood 200 feet from a tree. Lightning hit it. An eight-inch red fireball fell from the tree and slowly bounced away, not quite touching the ground, slowly growing smaller until it vanished. This could not be burning pitch vapor because it slowly bounced away, not touching ground. Another time, I also saw several small balls (three inches and smaller) hanging on a barb-wire fence, but lightning did not strike nearby.

Branch tip fireball rolls down trunk. [544A Dr. Kouvenhoven, MD] I am a professor of Surgery and affiliated with Johns Hopkins Hospital in Baltimore, Maryland, and submitted this report to Oak Ridge National Laboratory at the request of Ruark. While in grammar school in the Flatwoods Borough of Brooklyn, we watched a thunderstorm through the school window. An orange fireball, a few inches in diameter, appeared in the outer limb of a tree, rolled slowly down the limb to the trunk, down the trunk to the ground, and disappeared.

Hesitant lightning, dead chestnut tree, yellow fireball. [485B Moretz, Jr.] Lightning struck a dead chestnut tree, but doing less damage than to other trees I have seen hit. The lightning stopped or was hesitant at a point of the tree, ending in a deep yellow fireball, which appeared in a fork of some of the larger branches.

A dozen balls fall from tree. In a rainless thunderstorm and amidst a great wind, a series of over a dozen two-inch white-blue fireballs each of the intensity of a 200-watt lamp materialized in a tree and fell toward the ground but vanished first, each lasting only two seconds and covering 20 feet.

Buildings and Bridges

Because fireballs float up and down poles, along drain spouts and gutters, it is easy to appear on, and fly up and down, buildings and bridges.

Fireballs on Empire State Building. Technical observers who at 500 5th Avenue were watching the Empire State Building during the summer of 1937 as part of the lightning study saw ball lightning four times. When one of the engineers, who later became the chief technical executive of a large power company, saw a blue fireball slowly descend the 38-foot tower atop the skyscraper, he discovered that several others also saw fireballs on this skyscraper during storms. But all of their observations were omitted from the technical reports. The cameras and oscillographs had not detected the fireballs.

Since the fireballs officially did not exist, the engineers were told that the sightings could not officially be reported. Not surprisingly, fireballs have been reported on other tall structures, including the Eiffel Tower and even major bridges. Understandably, fear of adverse publicity, indemnification difficulties, and respect for negligent (and strict?) tort liability discourages reporting or acknowledging such fireball appearances.

Rare Drifters

Although most fireballs travel in controlled flight against and independent of even powerful winds, a few balls drift with a weak breeze or even float lazily. Some drift or float and then suddenly become controlled. None are known to be blown along by a strong wind.

Drifts with air current. [90 Lamphere] After lightning, I saw a two-foot fireball that quickly changed size, made sounds and lasted ten seconds before it suddenly vanished. It was airborne all this time and drifted with the air current. Surface tension appears to be more important to maintaining stability of the ball, but disruption results from changes occurring within the volume.

Small balls brighter, heavier; trio sinks to earth. [400 Sanders, 4802 Inskip Dr., Knoxville 18, TN MY1-4218 on 7/23/61 and 1953 sighting at ORNL Y-12 plant] During an electrical storm near the nuclear plant and prior to a downpour, a ten-inch deep orange fireball crossed from the top of one hill to near the foot of another some thousand feet away. A high-tension power line was over its path with towers on top of either hill. The ball followed directly below the power line and landed in a wooded area. There was no sound. The fireball seemed to float, without spinning, as if its density was low. As it approached the wooded area, it began to fall more rapidly and landed without any noise, smoke or fire. After six seconds, it was almost disintegrated by the time it landed.

The second fireball was different. Near the nuclear plant, I saw a four-inch fireball started from near a power pole and rapidly descend at a steep angle, about 45 degrees. It seemed denser and hotter than the earlier ball.

All three fireballs had several qualities in common—the larger fireballs were lighter or less dense and the smaller ones, brighter and shorter-lived. All were airborne near and parallel to the high-voltage line. They traveled in a north-to-south direction. The largest floated farthest and much more horizontally. Smaller ones, grapefruit-sized, were bright, shorter-lived, fell faster, and seemed denser. All sought the earth. There was no damage, noise, smoke, fire or fragmentation when they reached the earth—they simply disappeared.

Follows draft down hall. Before a storm arrived, an 18-inch blue fireball quickly floated down a hallway in the building of a military base. It seemed to follow a wind draft, but faster, before it left through an open window.

Flies Against Wind

Flying against a breeze or wind is commonly reported. Why and how? Flying against the wind takes energy. For a fireball without any apparent means of propulsion, this is a mystery, and was why scientists doubted ball lightning. According to existing physics, there is absolutely no explanation for how fireballs can fly, let alone against the wind.

Ball flies 20 mph against gale wind. [Spencer] It was 9:30 a.m. during a severe June storm in 1963. I am an engineer, an MIT graduate and a licensed pilot. I now live in Massachusetts but in 1963 was living on Phoenix Drive in Fort Worth, Texas. The house stood in the residential section, but had adjoining fields. We stood at the East window watching the hailstorm. A severe June storm struck at 9:30 am. It was a frontal storm, with an alleged tornado that turned over trucks half a mile away, blowing out car windows, and damaging our house. Noise was intense.

I didn't see the fireball materialize, the rain and hail were intense, and we could see little—just a bright-but-not-dazzling white fireball. I think it would have been dazzling in absence of the storm. At first I thought a power line blew down. The fireball was near a cotton wood tree. After the storm, I went out and found no damage. Then I realized that the fireball's drift had been to the south, against the wind.

Although I cannot say for certain it was a sphere, it seemed two feet in diameter. We were 30 feet away from it. It fell to ground from fifty feet up at a falling speed of 20 miles per hour in a curved path, flying against an intense wind. I have heard eyewitnesses describe ball lightning rolling in the Colorado mountains..

Floats like feather in gale. A storm lashed Wellfleet Harbor on the northeastern shore of Cape Cod in Massachusetts. Rain slashed. Wind raged. Two bolts of lightning flashed briefly brilliant white. Brilliant orange, a third flash lasted longer and struck the top of a telephone pole, but left in the air a "dark shadow" of itself. A glowing orange fireball atop the pole floated slowly off and drifted downward, and after ten seconds it vanished in a loud bang, just like a shotgun blast. The witness, a police officer, reported no damage.

Army jeep's fireball ignores fierce wind. Rain was imminent. To the right of a military jeep in a convoy, a bright orange ten-inch fireball the color of a sodium lamp appeared between an opening in the wire fence and then floated parallel to the jeep, staying twenty feet from it, about five feet above the ground flying at thirty miles per hour. It glided deliberately despite a fierce wind and vanished after fifteen seconds.

Fireball and remnants ignore gale wind. Stanley Singer in case 171, on page 43, of *Ball Lightning* reported a case in 1949 from a West German encounter. Despite a powerful gale, a dazzling white fireball floated in the air ten feet above the ground and 40 feet from a house. The size of a full moon, it emitted rays. The ball broke up, leaving a glowing crescent, like a new moon, but curved downward. A vibrant discharge shot to ground and a spark in the air. Still floating at ten feet in the air and impervious to the gale wind, the original fireball or its remnant emitted many one-foot red-yellow sparks. This phase lasted four seconds. A fist-sized glowing mass hit the earth and sprayed out red sparks.

Author—In a hurricane-like wind, the original fireball—and note this—its glowing remnants floated in the air as if there was no wind. This is no exception. There are other cases where the sparks and remnants of a disintegrating fireball were unaffected by a strong wind.

After contemplating the enigmatic paradoxes to physics that the prior cases present, you might not be as surprised to discover that fireballs appear aboard airliners and that many pilots and crewmembers have seen them or know others who witnessed fireballs aboard airliners.

4

Airplane Crashes

Is there a legal veil of silence—quasi-censorship when airliners encounter ball lightning? Probably not, but no one knows the answer. Lightning balls are seen aboard or flying beside aircraft more often than airline officials and aircraft manufacturers care to discover, want to know, or wish to acknowledge—all for understandable reasons of consequential restrictive legislation, adverse publicity, and tort liability.

From a legal perspective, reasonable ignorance of facts with proximate causality cannot lead to a tort of intent, which is difficult to prove, because no one readily can prove the *mens rea,* or mental intent of the airline executives, or that they did know or should have known, without express evidence to the contrary. If the executives do know—and I suspect that some secretly do—they want no fact pattern to ever emerge later in depositions propounded from the plaintiffs to the defendants that they did know.

However, if they by their action or inaction, and benign suppression of research and their failure to order all personnel to report and document every case, they personally become liable, and insurance firms may refuse to later indemnify. I hope that this book will create publicity that alerts scientists, the press, legislators and lawyers of this legal need for accurate reporting. If one airline disaster is averted because of this in the next decade, our effort will be worth their saved lives.

Unfortunately, today, benign quasi-censorship plagues the industry. A pilot recently related an interesting story. He works for a major airline. Inside the aircraft he was flying, a fireball appeared and floated past a screaming stewardess and many terrified passengers. The airliner made an emergency landing, but there was no damage. He reported the event. That major airline apparently downplayed the incident, and no news stories ever appeared. From conversations with many airline pilots and crew, he discovered that many airline flight personnel have witnessed the phenomena aboard or near their craft or been aboard and knew of the event.

And if wealthy nations that sponsor global terrorism ever learn how to artificially generate and intensify—or develop weapons based on the principles—of fireballs that reliably endanger airliners, then there will be hell to pay.

There are many cases—some included in this book—where fireballs destroyed semiconductors—microprocessors, LEDs, triacs, SCRs, and power transistors—devices so critical to airliners.

However, from a physics perspective, these encounters present a serious paradox. When fireballs appear close to an aircraft, it is equivalent to such balls flying—without deformation or distortion—against a 400 mile-per-hour wind. This proves unusual confinement, control or navigation, and propulsion. Can you adequately explain this? Neither can the leading physicists. And at least one leading airline manufacturer wanted to know more.

Major Airline Admits Danger

While an electronics journal editor, your author received a call in early March of 1977 from a scientist, Dr. B, with a major aircraft maker. Assigned to stop fireballs from appearing inside their airliners, this man read everything on the subject and came across a technical article in a journal that I wrote on ball lightning. I told him that metal-screened boxes and Faraday Cages will not stop them, and that they squeeze through keyholes and often ignore ground wires. Excerpts from his taped conversation went roughly as follows.

"I talked to a colleague," said Dr. B, "and he told me you wrote an article in *EDN* on ball lightning. Last October, ball lightning came in through a double window into the bar on the deck of a large airliner flying near a lightning storm. This is not the first time. I can't mention the name of this major airline, but I can tell you some details of the story. Saint Elmo's Fire was running in front on the window. Electrostatic charge is bled off, but this doesn't work in all cases, but it's harmless. Ball lightning entered the left-hand side passenger cabin window number three, floated across the cabin at about head level, and merged into the right window, either nine or ten.

I don't know how familiar you are with this aircraft. [Dr. B goes into a detailed technical description.] We are attempting to stop it next time and want to know if these balls penetrate plastic or metal. We are considering making all windows highly conductive."

I told him that this would not work—some fireballs slip through metal screens. But sometimes they melt a hole in metal, plastic and glass, but at other times they slip through without damage. However, to my knowledge, to this day, this manufacturer did nothing to mandate reporting encounters or to stop fireballs from forming aboard or near its airliners. And I suspect that the insurance indemnifiers, the manufacturers, and the airlines kept this subject verboten and out of the news and away from plaintiff's lawyers. I cannot prove it is a conspiracy of express silence and tort of intent, rather than an act of benign neglect and legal caution. Since then, several unexplained airline crashes could be attributed to fireballs, but the plaintiffs' law firms may never discover it.

Too many airline disasters, such as the Boeing 747 TWA Flight 800 on June 17, 1997, occur that are never explained. It blew up minutes after leaving Kennedy International Airport for Paris, killing all 300. After NTSB and FBI investigations, there are theories. While the cause may still be a mystery, and ball lightning probably was not involved, it is a risk to jet fuel.

The Leach case (described elsewhere) proves that exploding wires can instantly melt a hole in metal can also spit out a hot fireball. There is also some evidence that fireballs form and fly in vacuums. If ball lightning forms inside airplane fuel tanks and cannot ignite jet fuel in the presence of inert gases, then the jet fuel will squirt out through the hole and explode in the presence of oxygen.

For that reason, I authored a paper suggesting the FAA to legally require that all aviation cases be reported. I took law classes and knew legal implications, as would the airlines and their attorneys.

After brainstorming and coming up with a legal angle (I took many law classes), I wrote the following paper in 1999 at the request of Golka. He then gave it to Dr. Stanley Singer, who read it at the Fifth International Symposium On Ball Lightning held on August 26-29, 1997 in Tsugawa Toun, Nilgata Prefecture, Japan. Chairperson was Professor Yoshi-Hiko Ohtsuki. Mr. Hideho Ofuruton, Tokyo Metropolitan College of Engineering. 52-1, Minamisenju S. Akrakawa-ku, Tokyo 116, Japan. This brief paper warns of serious safety and legal issues, as follows.

"It is recommended that the Federal Aviation Agency and all aircraft manufacturers, airlines, and regulatory bodies mandate and legally require that all pilots, radar operators, and all military, private, commercial and airline personnel to formally report all observations of ball lightning, fireballs, and related phenomena near or inside aircraft.

The electrical nature, high energy content, explosive action, life span (up to 20 seconds) and high temperatures exhibited by ball lightning or fireballs make it a potential danger to the structures, engineers, jet fuel and electronics aboard aircraft.

Ball lightning researcher Golka interviewed scores of pilots and airline personnel. Approximately 40 percent were aware of ball lighting and six percent had witnessed it at least aboard or very close to their aircraft. Historical cases of encounters inside or near their aircraft have appeared in the scientific literature since the mid-1950s. Some external collisions have caused verified damage. In the past at least one major airline manufacturer (and there are reports of others) had assigned a scientist to research and stop ball lighting that formed aboard its airliners. It was alleged that this manufacturer was eager to cover up reports aboard its aircraft.

If in the future, ball lightning can be reasonably linked in civil action to the cause of airliner crashes, damage and personal injury, and airlines and manufacturers exhibited negligence or intent in failing to investigate this potentially dangerous phenomena, then exposure to civil torts, exists. Intent also could constitute *ultra vires* acts, permitting plaintiffs, and even allow criminal prosecutors, to pierce the corporate veil, thereby exposing corporate officers to personal liability and possible criminal prosecution in the case of an airliner crash. In such case insurance indemnifiers may not wish to protect the airliners or its officers. The Justice Department and individual states could then seek awards and punitive damages.

The following characteristics make this a potentially dangerous phenomena for aircraft: unpredictability, explosive behavior of some fireballs, intense heat, high energy density and content, apparent stable-plasma mechanism of unknown nature, ability to penetrate or materialize inside Faraday Cage structures, apparent unknown propulsion physics, and electrical nature. Fireballs can ignore or merge into conductors to generate intense electrical surges.

To fly, fuel-laden aircraft have become very dependent upon semiconductor-based electronics and fly-by-wire, which are protected from lightning, but not from ball lightning."

Terrorist Implications

The above paper that your author wrote caused quite a stir. However, unfortunately nothing came of it—and so the dangers continue. Today, these dangers worsen. If nations that sponsor terrorism ever discover how to create airliner ball lightning, there will be hell to pay. Since no one knows how to create fireballs, nor can they explain how it "defies" gravity in flying beside airliners and how it enters them, nor how the balls can explode so violently, then there is potential for dangerous encounters with global terrorist nations, with their unlimited financing. There are many cases that suggest the dangers to aircraft.

Dangerous Aircraft Encounters

Potentially dangerous encounters with fireballs occur not only near and above storms, but also on clear weather days. The following mix of cases is both new and historical. Many such cases exist, but most go unreported.

"RF balls" fly within USAF tankers. While a circuits journal editor, your author visited firms and heard interesting stories. An electrical engineer, Tim served in the Air Force and now was a manager at a micro-networking firm in Worcester, Massachusetts. He described how he saw a two-foot orange fireball aboard a huge jet tanker carrying jet fuel, used for mid-air refueling of other aircraft by means of a boom and fuel line. This fireball floated out to the rear of the aircraft and off into the air. This thoroughly terrified the crew because they were carrying a full load of volatile jet fuel.

Other Air Force crewmen also described fireballs. Robert C, a former director of engineering instruction at Centronics—the once-world-famous computer printer-manufacturer in Nashua, New Hampshire—previously served in the Air Force. In 1977, he described to your author the fireballs seen by air force crews flying jets and tankers. He himself saw a fireball roll on a New York hill.

Red ball on USAF flight. [309 Lyell] I am now with Oak Ridge National Laboratories. While in the Air Force, during a flight, I saw a red-orange fireball on a flight. It followed conductors for several seconds, before suddenly vanishing.

Inter-storm fireball briefly visits military flight. [145 Bowen 1946] While flying between lightning storms, I saw a blue-white fireball float through the aircraft. It disappeared after four seconds. Another crewmember also saw it.

USAF and Saint Andrew's Light. [471 Alnut] I saw what we in the Air Force referred to as "Saint Andrew's Light," or a glowing fireball moving at a constant speed at high altitude. There was no concurrent lightning.

Ball chases flight attendant. While working the day shift—there was considerable thunderstorm activity—meteorologist-in-charge of the National Weather Service Office in Rochester, NY, Peter R. Chastor, was meeting an acquaintance who was a commercial airline pilot, who just had a disturbing encounter. While the airliner was on decent to the airport during the storm, a basketball-sized "ball of sparks" entered through the engine intake. After moving into the fuselage, it chased a screaming flight attendant down the aisle. It vanished before it struck her. This upset the passengers.

Author—How did this fireball survive being sucked into a hot, spinning jet turbine? Did it morph? This is really weird.

Orange ball explodes, rocks airliner. A commercial TWA airliner at two-miles altitude flying from Paris to Cairo passed through a cloudy region. Startled passengers watched as a softball-sized ball of orange-yellow fire rose from under the cabin. The bright rotating ball, surrounded by a thin bluish mist, left a small trail as it moved next to the fuselage. It was only a few feet from the window as it flew alongside the plane. The passengers watched the fireball when it burst into a giant red "spray" that momentarily shot forward for ten feet and then silently vanished. A loud explosion rocked the plane.

Fireball burns captain, explodes. The former Deputy Director of the British Meteorological Office was flying during summer to Iraq through dense nimbostratus clouds over the Toulouse Gap at 8,500 feet. A fireball popped through a cockpit window, burned the terrified captain, and glided through the forward cabin. The ball then entered the rear cabin, and while floating down the aisle, it exploded.

Fireball damages transport. A twin-engine Russian transport plane on August twelfth at 12:30 flew near Nizh, Tamboske, 55 miles northeast of Komsomol'sk and 400 miles southeast of Moscow. Dense cumulus and cumulonimbus clouds filled the sky. Pilots Dubinski and Sergienko had seen several showers and storms. At 12:45 they flew into a heavy dark cloud. A dark red, almost orange, ten-inch fireball shot straight for the port side of the cockpit. The fireball came within one meter of the nose, swerved to the port, and circled the running light. It hit the engine, exploded in a blinding white flash, and a flaming band passed along the left side of the fuselage. The explosion sounded like "the explosion of a torpedo in water, muffled and sharp." After landing, the ground crew discovered several melted rivets and two one-inch holes punched into the elevator's trailing edge.

Fireball vaporizes fuselage holes. In 1984, the Russian news agency reported that a four-inch fireball entered a Russian passenger airliner, flew above the passengers' heads and departed out the tail section, leaving two holes in the craft. It appeared on the fuselage, at the front of the cockpit. The ball disappeared with a prodigious roar, only to re-emerge seconds later, piercing through an airtight metal wall, entering the passengers' lounge. After it flew about the passengers' heads, it entered the tail section and split into "two glowing crescents." These two then joined and silently departed the plane. US National Weather meteorologist Peter R. Chaston noted that mechanics later discovered two fuselage holes at entry and exit.

Not only can fireballs squeeze through existing tiny apertures, such as skeleton keyholes, they can vaporize their own holes of entry and departure in metal. Ball lightning has floated down the aisle of commercial airliners and penetrated sheet metal and exited through the metal skin and then bounced outside along the wing, despite the tremendous air resistance, and then flew off.

Your author examined under a microscope and photographed a hole melted through a metal sheet by a fireball, and from the heat of vaporization, then calculated the energy (described elsewhere in the Lawrence Leach case) needed for such metal vaporization.

Blue fireball floats down aisle. It was five minutes past midnight early on March 18th, when the flight out of New York to Washington was struck by lightning. The entire airliner was engulfed in a brilliant flash, and the crash awoke sleeping passengers. Seated up front, a scientist from the University of Kent Electronics Laboratory in Canterbury of Kent in England described what the passengers saw.

After a glowing eight-inch sphere appeared at the pilot's cabin, the bluish-white fireball floated down the aisle, past—passing 16 inches from the scientist—and continued on a level path of two-and-a-half feet above the floor at a constant speed of three-to-five feet per second, to the plane's rear and silently vanished. Although the fireball came close to several passengers, none felt any heat. With an almost-solid appearance, the ball was optically thick and in perfect equilibrium.

Orange fireball hits jet's nose, destroys electronics. A jet trainer preparing to land at Moody Air Force Base in Georgia encountered lightning. To avoid the storm, they were told to proceed to Mobile, Alabama. When they rolled out onto a westerly heading at 13,120 feet, a big orange ball of fire hit the nose head-on. An explosion shook the craft with such violence the pilots were sure it was a mid-air collision. The ground crew discovered the electronics had melted.

Fireball injures passengers. In a well-publicized encounter, an airliner flying from New York to San Juan, Puerto Rico, was flying off the Florida coast near Jacksonville when the pilots saw a "big fireball advancing with tremendous speed." To avoid a collision, the captain shot his airliner upwards at a steep angle, injuring a stewardess and several passengers.

Bright ball slowly floats down aisle. Flying at 16,000 feet altitude, a Fair-child Metroliner III was hit by lightning on its right front nose baggage access door, did some other damage, and vaporized three pinholes on the nose baggage door. A bright fireball flew down the aisle and popped out. The vaporized metal, if that was the source—the pilot thought this unlikely—had to enter a pressurized area and pass through the sealed pressure bulkhead. But there was no evidence of this.

Fireball visits president's plane. The late James Tuck, a famous fusion physicist and ball lightning authority, formerly of Los Alamos—and friend of Golka—while speaking at MIT, stated that Harry Truman's Secretary of State Dean Acheson saw ball lightning on the President's plane while in flight, and that the fireball floated over the President's dinner table. Truman inherited FDR's plane but upgraded to a 1947 Douglas DC-6, with its four-prop, reversible-pitch propellers, with radar, and painted it to resemble an eagle, complete with a white beak up front and blue tail feathers, and named it after his Missouri hometown—*Independence*. Tuck wrote that he read this in Acheson's book, *Present At The Creation*.

B-24s generate fireballs. [Brown] I am an electrical engineer and former supervisor of the Component Part Engineering Section at Sanders in New Hampshire. I generated ball lightning while in World War II. I was flying at 1,100 feet altitude in a B-24 Liberator between Taiwan and Luzon P.I. over the North Pacific at AP0245 in 1945 during mid-day.

Carrying ten machine guns, this large—110-foot wingspan—four-engine B-24 was a low-level bomber with a speed of 290 miles per hour, a 2,100-mile range, and a 28,000-foot ceiling. When a B-24 builds up static charge by friction, this ruined our communications. So I would discharge our B-24 by shorting the plane's skin to a 300-foot trailing wire antenna. Doing this, I generated a three-foot round ball of red-orange fire.

The ball materialized ten feet from me, rolled down the antenna, and floated off into the sky. We moved away from it at over 100 miles per hour and saw it for 40 seconds until it got two miles behind us. The ball gave off no smoke, but did appear to roll and pulsate. It was dazzling, so much so that we could still see the ball from a mile or more. I accidentally generated such fireballs on other occasions in the same way while on other bombing runs.

Our tail gunner Larson from Pinge, Indiana and our waist gunner Sterken from Pottstown in Pennsylvania witnessed the fireballs. This is not the first time I have witnessed ball lightning, since when children we saw them in Oklahoma often when twisters were nearby.

Author—Trailing wire antenna has capacitance and the metal skin of the aircraft, partly through friction, acquires an electrostatic charge that was bled off through the trailing wire. Gauss' Law describes mathematically how the charges are spread out over the skin. From the aircraft dimensions, and assuming the sudden contact of the wire to the skin created an RLC series resonant equivalent, your author calculated a damped sinusoid of 2.62 MHz.

Fireballs damage navy planes. [570 Martin] The Navy has records of fireballs appearing on aircraft flying through precipitation. I am familiar with one incident in which a fireball appeared in the Plexiglas nose dome of an aircraft, leaving it at the tail. It did some damage to radios and antennas. I can obtain the aircraft number, date of incident and names of pilot and co-pilot.

Exploding fireball hits tail fin. When Minneapolis International Airport closed for a lightning storm, on June 11 of 2001 at 6:30 p.m., numerous passengers saw a yellow fireball appear atop the tail fin of a Northwest airliner that was docked at Gate G16. It floated down to the fin and exploded loudly in a shower of sparks.

Fireball gets in the face of a Boeing 757 pilot. Lightning hit a Boeing 757, flying into the Dallas/Forth Worth airport, outside the copilot's window, burned each door hinge on the right side and departed from the right aft tip of the horizontal tail. It also disabled the automatic brake system. A bluish-gray basketball-sized fireball appeared 18 inches before his face, hovering motionless for half a minute. Unable to view the instrument panel, he twisted his body to see around the glowing sphere, which slowly and noiselessly faded.

Exploding red fireball damages military aircraft. A large military research aircraft trailed (from its reel) a five-mile-long wire antenna out the rear. Into it a transmitter fed 10 kW at the rather low frequency of 50 kHz. Running for 23 meters through the center of the fuselage, a one-meter-wide passageway was lined on both sides with equipment racks of steel. This ran to the cockpit, which was separated by a closed and painted metal door. Flying at an altitude of 6,096 meters or 20,000 feet over the ocean, the plane encountered a lightning storm, but continued to transmit.

Then intermittent corona glowed on the antenna reel, which began to loudly hiss. The transmitter automatically short-circuited from an overload, but the hissing continued. But when serious circuit problems started, an engineer threw a "grounding stick" (a heavy conductor sometimes called a "crowbar") across the antenna cable exit.

From the explosion came a deep-red fireball with a four-centimeter core surrounded by a bright corona, all having a 1.5-decimeter total diameter. It detached from the reel and slowly (2.0 to 3.0 dcm/sec) floated between the equipment racks at five feet altitude toward the cockpit. The four engineers safely squeezed their bodies between the racks but felt an electrical effect on their body hair.

Then, when it hit the cockpit door, the fireball loudly exploded without damaging the metal, but burned the door paint in a 4.0-dcm circle. An odor lingered of ozone and burned insulation. The pilots turned and flew back to base. But when they reeled in the one-ton "bomb" attached to the cable, it was gone due to the melted cable and there were many metal weld spots throughout the cable. It failed to damage the electrical systems and the independent tail-mounted generator used in the experiment, but the transmitter was badly damaged. They documented this to military authority.

Passenger sees parabolic dazzler at 80 miles. [568 Overton] I am with Union Carbide Nuclear Company at Oak Ridge. I saw a fireball last night, August 7th of 1961, from the window of an airplane flying in a northeastern direction over Madisonville, Kentucky, at 8:56 P.M.. We were flying in still-clear weather with a star-filled sky and few clouds. On the horizon from east to south, frequent lightning occurred. It was a medium glow and at first I could not see the separate strokes. Thunderstorms were occurring along a wide front across the middle and eastern sections of Tennessee and Kentucky. Most of the plane lights were out and we flew at 3,500 feet.

Out of one of the lightning flashes or glows came a bright point of light. Its trajectory was parallel to the horizon, from south to east. However, its direction could have been from south to north if the it was longer than it appeared. I estimated it was 80 miles away and the arc was subtending an angle of two to eight degrees. The lightning ball lasted a couple of seconds and may have been flying at supersonic speed. The dot or ball was extremely bright and its was not like a shooting star. It had a long gentle curve with an upward twist similar to a long sine curve. It disappeared suddenly.

Orange fireball materializes between pilots, floats down aisle. [561 Sverson] I was a pilot for [deleted] Airlines flying a DC-4 in 1951. Lightning struck near the nose or engine and a fireball materialized between us pilots and rolled slightly toward the bulkhead door separating the cabin and cockpit area. The basketball-sized orange fireball stayed suspended above the floor, rolled in air and disappeared through the bulkhead door after two seconds. Later, after I spoke to people in the cabin, several of them said they had seen the ball roll down the aisle and disappear toward the rear. Other than lightning, there were no other unusual noises.

Cockpit fireball roams aisle. [555 Clarke, Reactor Division; TWA Flight 84 April 4, 1958] I saw a fireball while flying from Albuquerque to Chicago. Halfway to Chicago, lightning struck the airliner and a basketball-sized, red-orange fireball traveled from the cockpit down the center aisle, for five seconds. I felt its heat.

Fireball fries airliner's electronics. [546 Marsh] I flew on a Lockheed Constellation of Eastern Airlines on April 5, 1958. I am Supervisor of AC Variable Speed at Allis Chalmers in Norwood, Ohio. A colleague, Denine of my firm, also saw it. Clarke of the Reactor Division made the report. Over Peoria, Illinois, on a flight from Albuquerque to Chicago, our airline was struck by lightning. An orange fireball rolled down the center aisle. I was sitting in a front seat, so I did not see where the ball went. An Oak Ridge colleague also witnessed it. No one was injured. The lights remained on, but all radio and electronics failed.

Fireball follows aircraft. [141 Sloan] During a flight near a thunderstorm in 1945, our aircraft encountered a fireball, which closely followed the wing. The ball was not associated with any lightning stroke. It caused magnetic and electrical effects. I suggested that other military and commercial flight crews be questioned.

From pilots' quarters, rolls fireball down aisle. [159 Harlan] Our plane flew through heavy rainsqualls and an electrical storm. An orange fireball with bluish tint, between six-to-eight inches in diameter, came from the pilots' quarter, rolled down the aisle floor and in one second disappeared through the tail of the plane.

Follows Leading Edge Of Aircraft Wing. While flying in a monoplane inside a storm cloud—there was no nearby lightning—we saw distant lightning. Rain began. Glowing green fireballs appeared on the leading edge of the starboard wing, close to the aluminum fuselage, then traveled along the front edge to the wing tip, then moved straight back in the air.

Wing-tip fireball explodes. A high-temperature chemist on a Boeing 727 that was landing during a thunderstorm in Dallas said that when lightning struck it, a soccer ball-sized fireball appeared on the wing. After 15 seconds it exploded.

Blue purple fireball traverses P-2V7, does damage. The Plexiglas nose of our Lockheed P-2V7 Neptune, used in anti-submarine patrols, while in flight, built up a large static electrical charge and glowed purple. A tremendous explosion rocked the aircraft and an eight-inch light bluish purple fireball jumped out from the nose wheel well. At a slow walking speed the ball bounced through the plane, set the magnetic gear in the rear afire, went up inside the vertical stabilizer and after its 20-second lifespan, apparently exploded, blowing the rotating beacon off the craft. We aborted the flight and landed.

Once again, here is a sobering thought—If ball lightning can destroy airliners, might terrorists adapt the underlying physical laws to weapons of interest to wealthy governments that sponsor global terrorism?

Klass Hypothesis

Ball lighting explains some UFO sightings. Others disagree. Writing in 1959, noted science writer Arthur C. Clarke in *Challenge of the Spaceship* suggested that ball lightning might explain some UFO sightings; but then in 1968 Clarke wrote on page 341 in *The Promise of Space*, that this theory had "recently been revived, but it can account for only a few cases."

In the late 1960s, mistakenly using potential and voltage interchangeably, aeronautical editor Klass suggested that fireballs were pulled or pushed by aircraft because of electrostatic force. He suggested that flying saucers are caused by ionized air or plasma—which is created by electrical discharges and high-voltage power lines—and that static accumulation of charged air, caused by airliners in flight, creates flying saucers that trail the airliners. The staff of the USAF-sponsored University of Colorado Condon Study, although they rejected flying saucers, also prematurely rejected Klass' explanation for several wrong reasons.

Easily calculated, the Coulomb force is infinitesimally negligible. First, Coulomb's law is an inverse square equation, and attraction or repulsion falls off rapidly with distance. Second, it is impossible to separate the required charges. Third, it is not possible to separate the electrical neutrality of large objects by much. Even our best capacitors are anemic. Fourth, if through some paranormal power such massive charge separation were possible, then the fireballs would ionize air around them violently—creating massive detonations rivaling nuclear explosions—or, ignoring that, they would be attracted strongly to and fly into everything around them, creating great damage and huge, billion-amp current surges. This would require an extraordinary upheaval in physics.

Klass' hypothesis was correct, but as then carelessly postulated, it was a mystery to physics. To his credit, Klass apologized for his error and went on to become a founder and an important UFO investigator for CSICOP.

The late Dr. James McDonald, former senior atmospheric physicist at the University of Arizona, prematurely dismissed the UFO and ball lightning connection because Klass attempted to explain away too many UFOs as ball lightning to the USAF-sponsored University of Colorado Condon Commission. Then, in his book *UFOs? Yes!* the former Condon team member Dr. David Saunders wrote that Klass was so persuasive that some team members would jokingly chant, "Klass dismissed." They were wrong.

Saunders prematurely dismissed Klass' hypothesis, writing that, "It seemed incredibly naïve to me for him to have based such a complex theory on so few cases." Condon team member Dr. Martin Altschuler, a solar physicist at NOAA, and the Condon Project's ball lighting specialist confirmed that ball lighting explains some cases.

However, after three decades, we have better insights. Phillip Klass was correct. Mystery lights and sky-borne, fair-weather fireballs are mis-reported as UFOs. But could monster fireballs cause UFO reports? Yes. From the Oak Ridge cases that your author studied, the answer is affirmative.

Klass is a an aerospace expert, editor and author, a UFO-debunking skeptic, and founding member of the *Committee for the Scientific Investigation of Claims of the Paranormal* (CSICOP). Klass questions the medical and anatomically viability of alleged aliens' body structures, including their thin neck and body musculature. They anatomically cannot exist. He disproved radical claims about Area 51 and MJ-12, questioned crop circles, and proved crashed desert saucerians were frauds.

Since fireballs appear near and within airliners, often traveling at hundreds of miles per hour—and since air at higher altitudes is more subject to ionization—this should make us ask if fireballs occur more within clouds and atop mountains. As we will next see, this may be true because of ionization and greater electrical activity.

5

Clouds and Mountains

Electrical activity is greatest on mountaintops. Rising from the earth, mountain ranges are sharp projections possessing electrical potentials close to that of the surrounding valleys and flatlands. But the air near them is at a much higher potential and is more easily ionized than air in the valleys. Thus the electric field strength or potential gradient in volts per meter will be greatest between the clouds and the mountaintops than to the valleys. These intense electrical effects create electrical phenomenon not experienced at lower altitudes.

Swiss fireballs upstage Oppenheimer. Ball lightning has puzzled eminent physicists. In his book (pp. 286-7), *The Quark And The Jaguar: Adventures In The Simple and Complex*, co-discoverer of quarks and a founder of the Standard Model of Quantum Mechanics, the Nobelist Murray Gell-Mann describes ball lightning, attempts at explanations by eminent physicists and Nobelists (such as Russian Pyotor Kapitza), and how all existing theories of physics cannot explain it.

Gell-Mann writes that around 1951 Harold W. Lewis was delivering a final paper co-authored by Robert Oppenheimer at a seminar. This was Oppenheimer's last research effort in physics (on meson production in proton-proton collisions) before becoming Director of the Institute for Advanced Study at Princeton.

The discussion led Swiss physicist Markus Fierz to describe ball lightning theories and how the Swiss government provided a railroad car and funds to research mountain fireballs in Switzerland. Eminent physicists in the audience began a lively discussion of ball lightning theories, wrote Gell-Mann, but concluded that physics could not explain it.

Pike's peak ball hurts professor. On page 341 of his book, *The Promise of Space* (1968) famed author Arthur C. Clarke described mountain fireballs. "There is evidence that ball lightning is more common at high altitudes than at sea level, and Professor J.S. Haldane is said to have encountered it when he was studying low-pressure physiology. So I once asked his much more famous son [the great geneticist and science popularizer], 'Is it true, Professor, that your father did some work on ball lightning when he was experimenting at the top of Pike's Peak?' 'No,' said John B.S. Haldane, 'ball lightning did some work on my father.' That was all the information I could get."

Alpine mountain fireballs roll down gorges. Lightning authority, Sir Basil Schonland interviewed observers in the Austrian Vorarlberg. In three of the five cases, there were several witnesses. In most cases there was a near-simultaneous lightning flash to ground, often nearby. In three of these Alpine mountain cases, the fireballs rolled down a fissure or gorge. All witnesses rejected St. Elmo's fire or retinal after-effects.

There are several reasons for mountain fireballs. The air is thinner and ionizes more easily. And at high altitudes there is a tendency for exceptionally powerful lightning flashes. Meteorologist Berger's measurements in the Swiss Alps confirmed that exceptionally giant flashes occur on a mountaintop every few years. The largest have very destructive currents exceeding 200,000 amps. By comparison, typical lightning strikes are only 30,000 amps, with 90 percent exceeding 20,000 amps and only 10 percent exceeding 60,000 amps. Reports are few because few witnesses live near mountaintops, and even fewer venture out during mountain lightning storms.

Mountain electricity intense. As one example of these intense electrical effects, two mountain climbers in Ecuador stood atop a high summit, 400 meters in diameter and 6,266 meters altitude. Clouds descended. Distant thunder rumbled. Granular snow pellets or graupel fell. They felt mild electrical shocks and a buzzing in their hands, their aluminum glacier goggles vibrated, their hair stood up. They dived into the snow. It stopped. But each time they raised their heads 20 inches above the snow surface, the frightening phenomena returned. After half an hour, they crawled 400 meters on their bellies and descended 60 meters. By then the graupel and thunder ceased, and they safely stood up.

Swiss fireballs puzzle Nobelist. The late James Tuck, eminent fusion researcher of Los Alamos and founder of Project Sherwood, wrote in a paper delivered at MIT that the Nobelist and former Director of MIT's physics department, Victor Weisskopf, reported two cases of mountain ball lightning at his Swiss cottage, where he would spend some summers. In a letter to your author, Weisskopf wrote that a neighbor living 300 yards away twice saw it. He talked to Golka on many occasions about generating fireballs.

Ten noisy mountain fireballs. When several mountain climbers reached the 3,900-to-4,000 meter altitude of the European Albus, intense electrical activity plagued them. When they raised their ice axes, sparks flew off. They climbed higher to 4,200 meters and took refuge from the storm in a house. A 25-meter-per-second wind gusted to 50 m/s. Looking out that early morning at two, one climber saw various-sized electric tongues of light blue fire hissing from every sharp steel protrusion of a building that lay in ruins just below the site. The taller protrusions had larger flames. Below the wreck, at 4,000 meters, lighting flashed. Ten large basketball-sized balls of orange fire flew randomly, despite the wind, in all directions and hissingly burst. All this they heard above the moaning wind.

Clouds Spawn Divers

No one knows how often fireballs form and behave inside cumulonimbus thunderstorm clouds. The bottom or floor of these clouds is typically 3,000 feet; the top ceiling of their anvil head, 27,000 feet. Few planes fly into these clouds.

How high fireballs occur and how often, and if they are different from ground level balls, is unknown. Observers are more common on the ground, and what occurs inside clouds, or higher, is uncertain. However, there are sightings of fireballs moving between clouds or falling from them and it seems likely that the intense electrical activity would generate many more in-cloud fireballs than anywhere else. What effect, if any that these fireballs have upon thunderstorm clouds remains conjecture.

Falling fireballs sometimes plummet faster than a falling cannonball. Upon reaching ground level, they can break up, level out into a slow horizontal flight, or rapidly bounce back up to the clouds. Occasionally, a falling fireball will split into two or more smaller fireballs that float above or roll on the ground.

Let us look at the atmosphere. Above 260,000 feet or 79.3 km (49.2 miles) in the thermosphere, temperature rises above -130 degrees F and there are tenuous noctilucent clouds made from water vapor that is the byproduct of the breakdown of atmospheric methane, which condenses on tiny meteoric particles. The air pressure drops from 1013 mb at sea level to below a mere 0.01 mb at 80 km.

Inter-cloud leaper over softball game. [463 Stansbury] At about 7:00 P.M. on a July early evening in 1960, I was playing in a softball game at one of the fields along the Oak Ridge Turnpike near the high school. Thunderheads of an approaching storm from the northwest were moving in fast. I saw a ball of fire near the front of the thunderheads almost directly overhead. I estimate the fireball was five feet or more in diameter. The fireball leaped from one section of cloud to another. One other person also saw it. The storm became a bad one with heavy winds.

Ball falls onto field, bounces thrice, winks out. [436 Franklin] A late afternoon storm came after a very hot afternoon: the temperature exceeded 95 degrees. Lightning flashed every 30 seconds. I watched the storm from a shelter and could see half a mile across an open field. The falling glowing white fireball seemed to hit the ground and then lift. It bounced three times and then winked out. The hissing ball traveled 1300 to 2600 feet in five seconds. It was a quarter-mile distant and the diameter of two basketballs.

Cloud drops red fireball at squirrel hunter's feet. [320 McClellan] While on a squirrel hunt in 1948 near my home in southwestern Virginia, light rain began to fall. Lightning flashed. I took shelter near a cleared field of ten acres. After rain stopped, I started across the field. After walking a short distance, I saw a round, red fireball fall fast out of the clouds. The size of a balloon, it hit the ground close in front of me. It left no signs.

Randomly-curving inter-cloud dazzler. [104 JRM] The sky was clear except for one area encompassing a solid angle of about 30 degrees. This area was extremely active and produced a series of "hair net" lightning patterns. An intense white, large fireball moved rapidly along a randomly curving path between two low-cloud formations.

Low skyball illuminates earth. [157 Logan] An orange-and-golden 18-inch fireball moved in a continuous zigzag horizontal direction at 50 feet above the ground, emitting light sufficient to render small objects such as baseball-sized stones and vegetation visible. When it approached the crest of a low hill it vanished with an almost inaudible crackling sound.

Cloud drops silent supersonic fireball into swamp. [16 Long] During a 1926 summer thunderstorm centered half a mile away, I watched frequent cloud-to-ground discharges, when a white fireball descended rapidly to ground in one second. There was no lightning before or during the fireball. It fell behind a hill into a wooded swamp. A glowing trail lingered for less than a second. As it fell, it did not change appearance or size and I heard no sound.

Wheel-on-fire above trees. [124 McNew] It was a late summer evening in 1957. The fireball was in the east from where I live on Watt Bar Lake near Eagle Lodge Boat Dock. It resembled a fire rolling through the air, not far above the trees. Fire seemed to come out all around it.

Red skyball afire in clouds. [165 Vaughn] I am with the ORGDP Barrier Manufacturing Department of Union Carbide Nuclear. On an early 1958 summer evening, returning to Oak Ridge from Knoxville, my family and I were watching an electrical storm and the peculiar lightning flashes. The oddest was following a flash or streak, when a ball of white-to-rosy-red, then diminished in color to a flame, seemed to ignite in a black thunderhead. The fireball seemed to be afire deep in the thunderhead and disappeared slowly.

Swelling orange in-cloud skyball turns pink, vanishes. [333 Martin 1959] While looking into the skies and clouds, at a regular cloud distance, a fireball in the cloud suddenly illuminated to a ball shape, partly orange and yellowish red. It quickly swelled up and suddenly disappeared. Just before disappearing, it moved away from its center and simultaneously turned pink. There was no lightning.

Yellow tails follow two fireballs to earth. [334 Allan] During a night electrical storm, I saw two unusual orange-yellow balls that went down in a perpendicular path, at two miles distant, followed by a yellow trail. They came to earth, splashed and disappeared. I am not sure if the two fireball flashes were simultaneous or if one immediately followed the other. They were about a quarter to half a mile apart. There was no unusual thunder.

Shoots up "like a rocket." [550] I was in Missouri during a storm. A family of fireballs one at a time ran 40 feet from house to cellar. This reddish eight-inch fireball cracked, or appeared to, as I ran—about ten feet in front of me, above waist high. It took off northward, upward at 45 degrees in a straight line like a rocket.

Fireball slowly descends from clouds, explodes. [W.F. Smith of Kent, UK; "Ball Lightning"; *Nature*; July 22, 1880; pg. 167] At nine on the Saturday evening of July 17, sheet lightning flashed and thunder rumbled distantly. At midnight the dense mist lifted. A quarter mile ahead a large fireball slowly descended from the clouds and after a six-second descent, disappeared upon approaching or touching the ground with two slight-but-quick concussions—perhaps an explosion and echo.

Cloud drops ball, rebounds up. [Barnhouse] I am a veteran and auto mechanic living in Braintree, Massachusetts. I saw a fireball fall from the clouds on September 14, 1954 near Boyds, Maryland, where I once lived. I was hunting and crossed an open field at night. I looked up and saw a reddish fireball come down from the clouds, curving to the right at 45 degrees as it fell. When it reached the ground, it whipped out, and bobbed back up into the clouds. It was a thousand feet away and was bright, but not real bright. It never stood still and was moving faster than a man walking. The entire flight lasted 35 seconds, and my father also saw it.

Fireballs are a frightening experience if you don't know what it is. I have seen it twice in my life, both times in Maryland where I would go fishing and hunting quite a bit. I have also seen something else I don't think too many people have seen, except for old timers and late night hunters—wil-o'-the-wisp or jack-o'-lantern. It is a form of loose fire that some say is methane gas oozing up and burning coldly that forms these ghostly lights in the swamps at night. My mother, aunt, and grandfather saw these strange lights.

Yellow-orange skyball explodes. [Wilbur of Pensacola, Florida] I am a Navy Tech Librarian stationed at Pensacola, Florida. During October of 1973, I saw a yellow-orange ball of dazzling light, apparently not a bolide (meteor), move over Route 44 at the Massachusetts Raynham-Middleboro line. I saw a large ball of fire going through the air, so bright that I couldn't help but notice it. It grew bright, and when overhead seemed to slant as it moved to the northwest. Its motion was steady and it disintegrating without exploding.

French sea-clouds hurl inter-cloud fireballs. An intense summer hailstorm with thunder lashed the French coast. The sky cleared, the hot sun came out. Just after sunset, a fast-moving black cloud appeared in the east. As it approached, numerous lightning strokes without sound were seen, and fireballs shot out in every direction. Fireballs of red and yellow fire came from the cloud's center, and moved horizontally. An ordinary hail and lightning storm soon followed.

This is not unusual for this region and other similar areas. Certain storms along the French coast encourage intense storms and fireballs. High-level summer storms affect southern England, the Lowlands, and northern France. In the heat of day, France originates these storms that then move across the English Channel. Gray puffy masses, altocumulus clouds are middle clouds with bases between 6,500 to 23,000 feet. One part of the cloud is darker, with rounded masses or rolls.

The first indications in this region on a sultry morning are easterly winds, near-fog, and altocumulus (alto) clouds. These clouds look like "little castles" (castellanus) with turret-tops and small round, chaotic globules of floccus (resembling wool). This may forecast a thunderstorm, depending on how they move surface wind. These castellanus indicate rising air at cloud level. On such days, these clouds appear on a warm humid morning, which forecasts thunderstorms by late afternoon that form in lines along the wind in this section of coastal France.

If the alto-cloud-level wind crosses the ground wind, thunder will occur. But if the alto-wind stays in the same direction as the surface wind, there will be no thunder or the storm will be delayed into the night, in which case inter-cloud ball lightning is more visible. High temperature and atmospheric moisture create thunderstorms. High surface heat increases evaporation and relative humidity over 75 percent, and raises lower-air buoyancy. These thunderstorms will only last as long as their water supply lasts.

Night skyballs zigzag into ground. In a western house atop a hill that was hit several times each year by lightning, a witness watched four-inch white-blue fuzz-balls zigzag slowly down from the sky and vanish into the ground. Then the fourth fireball moved rapidly horizontally above the earth, followed by a large white fireball double its size. Numerous golf-ball-sized white balls fell.

Multicolored lightning and mother ball spawn zigzag skyballs. [466 Curtis] It was in late summer in 1938 and I was on the river near where the Tennessee Central Railway trestle crosses the Emory River, and was on the Harriman Tennessee side of the river. At this location there is a sharp rise of low hills just across the river and to the southwest. To the north, Walden's Ridge shields Harrinman like a long protective arm and is quite high. The river flows from the north through a break in this ridge.

While we were at the river, a storm blew up from the southwest with many dark clouds that were not connected, that is, there were spaces in-between when lighter-colored clouds showed through. I was with my brother Tom. We decided to go home.

It was comfortably warm despite the moderate wind. Lightning cracked loudly. It looked peculiar because it was not all the same color. We reached the Tennessee Central grading at the Christmas Lumber Company on the Tennessee railway trestle. Approaching the trestle. we looked up at the lightning. There were many bursts from one cloud to another, some which were the familiar heat or sheet lighting that makes no noise.

There were other inter-cloud bursts that varied from bluish green to red and orange. Most of the activity was hidden in the clouds. But one particular burst jumped from a cloud out into the air. It began as a particularly violent burst, blue-green, from a very dark cloud and jumped out into the fringes of a very thin cloud. There it moved more slowly with time until we could be see that they were rapidly moving balls and not a streak of lightning.

The mother bolt came from a dark cloud directly overhead and moved to the northeast at a 60-degree angle and flattened off to a 20-degree angle and was over the area near the junkyard in Harriman, a little to the southeast of it. The balls continued in a course fan-wise nearly parallel with the angle of the fan at the point from which they were ejected spread out like a fan. And from the radius of the fan, a different color group of bolts, perhaps six, jumped out. These red or reddish-orange bolts moved swiftly at first in a random zigzag path and estimated the balls to be up a quarter mile, and maybe slightly higher than Walden's Ridge, but not much, that is, not a hundred yards or so higher.

They changed color more from red to orange as they traveled. They did not all disappear at the same time, but did so a second of each other, and probably two seconds of the time they were beorned [sic, undecipherable handwriting] in the fan-shaped flash. They disappeared in the same fashion that a hot coal in the fireplace hearth wood, by cooling down. Although the mother lightning bolt was followed by a violent crash of thunder, there was little if any noise associated with the balls. They just glowed and later disappeared. The storm did not produce any rain, but later the sky became very dark. After the storm, the 3:00 P.M. afternoon heat of that late August day turned cool.

White tumbles inside orange, rolls to edge, vanishes, repeats. [111 Jones] I am a chemist with the ORNL Thermonuclear Experimental Division. One night during a storm, lightning flashed across the sky from one horizon to the other, a stationary source of light appeared inside a cloud (a mile distant) and lower (200 to 300 feet altitude) than any of the other clouds. The light slowly pulsated during the fifteen seconds I saw it. The orange light was mixed with white, which tumbled inside the orange. The white light rolled from the inside to the outside. Upon reaching the edge, the white would either fade or else was covered up again by the orange light. The orange color faded after about 12 seconds, then the white light faded and went out in the next few seconds. The size seemed to change, perhaps due to the two colors mixing and the light pulsation. A soft rumbling noise that varied in pitch accompanied the light. I am a chemist in thermonuclear research.

Supersonic fireballs or sky streaks. Before an early evening summer thunder storm in 1989, an engineer saw from the west lower door of his home, a round glow with fuzzy and indistinct edges—if a fireball it may have been semi-cloaked by clouds—about nine degrees "diameter" (a fist at arm's length) quickly rose from a 20-degree elevation in the west to beyond overhead and vanish, only to instantly repeat 50 times per minute silently for 20 minutes. There was no sound or rain, no lightning or thunder, and the air was still. After awhile, a regular thunderstorm arrived. With trigonometry, and assuming various cloud floors, he calculated that the phenomenon was supersonic, despite the total silence.

Three supersonic fireballs. Hail, lightning, and a tornado lashed Oklahoma. In a lull, lightning flashed and two white fireballs came from it and shot east, diverging. Another fireball came from the west towards the other two. All three traveled in almost straight, but slightly arcing paths across the sky that left a luminous trail. They traveled at supersonic speeds, but oddly without making any noise.

Christmas cloud displays multi-colored fireballs. Just before nightfall, a large cloud flashed for half an hour: within it, fireballs of blue-to-green and red-to-yellow shot randomly—usually two or more visible at a time.

Big skyball trails white trail. A two-yard-wide bright orange fireball up at 90 yards altitude floated slowly horizontally at a steady speed, with only a slight zig-zagging, for 3,200 yards, trailing white smoke that vanished after several seconds. After one minute, it suddenly vanished.

Floats down and up from a cloud. A bright white-yellow basketball-sized fireball popped out of a cloud bank and floated down to about 12 feet above the ground and floated horizontally, then slowly moved up and down, then ascended back up to the cloud floor.

Green Skyballs

Green is not completely uncommon a color for fireballs, but bright green is rare. So, to see green fireballs flying overhead, and only at certain times and places, is most unusual.

Moonlit green skyball disrupts card players. [7 Gambill] At about 9:30 P.M., three adults and I were playing cards on an evening during the fall of 1955 in a screened-in porch in South Harriman when one of the others sighted the fireball passing over a hill on my right. It was a perfectly-spherical green ball of fire, and the size of a quarter held at arm's length. It silently floated horizontally, with no apparent rotation. It was in our view for ten seconds, and after passing over the shallow valley in which we were located, it disappeared over a hill on my left. The night was warm and clear, with some moonlight. I remember seeing reference made to Soviet work in ball lightning in an aeronautical journal.

Green fireball emits horrific noise, flies fast. [82 McSpadden] A greenish basketball-sized sphere swished by with a horrific noise. It moved randomly in the air completely free of any objects. After 15 seconds, it completely faded very slowly away, accompanied by a railroad train sound. It appeared after a lightning stroke.

Green skyball, clear night. [5 Lewis] The weather was hot and dry, a summer night without clouds, only a haze. The light green fireball moved rapidly for a short distance high in the night sky, vanishing after several seconds. I am with the Thermonuclear Experimental Division at ORNL.

New Mexico's Green Meteors

Green meteors occasionally appear. They cause no alarm. But when they appear the size of a full moon and each night slowly fly across the night sky in straight lines, they aren't meteors. It began in late 1948 in the American Midwest. I first learned of them through an authoritative book, *The Report on Unidentified Flying Objects*, by Captain Edward J. Ruppelt, an aeronautical engineer and the founder of the Air Force's Project Bluebook. Since his account is the best known, my report is totally rewritten and adapted from his book, and with my own additions.

People near Albuquerque, New Mexico, saw the first green fireballs zooming through the clear night skies in late November of 1948. Scientists wrote it off as surplus war flares. It would pass. But days passed and the reports got better. Those green streaks seen low on the horizon now became great balls of green fire. Since there was no publicity, the reports were not psychological.

Finally, the first well-authenticated case on December 5, 1948 received worldwide attention. While flying an Air Force C-47 transport at 18,000 feet ten miles east of Albuquerque at 9:27 P.M., Captain Goede, his copilot and engineer saw a light rising low from Sandia Mountains' eastern slopes, a green fireball too large to be a meteor arch upward and level out. They had already seen an identical fireball twenty-two minutes earlier near Las Vegas, and decided two was too much and radioed Kirtland AFB control tower.

Ten minutes later, the captain and copilot of Pioneer Airlines Flight 63, on a westerly heading, saw a light ahead and a little above them. On a too-low and flat trajectory to be a meteor, the fireball shot straight for them on a collision course. As it rapidly grew in size, it changed from orange-red to green. The DC-3 swerved in a tight turn and the green fireball, now larger than a full moon, shot past and grew dimmer as it fell. Upon their landing, USAF Intelligence officers debriefed the crew.

By next morning, reports phoned in from all over northern New Mexico triggered a full-scale investigation. The green fireballs were near two key installations critical to the fledgling atomic bomb program—Los Alamos Research Laboratories and Sandia Base—while other installations, such as the White Sands Proving Grounds and fighter-interceptor bases, just as essential to our national defense in the early Cold War, were nearby. Nervous, the intelligence community called in famed meteor expert Dr. Lincoln LaPaz of the New Mexico Institute of Meteorics.

When intelligence officers showed him field reports, LaPaz said that the fireballs might be meteors. And since he had successfully plotted meteor paths before, it should work here. First, he interviewed witnesses and discovered eight separate green fireballs were seen that night. The largest flew form west to east. From the many reports, LaPaz plotted the exact location of impact, near the western border of Texas. But when the team searched that area with steel rods and detectors, they discovered nothing and returned, baffled. Each night brought more green fireball reports, with thousands of witnesses.

On December 8th, two Kirtland intelligence officers flew a plane around north of Albuquerque after dusk. While 20 miles east of a Las Vegas radio range station on a 90-degree heading, they saw a brilliant green fireball ignite 2,000 feet above their craft. The dazzling giant green fireball flew level straight at the plane from the left. After several seconds, it rapidly fell, then flamed out.

Nightly reports poured in. By mid-February, a scientific conference was held in Las Vegas and was attended by leading scientists. Dr. Edward Teller, father of the thermonuclear age, Dr. Joseph Kaplan and later director of the worldwide IGY in 1957-58, Dr. LaPaz, Dr. James Tuck who later headed the Los Alamos fusion program, and other luminaries attended. Almost without exception, each had personally seen at least one green fireball in the last three months.

After two days, the conference concluded with recommendations that the USAF Massachusetts Cambridge Research Laboratory set up Project Twinkle to study green fireballs. The Air Force revealed a case where they tracked a fireball at 14,000 miles per hour. Budget cuts and the Korean War ended the project before it began. Shortly after this, the reports of these oddly silent fireballs ended as mysteriously as they began. In the excitement, few knew that green fireballs were seen over several other states at that time and later, and in some deserts.

In a related case, on the December 12, 1951, Dr. Lincoln LaPaz investigated an unusual case. The four persons near a thirty-foot water tank were killed in Tucumcari, New Mexico, when a fireball plunged from the sky toward a 750,000-gallon, ground level tank. As horrified witnesses watched, the tank buckled and collapsed, demolishing twenty buildings. The team could find no meteoric remains. Since meteors leave broken fragments when they hit, and from other data, LaPaz concluded it was not a solid object.

More recent proposed explanations include frozen metastable nitrogen, which undergoes a transition at 16 degrees Kelvin and produces blue-green radiation, or encapsulated micro-black holes, or anti-matter specks. None are satisfactory hypotheses. However, just because physics cannot explain a phenomenon does not mean that there is not a natural explanation, for there is, but we have not learned it yet. Some of the principles involved in standard fireballs could be involved in these green meteors.

Ionized particles scatter electromagnetic waves. Ions and electrons are generated by St. Elmo's fire, lighting, meteors, reentry spacecraft, ionospheric layers, the aurora, static discharges from high-speed aircraft, and probably ball lightning fireballs. These phenomena can return radar echoes. Lightning echoes have been detected on the radar and plan position indicators exceeding 100 miles in length, the horizontal flashes. Schonland had previously listed the maximum record at 50 miles. (Ligda, 1956; Atlas, 1958a). Longer wavelength radars are more likely to detect lightning, in part because they detect less precipitation. Radar can detect lightning that is not visible.

Although most meteor echoes last under a second, some last several seconds. Although only lower-frequency radar will detect meteor ion trails, if the meteor survives entry, they produce spectacular displays and are detectable on any frequency at 120-mile ranges and 20,000-mph rates.

Clear Weather Fireballs

The term *ball lightning* is a misnomer and misleading. *Fireball* is a better word. True, most appear during or near lightning storms, but do not occur near a lightning stroke. Although Stenhoff reports that nine-in-ten fireballs occur during or near a storm, some do not, and some do occur during sunny days. These clear sky and sunshine fireballs need an explanation, but there is none. The rare clear sky lightning ("bolt out of the blue"), in which lightning from a distant storm travels horizontally through a clear sky, and is weaker, is not the culprit.

Did radioactive fireballs tear up steps, invade girls' dormitory? [562 Guion] There was no lightning. A blue-orange fireball came in through the open front door and rolled left across the floor to the radiator against the opposite wall. The pine tree in front of the steps had a long jagged scar and the right front edge of the concrete steps was partially torn off. There was a strong acrid smell of sulfur and the room was filled with bluish smoke. None of the girls in the room was harmed. Shortly after this I became seriously ill—probably from sunburn [and radiation burns?] I received that day, the Fourth of July in 1931, while on the river.

The next Sunday while in chapel, severe back pain struck and I could not straighten up. I was so ill in bed for two weeks that I thought I would die [radiation sickness?]. Prior to this event, one girl told me that during storms, lightning ran along in the dormitory, located on the second floor in Farner, Tennessee, where the school also was located, about 20 miles from Copperhill, to the west, and about 100 miles north of Atlanta. If you wish to corroborate this, or need further details, I include the names of thirteen girls as witnesses.

Clear sky lightning hits walnut tree, fireball descends, follows roots. [522 Guetlner] It was very dry. There was a very small cloud. There was one clap of thunder and lightning hit a walnut tree. A two-foot dark-red fireball hit the ground and followed the roots of the tree, which were a few inches above the ground and exposed, running uphill. There was no rain.

Rainless night skyball illuminates house. [342 Deal] Living on Illinois Avenue in Oak Ridge in 1955, my wife and I were in bed. A light woke us up. It was like sunrise. Then a four-foot light-yellow-orange fireball moved east to west—not fast, not slow—at tree height. After the fireball passed, it got dark again. It had not rained, and it did not rain afterwards.

Foggy moonlit power line fireball. [No assigned ORNL number; received 11-30-66 from Stressinger] The event date was 11-22-66 and the time between ten and ten-thirty before midnight. The sky was slightly overcast, with some moonlight and a slight fog. It was a cool temperature about forty. Near a high-voltage (60 KV) power line a quarter-mile from where I stood, an airborne soft-ball-sized greenish-orange fireball fell several feet and in three seconds suddenly vanished.

Sunny-day red fireball explodes. [275 Schrader] I was riding turtle back of a car driving over a dirt road, with woods on both sides. The sky was moderately cloudy, with intermittent bright sunshine. It was slightly muggy at 85 degrees. A slightly red softball-sized ball of fire flew in a straight line, five feet above the ground, across the road. After several seconds it exploded, sounding like a 30-30 rifle.

Sunny day skyball falls onto log. [280 Ballard] There was not a cloud in the sky. I stood in our backyard. An eight-to-ten-inch red fireball came down at a 45-degree angle and hit a log about 200 feet distant and exploded, sounding like a small firecracker. It made a white spot on the log, which remained until the log later rotted.

January, tailed-blue-duo turn red, vanish after half-minute. [567 Woody and Love] Both sightings took place within a three-mile radius of Widows Creek Steam Plant. On January 6th, while traveling by auto at 1:45, we saw a fireball travel at a 50-degree angle north to south. On January 23rd, the second fireball traveled at a 15-degree angle from east to west. Both were basketball-sized, with blue that turned red, and descended trailing a tail. After 20 to 30 seconds, they slowly vanished. There was no lightning.

Sunny day fireball grows brighter. [274 Easley] It was a clear day and the sun was shining. We heard thunder and saw a fireball, which grew brighter and moved randomly.

Cloudless ball sails under bridge. On a night clear and cloudless, a soccer-ball-sized fireball sailed before a car and then off into a gully, beneath a small bridge the car passed over.

Ball appears, lights fade. During a clear night in early June, on a peninsula, the house lights alternately dimmed and brightened (with a humming noise), apparently associated somehow with a brilliant red-and-orange sparkling fireball above a neighbor's house.

Sunny Multi-Balls. A few clouds floated high during a sunny summer day. One after another, basketball-sized fireballs appeared 20 feet above the ground and smoothly drifted downward to the ground, 300 feet away. Many reached earth, but many did not and sizzled out in mid-air.

Three rising balls dazzle. The house lights flickered. Very intense light out-side shone through the window drapes and a deep low electrical hum and crack-ling lasted for several seconds. It occurred again, coming from a dazzling flash behind a house across the road, and a fireball rose up 40 feet and attached to a power line. It shrunk and rapidly followed the line and disappeared. A third time, a fireball arose from the ground. No evidence of the three fireballs was ever found and the electrical utility company had no reports of any abnormalities. Although the event took place in mid-August, there was no lightning.

Sunny day fireball dances above rug. On a sunny mid-afternoon, a silent basketball-sized flame-colored (orange and red) fireball appeared and danced in one spot, half a foot above the living room rug, and after a minute it silently van-ished. Full of "activity" and sparks and fire, the ball did not warm the rug beneath it.

Green glow mystery light. Lightning struck near the house and a loud hum that changed intensity and a green glow came of the bedroom, then after four sec-onds, faded, only to repeat itself after thee seconds.

Sunny day fireball climbs tall tree. During a late March sunny afternoon, a brilliant reddish orange, soccer-sized fireball appeared near a tree. Its center faded to dark brown. After several seconds, it ascended a 100-foot tree and floated to a second tree, then to a third.

For those who feel high voltage discharges such as lightning are needed to create fireballs, these clear weather fireballs remain truly enigmatic. They account for five percent of reported fireballs. But even more enigmatic are the so-called "mystery lights."

6

Mystery Lights and Methane

Do "mystery lights" exist? The late Isaac Asimov once wrote how Nobelist Svante Aughust Arrhenius almost failed to graduate. His doctoral thesis in 1884 explaining acids, bases and ionization was considered too incredible, and the committee was skeptical. But later in 1903, for these criticized theories, he was awarded a Nobel Prize. Today, zealots and crackpots point to him and similar scientists like Galileo to scold and castigate scientists for being such hard-nosed skeptics.

But, we are skeptics for good reason. Isaac Asimov pointed out that many strange and inaccurate theories are proposed each year—and, not just by doctoral students, but also by misguided mainstream chemists and physicists, who mistakenly let their enthusiasm guide them. Doubt and reason guided by healthy skepticism guide best.

Many scientists—including your author—once doubted odd lights, fog fireballs and mystery lights. But people observed and reported unusual nocturnal lights, such as the Marfa Lights, for two hundred years and probably since the dawn of time. Unlike ball lightning, these night-lights are confined to particular remote localities, such as a dusty road or deserted railroad track. Some are multicolored lights that dance in the distance and recede if approached. Others will disappear and reappear on the other side of the observer. They seem unrelated to electrical storms and ball lightning. However, since science cannot explain these lights, then could they and ball lightning share underlying physics?

It is tempting to alter facts to fit proposed theories—scientists do it all the time (and who can blame them?)—but if it is not at least scientifically dishonest of us, then it is our prejudice that blinds us to the possible existence of things that we dismiss. We all do. Even Einstein's scientific prejudice blinded him to the existence of black hole singularities, even after Schwarschild proved their existence in 1916 with Einstein's own equations! Einstein's reaction intimidated Oppenheimer and Snyder out of vigorously pursuing their research into collapsing black holes, thus delaying research for decades. And, like a Don Quixote,

Einstein attacked subatomic physic's foundation. But, quantum mechanics was no mistake—Einstein's prejudice was. Einstein dismissed Kaluza-Klein higher dimensional physics, thus strangling further research for a half-century. Its partial successor, string theory, incorporated some of the Kaluza-Klein concepts.

So, cautious skepticism is better. Never dismiss totally; periodically re-examine rejected data and hypotheses. So, rather than ignore these cases, I highlight them. We learn from what we cannot explain, not from what we know. I now know that these mystery lights and strange glows do actually exist. Even the scientists at Oak ridge National Laboratories saw strange lights and reported them.

Roaring monster sphere lights golf hillock. [85 Lewin] In 1938, I was caddying on a hilly golf course in the midst of a thunderstorm. It rained heavily. Lightning frequently discharged all around, from cloud-to-cloud and to earth. Lightning hit nearby—the flash and crash were simultaneous—but this became a continuous roaring. Simultaneously, the lightning flash lit up again at a neighboring hillock. The light grew much brighter and assumed the shape of a thirty-foot sphere. The roaring sound and light lasted for over five seconds, and then slowly disappeared. There were clouds of steam at the little hillock where the giant sphere had been. At all times the ball of light remained suspended in the air directly over the hillock.

Lightning-colored cloud rolls down hallway, out office window. [396 Brogden] Lightning struck a transformer on the fifth-floor where I worked. Resembling a "lightning-colored cloud" with a three-foot shape that changed slightly, the fireball rolled into-and-down the hallway, then out through an office's open door and out a window. I stood at the extreme or dead end of the hallway and not too close to the ball.

Hot surface-glow scorches 50-by-5-foot wet field. [117 Fuqua] For one hour, a surface glow that did not change size followed the wet ground, going slowly. I felt considerable heat and it was bright enough to burn my eyes, which I had to close, but suffered no injury. It was accompanied by a deep roaring sound [at disappearance?]. It scorched and parched the wet green clover field for a 50-by-5-foot area.

Author—This and other odd metrological events are rare but explainable, and was reported by meteorologist Peter Chaston. In some heat bursts and related phenomena, temperatures exceed 140 degrees and scorch crops and kill livestock. The first authenticated case was on June 20, 1989 in Pierre, South Dakota. Fierce dry snow and desert sandstorms sometimes create tremendous charge separation, strange electrical effects, and localized lightning.

Occasionally strange electrical lights and effects also are seen during hurricanes. During the Labor Day Hurricane of 1935—which recorded the lowest barometric pressure ever recorded—so fierce was the wind, it blew sand, and friction created a giant Van de Graff generator. One terrified survivor exclaimed, "I thought that I had died and gone to hell, because the sand was sparkling electrically with an eerie electrical fire!"

Giant flowing glowing fog not St. Elmo's fire. While standing outside his car, in a Yellowstone Park marsh during a nice September day, a man saw a fuzzy fog of blue light—150 feet wide and 1,000 feet long—fly at several feet per second towards him. When the glowing fog enveloped him, his scalp tingled and he felt tiny sparks snapping. He felt a sudden temperature drop and smelled ozone. The strange pulsating light enveloped him and the objects near him in a glow. When he touched a shrub and his car, he felt no shock.

A 300-feet-wide glowing "fan" towers 600-feet high. During a rainless summer night, humid and windless, with distant glows of heat lightning, a red glow quickly arose from behind a ridge and soon became pink-orange and then orange-yellow. It silently grew brighter and became white with a bright red halo. So bright was it that its light reflected off the low clouds. After one minute, the glow faded. An Australian population biologist, a ranger, photographed the glowing fan. By triangulation with another observer, he pinpointed the fan's location at 3 kilometers to his east. They performed an extensive investigation, but discovered no explanation.

Spider-web lightning fills electrical sky for hours. Driving eastward, with a beautiful sunset behind his car, a meteorologist was puzzled by the lengthy sunset. The entire sky emitted a beautiful and purple-pink that looked the same in every direction. After one hour of driving, he realized something was drastically wrong and pulled over to park at a rest stop, got out and heard something eerie—the glowing clouds above crackling.

Coating these clouds, fiery spider-web strands coated the sky and glowed pink. The discharges lined up in distinct east-to-west bands with clear spaces between. Each electrical "tree" had a trunk pointing south with its V-shaped branches pointing north. The group migrated northerly, with each band sprouting new branches. When several of these brightened, simultaneously the trunks blinked out and split that electrical tree into two or more smaller trees. Each structure steadily moved upwards, but grew new feelers out of the V's, and the trunks popped off.

After 45 minutes, he drove on, and then after two hours he drove out from under the sky. There are meteorological phenomena besides ball lightning, mystery lights and green fireballs for which physics has no explanation, but through which there might run a common thread.

Mystery Lights Haunt World

Mystery lights both resemble and differ from ball lighting and fireballs. Is there a common underlying physics? Although mystery lights do exist, physicists remain puzzled. With that in mind, we briefly examine them.

Since earliest times, people have seen mysterious glowing lights that reappear at night. Many are famous and were seen by the earliest settlers. Others are less well known. But a few are local, appearing occasionally. They are often only known locally, and are even unknown by most local residents. For example, one appeared near Attleboro in Massachusetts, but only infrequently, and has not been reported recently. Methane gas is not the cause of mystery lights, but it creates its own eerie glow at night in marshes and cemeteries, but much remains unknown.

The world-famous **Marfa Lights** of Texas—unidentified floating balls of yellowish-green floating lights—were seen for the past 150 years at night on the horizon around the Big Bend area, usually southwest of Marfa between Alpine and Marfa and the Chinati Mountains. First reported by rancher Robert Ellison in 1883—but they were probably seen by Indians long before that—these lights move and change their color and brightness or intensity.

Established in 1881, Marfa lies on U.S. Highway 67/90 in Presidio County just 26 miles west of Alpine. Appearing over the flat prairie north of Cuesto Del Burro Mountains and near Twin Peaks. These spheres appear in a region southwest of Chinati Mountain near Twin Peaks on Mitchell Flat. Today, a town that once saw better days, this small western Texas ranching town of 2,000 citizens near the Big Bend National Park, has its claim to fame: The classic movie "Giant," which starred James Dean and Elizabeth Taylor, was filmed here. Eight miles east of Marfa on U.S. 90, there is a roadside plaque that commemorates these lights.

These glowing spheres appear each night south of Marfa, and each night dozens of people park at the viewing area that is on Highway 90 just nine miles east of Marfa and 16 miles east of Alpine, on Mitchell Flat southwest of town, and just east of Marfa and west of Alpine on a large plateau.

The lights vary in color, are round, are baseball-to-basketball-sized, and move quickly and erratically. The lights do not pulsate and are constant. Some witnesses saw the lights up close. The lights chased some witnesses. Ten miles east of the town at the official observation side, the plain is covered with short cacti and small bush. Before sunset, a dozen or more cars pull in and park to observe the lights.

Although not visible in daylight, an hour after dark the radio tower is visible, and the Marfa Lights appear near the tower. Some Marfa Lights float clockwise along the mountain and a few appear to the northwest. Though an eight-inch Celestron telescope, the lights resemble fuzzy spherical balls several feet in diameter at the mountain. They appear to blink, but this is because the light passes behind rocks. But other smaller rocks were briefly silhouetted. Through telescopes, witnesses see the light illuminate the rock wall behind it and the ground below.

Sometimes, half a minute after a ball materializes, a second ball appears next to it. Although the human eye cannot distinguish the pair, a telescope shows that they are separate. Through telescopes, witnesses observe that the twin balls may alternately dim and brighten for up to ten cycles, then often separate so that the unaided naked eye could resolve the pair. When they begin their strange nonlinear dance, they look like car headlights on a mountain road, except that there are no mountains there, and no roads.

Some witnesses report that the lights appear almost constantly for four hours—sometimes up to five lights are visible simultaneously—but taper off after midnight and stop. The lights were seen ten miles away. Some motorists report that the lights flew beside their cars, just 20 feet to the side and off the road, and flew beside them at speeds up to 60 miles per hour.

During World War II, pilots training at the now-abandoned and overgrown Marfa Army Air Field, located just several hundred feet from the viewing site, would see the lights. In their attempt to solve the mystery, they flew at night and dropped bags onto the lights and returned by jeep next day to see what was there—nothing but broken flour bags. Today, the town of Marfa sponsors an annual Marfa Lights Festival with entertainment, dancing and events.

The **Brown Mountain Lights** are equally as famous as are the Maco Station Lights and almost as famous as the Marfa Lights. Observed first in 1850, these spectacular lights do not appear regularly. Green, blue, white, yellow and red, these 18-inch spheres sometimes sizzle. The Brown and adjacent Chestnut Mountains are in the northwestern corner of North Carolina. The Grandfather Mountain Tectonic Fault runs very close to these mountains. On some nights in the gorge, the lights appear with no pattern. They begin as blue-white or yellow lights that weave in and out of trees. When they reach the river, they vanish and reappear elsewhere. They last two to twelve seconds.

The lights are eight miles northwest of Morganton in North Carolina and are observed from Wiseman's View or Linville-Grandfather Mountain. The lights are best observed in good weather, but also have been observed in heavy overcast and in humid weather with mist that hides the mountain.

Resembling pink-orange or red toy balloons whose size may vary, they rise—sometimes slowly shrinking in size—then hover and fade after another one to fifteen minutes. Sometimes up to three widely spaced spheres appear. They are bright and easily seen at Blowing Rock at twenty miles to the northeast. Like the other Mystery Lights, the Brown Mountain Lights are also shy, and these bobbing lights refuse to let anyone get too close. There are exceptions. A gas station owner and several friends climbed the 2,600-foot summit. At dusk they left and a sizzling, three-foot light formed and hovered above them. A strong light, it lengthened and shortened several times.

The **Maco Station Lights**, first reported in 1862, appeared close to a trestle span of the Atlantic Coast Line railroad tracks over Hood's Creek. The lights used to weave back and forth just three feet above the racks. Sometimes there were two lights, one below the center span and the other near the end of the trestle.

Because the engineers mistook the lights for the lights of oncoming trains or flares, they switched to two lanterns—one white, the other green. The lights were mostly seen in desolate areas of the tracks and in the adjacent woods. Several lights chased a soldier during World War II down these tracks. In 1977, the tracks were torn up and the lights ceased.

The **Ozark Spook Lights** are also known as the Hornet Ghost Lights. This famous nightly light appears ten miles southwest of Joplin in Missouri near the Oklahoma state line. On a hilly road along a two-mile stretch appears during clear night. About twelve miles southwest of Joplin along State Highway 43, signs direct visitor to the Spooksville Museum that is on a gravel road that runs west across the Oklahoma state line and leaves for the state highway two miles southwest of Hornet, a small village.

The color varies from golden yellow to orange or red, and they can be so bright that a glow reflects from the road surface. Every color of the rainbow has been seen. Seen from two hundred feet to a mile away, the lights expand and contract, changing in size from an apple to a bushel basket, and move left to right and back, from ground level to tree-top heights.

Although moving randomly, the lights never stray too far from the road. When they appear or disappear, they blink on and off abruptly. When witnesses approach them, the lights recede, vanish and reappear behind them. Sometimes there are two lights, one above the other. At other times the upper light is tilted off at an angle, about forty degrees from the vertical. Through telescopes, observers saw up to sixteen distinct lights that continuously changed. They had indistinct glowing edges. Alternately growing larger and smaller, brighter and dimmer, these spheres of diffused light disappear within a minute.

This light is sometimes seen in daylight. At night, witnesses have seen it as a glow over the hills and as a yellow-green five-foot glowing sphere that enveloped witnesses. Ostensibly, some investigators once demonstrated on one clear night that light shining from a car headlights is refracted or bent by the lateral differences in atmospheric density. This is because the highway is 200 feet below the gravel road, with a range of hills in between. These variations create differences in near-surface temperatures through ambient cooling caused by the unevenness in convection and infrared radiation.

Refraction or bending of light rays from man-made lights cannot explain the Ozark Lights—or, for that matter, cannot explain other mystery lights—because they do not resemble the lights, and these lights pre-date cars and other electrical lights. Nor does the phenomena of refraction and reflection create the phenomena in other mystery light locations, even where the conditions are better.

Independent of weather, the Ozark Spook Lights appear in the two hours before midnight. Although there may be many coalescing lights, these sometimes split into two or three intermingled lights, and occasionally emit sparks. These lights can do acrobatics and have appeared atop automobile hoods. Sometimes there is a trail of lights. Some appear transparent and others are revolving balls. They never fade out, just wink out. They have floated onto porches.

Hessdalen Valley Lights were never true mystery lights. Some mystery lights cease, such as the Maco Station Lights, when something in the local environment is altered. In the case of Maco, the lights abruptly stopped in 1977 when the railroad tracks were removed and they never reappeared. On the other hand, some new lights suddenly begin to regularly appear, usually for no known reason.

In the middle of Norway lies the three-mile-wide and seven-mile-long Hessdalen Valley. For no known reason, in December of 1981, strange lights began to appear three or four times each day all over the valley, but usually under nearby ridges, and sometimes near homes. The lights usually appeared at night and often in winter. Unlike most other lights, these are more than just spheres—some are cones and other shapes of many colors, although white and yellowish white predominate.

The Hessdalen Valley lights frequently appeared in groups of three or more lights in a triangular shape that contains one red and two white-yellow lights, or sometimes blue flashes. Witnesses described bullet shapes of yellow pointing down, a moving bluish white blight that sometimes flashed, and a pattern of different-colored lights that moved together as if connected.

With the assistance of the Norwegian government, scientists from the Hessdalen Project set up field stations of magnetometers, cameras with diffraction gratings, and other instruments. Events seen as lights were also occasionally tracked on radar, sometimes at great speeds. Magnetometer readings often correlated with appearances, and spectroscopic images yielded continuous spectra without absorption or emission lines.

To date, there is no explanation. Ostfold College was preparing a more sophisticated surveillance network of automatic measuring stations, but it was too late. In 1984, the sightings waned, and few were seen since 1985—only twenty per year.

Likewise, the **Min-Min Lights** are also not true mystery lights. They appear in Boulia of Queensland in Australia, the driest of continents. These lights appear on Aboriginal lands and are visible for miles. They are as bright as headlights but move erratically. They float about Alexandria Station—one of the world's largest ranches that cover 11,000 square miles.

A few witnesses report that the lights are hot. They appear in different colors but are often white and the size of car headlights. Since the original hotel from which the lights were first observed was small, and once the lights became famous, the government constructed the present Min-Min Light Encounter Center. However, whether the lights are named after the original hotel or just the reverse is now unknown. What is certain is that the original native Australian inhabitants, the aborigines, were seeing the Min-Min Lights long before the English settlers first arrived.

The light has followed travelers for long distances. Descriptions vary. Sometimes the light is large and brighter than a car's headlights, but also can be small and dull, fast or slow, oval or round, multi-colored or mono-colored, flying straight or erratic. Most often the light is never higher than several yards above the ground. Witnesses report that there seems to be intelligence in their behavior. They never appear before dusk. The phases of the moon and weather do not affect the sightings. Researchers have discounted some sightings as mirages, road reflections, heat haze and other natural explanations.

Some Aborigine tribes are terrified of the lights, but others are not. The Lights occur in many locations, most of them remote but not confined to roads, and are reported by ranchers, campers, drovers, stockmen and others who travel in sparsely populated areas. The first reported sightings date from the Cobb and

Company Coach years of the 1880s and from sightings near Ovens River, Victoria, in 1838, by Aborigines, who report that the lights appeared earlier.

The Light got its name from the infamous Min-Min Pub Hotel, dead house, graveyard, mail-change, now in decay that once stood in the Boulia District of Northern Queensland. The phrase in Aborigine has different meanings, depending on the tribe. The light mostly appears in "The Boulia Triangle." Today, Boulia is a small settlement of 300 people.

Civilization is encroaching on the outback and the road network is being upgraded, and tourism to Boulia keeps increasing. The government constructed a Min-Min Encounter Centre, drawing more visitors. The Light avoids congested areas. Only time will tell if this encroachment will diminish the appearances. Around Boulia, the reddish Cawnpore Hills resemble a Martian landscape. Boulia has the largest stand of the famous Waddi trees, which are so hard that they damage saws and axes.

One group of researchers set up camera arrays, monitoring equipment and a magnetometer in the remote Kimberlay region of Western Australia. At first brief blue white lights appeared, moving sporadically on the opposite slope of a small valley, 2,000 feet away. Then a bright light came from a hill before a ridge, moving slowly down to the desert and vanished. Another bright light appeared six miles on the other side and the magnetometer began to detect large field anomalies of 800 nanoteslas (p-p) on a two-hertz pulsation. Normal peak-to-peak amplitude is under one nanotesla. After five hours, the reading returned to normal.

Unlike the wandering Min-Min Lights, the **Bragg Road (Big Thicket) Lights** remain near Bragg, Texas. The Bragg Lights appear along the 7-mile-long dirt Farm Road, which runs from Saratoga to Bragg, now a ghost town. Although the road was an old railroad link put down in 1901 and torn up in 1934, and the residents saw the light near and on these tracks, they appeared before 1901. The light varies, from a point source and then a dull orange glow.

A nightly mystery light which appears near the town of **Ahoskie** in North Carolina, about sixty miles southwest of Norfolk, Virginia, on a railroad changes size, color, and shape as it runs up and down the tracks. In Saskatoon, SK Canada, there is gas prevalent in the **Langenburg** area in Saskatchewan, known as "taber lights" which seem to appear in cemeteries or areas where the soil was never disturbed by agriculture. Suffice it to say that we cannot go on simply because there are thousands of mystery lights worldwide, and their mechanisms are as mysterious as ball lightning.

Methane Gas

Is methane gas a nocturnal reality or convenient scapegoat? Natural methane or swamp gas was blamed for many things, from ghosts to aliens. Called jack-o'-lantern, foolish fire, ignis fatuas, fire damp, marsh gas and wil-o'-the-wisp, it is an inflammable gas burning with a faint blue flame—sometimes white and yellow, and sometimes faint blue or red—and is seen by rural folks in marshlands during summer nights. Placing their hands into these luminescent gases, witnesses feel no heat, copper rods do not heat, and dry tinder does not burn. Folklore blames elusive ignis fatuas or wil-o'-the-wisp for many things—from flying saucers to ghosts and elves.

To understand mystery lights, we need to understand methane. Organic chemistry textbooks tell us that methane is the first and simplest molecule of a chain-series of straight chain—that is, no appendages, as in branched-chain—hydrocarbons in the alkane (single bond) series. Each molecule of methane gas consists of one carbon and four hydrogen atoms, looking like a three-sided pyramid or tetrahedron. Natural gas from oil or gas wells is 85 percent methane, which is used as a fuel in many homes and factories.

Anaerobic (airless) subsurface methanogenic (methane producing) bacteria feed on and decompose a marshland's subsurface organic matter. There are many species, but all of them belong to the family Methanobacteriaceae. Their methane gas oozes up in slow bubbles from warm, shallow marsh bottoms. A colorless gas, when ignited in the presence of oxygen, methane burns readily, yielding significant heat to become water and fine carbon. In field visits to local marshes to verify this, your author captured bubbles of methane gas that displaced the water in an inverted vessel. Also, if you crush bituminous coal into powder and fill a glass funnel and invert it in a jar, then fill with water, and let it set for several days, the tube will fill with methane.

For years, your author searched the outskirts of our nearby Massachusetts Hockomok Swamp, the largest in the Northeast United States, but never observed wil-o'-the-wisp nor met anyone who had seen it. Our Southern New England has numerous swamps and bogs, but no wil-o'-the-wisp sightings. Your author suspected that it is due to the formation and type of wetlands. Our Northeastern summers are hot, our winters cool, even cold. Because the summers here are hot and humid, and the winters moderated near the coast (typically 10 degrees warmer than 90 miles inland), they create marshes and swamps of a different type.

Sections of ponds become choked with vegetation in their conversion from open water to dry land. In Southern New England, reedy marsh is a mid-transition—a semi-floating mat of sphagnum moss and leather leaf shrubs extending across the surface of the water. The floating mat becomes firm enough for vegetable debris and small trees, and then becomes heavy enough to sink and thicken. The mat starts at the edge of a pond, and then slowly extends across the surface, and in a century or two turns it into a bog.

Our cool bogs lack plant nutrients, so crucial for generating swamp gas, because the mat seals underlying water from the air, and vegetable debris decays slower and peat builds up. The small decay that occurs in the dark water below the mat consumes nitrogen and generates acid—making water absorbtion difficult for most plants. (Coincidentally, this problem is also faced by salt-resistant halophytes growing in coastal salt marshes and saline deserts.) Colloidal iron in suspension is washed into the bogs and swamps and precipitates. The soluble forms precipitate during summer heat, which can reach the low nineties. These precipitates form nodules of iron ore.

The early colonial settlers made precious nails by mining this bog iron. In Colonial days, settlers mined bog iron to make tacks, nails and even cannon balls for the Continental Army. Despite many historical records, which I examined at historical museums, few historical reports exist here of possible swamp gas phenomena. But in the southern marshlands, it is a different story.

Since the last ice age, lakes and bogs in the midlatitudes of the eastern United States possess more carbon in their sediments. Although aquatic vegetation grows faster in lower and warmer latitudes, its decomposition also is far greater. Thus, in colder latitudes—the New England and mid-Atlantic regions are best—vegetation both grows and decays slower, thus building up more organic matter in the sediments of our marshlands.

With little free oxygen below warmer marshlands, the biochemical reactions do not turn carbon into much carbon dioxide, but into methane, sulfur compounds of hydrogen sulfide, and nitrogen compounds. Methanogenic bacteria in airless stagnant marshes give up methane and phosphine that can ignite spontaneously and burn with a pale blue flame, visible at night. Other species of methanogens live in sewage sludge, shallow coastal waters, tidal mud flats, and permanent waterlogged soils. Russia's Lake Beloye bubbles up an 80 percent methane gas. Methanogenic bacteria are most common just below the surface of marsh mud. Bacteria concentrations of 250 cells per gram are not uncommon and generate 1.4 to 47 micro-liters per ton of silt per day. But methane is not only liberated on land.

Frozen methane is also buried beneath in some offshore mud of the seafloor and bubbles up to the surface. In fact, vast quantities of methane gas lay frozen and buried in sediments beneath the seafloor. Only recently was it discovered that there is a potential danger. If something goes wrong and the frozen methane is melted and rapidly released into the atmosphere, then such massive amounts of methane gas might create serious and sudden alterations to our climate.

At the end of the Paleocene, at 55 million years ago, when deep ocean temperatures suddenly rose twelve degrees Fahrenheit, a worldwide heat wave exterminated many microscopic deep-sea creatures and generated a sharp spike in carbon isotopes. Belching methane, the seafloor burped up methane hydrate that floated to the surface in huge blocks that then released a greenhouse gas with nearly thirty times the heat trapping ability of carbon dioxide.

Then, in a geologic blink of an eye, half of the deep-sea and bottom-dwelling creatures suffocated in the warmer water—all because of a decline in oxygen of the warmer water. Bacteria digested dead plants and animals, creating the gas that still lie entombed in 15 trillion tons of gas in cages of ice that bubbles to the ocean surface.

Methane has thirty times the heat-trapping abilities of carbon dioxide. Entombed in the crystalline cages of ice, sub-seafloor hydrate deposits in several locations, such as buried under a thousand feet of sediment off Florida's northeastern coast, melt, escape from the seafloor and bubble up to the surface. They then enter the atmosphere. Refrigerator-sized chunks of frozen methane chunks float up from half a mile below the sea off the coast of Oregon and then disintegrate.

At the end of the Paleocene, the five million year warming heated the bottom seawater to the critical temperature and the frigid hydrates decomposed in a sudden blast of massive amounts of methane gas that created a greenhouse effect which raised global temperatures. Deep-sea temperatures rose six degrees and even microscopic sea creatures died, producing a spike in carbon isotopes.

In the late-Paleocene disaster, as the spring temperatures suddenly rose sooner, the links between interdependent species were rapidly decoupled. And since timing is everything, few could adapt fast enough, and they shifted dangerously out of sync. Altering the relationships among species by weakening links in their interlocking food chains (as in birds that migrate and nest) created massive domino effects that branched out to affect many other interdependent microbes, animals, plants and insects. Many windows rapidly shifted, further upsetting the food-mating windows and multiple interconnections that dramatically altered migrations. Populations plunged.

Changing climates tore apart existing interconnections, and created new ones, but left no new havens for species that failed to adapt fast enough. If the ecosystem shifts are gradual, most species can adapt—too sudden, few can, and extinctions become massive. Certainly, new species constantly appear. But it takes time. Today, our Earth's temperatures are rising, with dramatic extremes in seasons, further harming interdependent species that are already gradually becoming decoupled. More species will become extinct.

But, imagine this nightmare—if because of global warming, the seafloor methane hydrates suddenly "break lose" again, then catastrophic changes will occur suddenly. Then the Greenhouse Effect will be magnified far beyond mere carbon dioxide and man-made pollution.

Man-made pollution is a two-edged sword. Pollution reduces greenhouse warming. It reduces rainfall by breaking large water drops into droplets that reflect more sunlight back into space and suppresses rain from tropical clouds. Huge carbon sinks, forests, grasses and oceans counteract greenhouse warming by absorbing carbon. Temperate-zone forests absorb massive carbon. Surprisingly, tropical and Canadian forests help little, although the warming northern tundra and longer growing seasons are growing larger brush and trees up north, thus absorbing even more carbon.

However, all is not so neat. Carbon dioxide is less soluble in warm water. Carbon cannot be removed unless with carbonate through marine plants, which sink. But carbonate-containing rocks (limestone) only replenish the oceans at a steady rate, so if the buffer precipitates out faster than rivers can replace them, then the oceans' carbon appetite will drop. Solutions at sea are being explored. As for resurgent forests, we must prune them to foster new growth.

Most atmospheric methane from the sea floor sediment is generated in low-oxygen, much by oxygen-fueled single-cell archaea microbes. But luckily for our atmosphere, some archae have a different metabolism—instead of oxygen, they use sulfate ions and consume methane. Two microbes—one consumes sulfate; the other, methane—together feed on methane. They form tiny (3.2 micrometers across) spheres of 100 methane-consuming archaea in their center, enclosed by a shell of 200 sulfate-consuming bacteria. There are 30 million of these microspheres per cubic centimeter in the top 5 cm of sediment.

Some methane beds lie off the coast of Oregon, so we know that these microbes grow and metabolize sulfate ions and methane in colder sediments. But most of today's methane is generated on the land.

Methane can only be produced under oxygen-free, reducing conditions because it is generated in a series of steps, creating unstable products that easily oxidize, as does the escaping methane, into carbon dioxide and water. Actually, there is always some oxygen present in marsh soil, which will form a minute amount of carbon dioxide, ammonium compounds (NH_4), nitrate (NO_3) and sulphates (SO_4). The methane that constantly escapes into the air breaks down in a series of steps to yield carbon dioxide and water. This explains why methane does not poison our atmosphere. The atmospheric oxidation of biologically produced methane produces 55 percent of the water vapor in the atmosphere, the rest by evaporation. Methane oxidation removes oxygen from the air, preventing the rise in oxygen, which would create uncontrollable fires, restructure partial atmospheric pressures and alter plant growth.

The bottom line is this—that even though all the above explanations are connected, something in our explanations is missing. Unfortunately, methane gas cannot explain fireballs or mystery lights, or many—if any—UFO sightings, for that matter. And, as much as we do know, we still cannot experimentally duplicate wil-o'-the-wisp. The truth is that our explanations remain glibly superficial.

But, returning back to true fireballs in the next chapter, we might learn something by examining long-lived fireballs that dazzle and hiss loudly.

7

Methuselahs, Dazzlers, Hissers

How long a fireball lasts is important. Long-life fireballs provide clues to their physics. Fireballs that last five minutes, even one minute, must store incredible amounts of energy—that is, must have high energy densities. But high-energy storage creates trouble for any theory. Or is energy fed into them? This, too, creates difficulties for physics.

Also, plasmas are unstable and need confinement mechanisms. Thus they were of great interest to controlled fusion researchers, and some physicists once believed that fireballs might provide clues to fusion reactors that will use deuterium and tritium for their fuel. Today, physicists feel this may not be a direct road, but since no one knows the mechanism behind true ball lightning, who can say for sure. In either case, such long life-timed Methuselahs really puzzle physicists.

Yellow ball slips through glass. [295 Kirkpatrick] A six-inch yellow fireball appeared at the window and passed through it, moved across the room and out another window. Its center glowed yellow, with a whitish ring around it. It lasted one minute.

Author—The witness implied that the ball slipped through glass. In many other cases, fireballs did slip harmlessly through glass without doing visible damage. However, in some events, fireballs melted holes in glass windowpanes and Leyden jars, and in one case, the fireball slipped through one glass pane but melted a hole in the second pane. Sometimes the holes are clearly punched through the glass, as discovered by Russian physicists, as a feat only duplicated by ultra-fast lasers.

Size-changing fireball last two minutes. [351 Hays] A five or six-foot airborne fireball gradually changed in size and began slowly disappearing after two minutes, lasting up to three minutes. There was no lightning flash.

Methuselah lasts 15 minutes. [296 Kent] For fifteen minutes, this one-foot yellowish-white fireball was airborne. I was a quarter mile distant. The size gradually changed and it slowly decayed.

Author—Did he mean 1.5 seconds? And did he estimate it or actually time it? There are rare accounts of mountain fireballs that lasted for fifteen minutes. But most disappear after several seconds, and few last a minute. Few exceed three minutes.

Mountain fireballs parabolically toss to and fro. In 1885, heavy snow and rain fell at 8,215-feet altitude upon two Swiss mountain climbers. A horizontal row of smaller yellow fireballs appeared on a ridge and jointed to form a large glowing mass that ejected red and blue fireballs, which kept floating downward and exploding. Another fireball materialized in thin air and began tossing left and right in a parabolic arc. It would disappear, then reappear, and continue its tossing back and forth again for several minutes each time.

Questions exist. Why did the horizontal row or "fireball necklace" ignore gravity, but then coalesce into one glowing mass? And why did it emit blue and red balls that fell under the influence of gravity? Why did the fireball toss back and forth? Why when in flight, did it seem influenced by gravity? What force tossed it to and fro at the same angle of elevation and with the same force so accurately that the strong wind did not affect its flight? What was its internal propulsion force and guidance? Why did the earlier red and blue balls float downward while this ball seemed like a basketball being thrown back and forth? If the ball was purely subject to Newtonian mechanics, then the time of flight is easy to calculate from simple physics. But not knowing the distances and times, we could guesstimate that there were at least a hundred tosses per appearance.

Hissing Noise and Ozone Odor

Many fireballs are silent. Most, if not all, bouncers are silent. Some crackle or hiss. Some are loud. When they emit rays or horns, they can make sudden sounds. They may vanish silently, explode mildly or loudly, or they may disintegrate with changing noise. A few fireballs roar. A few sound like a train. Nearby radios (AM only?) rumble static. If FM also does, this is no surprise, because FM also can detect static. You can demonstrate this by placing a heating pad near the FM radio. Some experts have tried to correlate color and intensity with noise, but there are too many exceptions for any clear rules.

Many witnesses—perhaps roughly one in three—close to fireballs report a pungent odor and compare it to ozone—that is, if they smelled ozone before—or compare it to the electric odor sometimes near a photocopier or to burning sulfur, to ammonia, or even to burning tar. However, many other witnesses report that there was no odor. Even for someone with an impaired sense of smell, the strong odor of ozone is unmistakable, so it is unlikely that they overlooked the odor. No witness ever compared the odor to anything pleasant, such as flowers.

Yellow-red dazzler flies crackling, slowly disappears. [176 Gosen] A two-to-three-foot crackling yellowish red fireball, which occurred after a lightning stroke and was airborne, disappeared slowly after several seconds, leaving an odor that was outstanding, and similar to a welding arc.

Red fireball tours nuclear bomb facility. [245 Smith] A very bright red two-foot fireball moved through the [deleted] room of a nuclear bomb-making facility, leaving an odor "like a cold furnace." It did not change size. There was an external nearby lightning flash prior to this.

Dazzlers

All fireballs glow—with the possible exception of the infrequently reported alleged gray and black fireballs, described elsewhere. Most balls are bright, some are dim, and a few are dazzling. Rare are those so intense that witnesses cannot look at them.

One-minute dazzler like sun. [208 Peterson] After a nearby lightning stroke, an eight-inch fireball gradually grew in size as it randomly moved. It was like sun-rays—an intense glare like the sun. There were no color changes, and after a minute it slowly decayed. After its disappearance, we had spots in front of our eyes, as if staring at a bright light or the sun.

Dazzler over house. [200 Lamb] There was a keen streak of lightning from the south. On the end of it shot a softball-sized ball of fire that flew to the north. It passed right over the top of my house. The fireball was very hard on my eyes and was just like looking at a welding machine.

Dazzling blue ball. [301 Sharp] Simultaneous with a nearby lightning stroke, a blue three-foot of dazzling fire appeared. Its appearance caused a glow like an arc from a welding machine. After thirty seconds, it burst into a puff of smoke.

Five red dazzlers slowly float. [18 Clark] Within three hours on a day in 1934, I saw five, five-foot, brilliant reddish-orange fireballs occurring after lightning flashes that floated up, changing in size and slowly disappearing after five to ten seconds. This took place five miles east of Crossville in Tennessee, fifty miles west of Oak Ridge.

Greens, Combos, Color Changers

Although multi-colored fireballs occur often, color-changers occur less frequently. What are the chemical reasons for such changes? No one knows.

Green fireball visits lineman atop pole. [416B Guice] As a lineman, I was restringing a three-phase line of 13,800 volts. It was a hot afternoon around three. I was at the top of an electric pole, which was on a slight hill tying in the line. A thunderstorm came up quickly and on the second strike, the bolt struck the outside phase, two spans away. A five-inch fireball came up the line in my direction—green, as in a prism spectrum, with a corona two inches around the ball. It traveled a span-and-half, about 300 feet, and disappeared abruptly. I received a slight shock from the line I was working on, which was the line on the inside.

Multicolored red-white-green-orange. [284 Hill] During a late damp October afternoon in 1934 at Cumberland Gap in Tennessee, I saw a green ball of fire come from a southwesterly direction, about 30 degrees above the horizon. It changed to orange, traveled in an arc to southeast. The light around the fireball was white, with a red center. After fifteen seconds, it exploded like a Roman Candle and broke up into several points after the explosion. This occurred in a rural area without power lines or conductors. On another occasion, last summer during a thunderstorm, I observed fireballs come down a power pole and hit the ground and bounce.

Sky-blue fireball rolls down roof. [539B Metcalf] In 1912, I lived on Second Street. in Elyria, Ohio. The fireball was the most beautiful sky-blue I have ever seen. It rolled down a roof slope like a blue streak. It appeared during a heavy downpour April 15 at midnight. The roof was very hot. There was no damage to the roof. It was in Elyria, Ohio.

Silver-yellow big ball rolls, explodes. [44 Hess] During a storm, I heard a bang in the front yard, where grow many trees. About 30 feet from where I stood, a two-foot silvery-yellow fireball rolled for about six feet and exploded after three seconds with a terrifying noise.

Blue-red color-changer trails orange. [392 Stinbrett] Lightning struck and this blue-green to green-yellow basketball-sized fireball jumped or arced away some fifteen feet. It got smaller as it traveled in the air. It began as bluish-green, changed to an orange-red, then went out after a couple seconds. It was pretty. A yellow-orange red streak followed behind like a vapor trail, which also went out when the fireball did.

Junction box tosses dancing chameleon balls. [394 Buchanan] In 1951, while working as a telegrapher for the Illinois Central Railroad, I witnessed a violent electrical storm near my office in Lowes, Kentucky. I think a bolt struck the Western Union wires adjacent to the tracks and followed these wires to the junction boxes inside my office. As this occurred, these balls of fire danced across the boxes and down to the floor and gradually disappeared. Like chameleons, they were at the outset a dark blue, slowly changing to bight blue, then to green.

Yellow fireball changes to blue. [343 Teal] During a lightning flash, a six-inch ball of yellow fire appeared, but within seconds it changed to blue and then slowly disappeared.

Deep rich green shell. On a cloudless and windless day at noon, a very bright eight-inch green fireball slowly descended with a deliberate purposefulness downward, and then at an angle, and disappeared silently behind some bushes. The brightest part of the ball, its core was white with a faint green tint and was surrounded by a transparent deep rich green.

Weird orange-blue-red encircles white ball. Twenty yards before them and floating 15 yards up, moving at 2.5-feet-per-second across the road as it gradually descended, the fireball with a bright center had a weird orange blue periphery and red boundaries.

But not all fireballs are beautiful Some nasty and deadly fireballs harm, burn and kill their victims. As beautiful as are these fireballs just described above, there are some other fireballs that their victims wished they never met.

8

Mild to Deadly

Physical effects include any change to a person, animal, plant or inanimate object. From electrical shock and explosion, fireballs do cause injury and death. They manifest themselves in mechanical, heat, electrical, magnetic, optical, x-ray, and radio static effects. There is some mysterious connection or underlying cause, yet unidentified, associated with fireballs that will cause effects.

Whatever this underlying cause is, it is somehow associated with many fireball appearances. For example, it is not uncommon for a ball to stay, say, ten feet from an object such as a telephone, and yet that phone may be burned out or ring strangely. And yet there may be no lightning storm, or at least no nearby flashes. Explaining this strange association is frustrating for physicists. At any rate, the physical effects and injury are visible for witnesses to later examine.

Injury to animals and humans is not frequent, and most eyewitnesses escape injury. Of those that are injured, most recover and death is very uncommon. Most fireballs are mild, others are dangerous, and most witnesses wisely refuse to touch fireballs. One child that did so once was instantly killed. But injuries are usually minor. And spooky confabulations like spontaneous human consumption are harmless—simply because they do not exist. Injuries divide into categories of severity.

Categories of Injuries

Injury and death from ball lighting are factual. The statistics are not. There is very little solid literature on the number and type of electrical and explosive injuries from fireballs. Most information comes from anecdotal reports. It is confusing and difficult and even impossible to separate the different causative components in fireball cases. Furthermore, most people are not looking at the victim when he is hit, so that some lightning victims that were injured or killed by ball lightning leave no witnesses. And because ball lightning was historically deemed not to exist, many cases of injuries were incorrectly reported as originating from traditional lightning. There are few clean cases.

In fact, even worse, there is no agency that requires reporting any lighting injuries. And because many lightning victims fail to seek immediate medical treatment, or even treatment at all, it is difficult to acquire solid statistics. There is even less medical literature on low voltage and low impact explosive injuries because most of these injuries go unreported and often untreated. This is unfortunate, because, excluding Hurricane Andrew in 1992 and flash floods, lighting each year has killed more people than any other natural disaster.

Historically, farmers were the primary victims of lightning and also frequently saw fireballs. But now that farming is mechanized and fewer people work on farms, there now are more recreational injuries. Because there are more groups of clustered people outdoors today, lightning accidents now involve more victims per stroke because the current splashes and spreads.

To understand ball lightning injuries, it is important to understand lightning injuries. With most high-voltage lightning injuries, histological research reveals coagulation necrosis not inconsistent with thermal injuries, and other damage such as changes in protein configuration that disturb cell functioning and damages cell walls. Electromagnetic and magnetic damage also occurs, but little is known.

High-voltage direct current causes a single muscle spasm that reflexively convulses the victim away from the source. But alternating current is three times more deadly because of tetany, which can occur even with very low currents. Because the muscles are stimulated 120 times per second, they are in continuous contraction. Also, because the flexor muscles of the hand and arm are much stronger than the extensors, the hand grips harder. When the current exceeds 8 mA, which is the let-go threshold, the victim cannot release the "hot" current source until he collapses, unfortunately exposing him to prolonged contact. In the Edison-Tesla dc-ac debates over safety, Edison was correct—at least on the safety aspect of dc.

Unfortunately, what is unknown about the patho-phsiology of current flow in humans and animals exceeds what is known, and medical science has much to discover, but much is already known.

If the impedance of biological tissue is low, current flows through with less opposition. Thus there is less damage because Joule's law states that power is current times voltage, or current squared times impedance, and so there will be less heating. Impedance is resistance plus reactance. Reactance does not consume power and has inductive (coil) and reactive (capacitor) components, which are either lumped or distributed. But, it is the resistance that opposes current flow and causes heating. Further confusing matters are the transient and ringing sinusoidal waveforms, or those that are composites as described by Fourier and Laplace Transforms. They are important to this field, but not to our discussion.

Unlike resistance, reactance does not consume energy—it merely shifts energy back and forth like a pendulum between inductance and capacitance. It is like hitting a bell and hearing it ring and slowly die out. In perfect theoretical (ideal) inductors or coils and capacitors connected in parallel, the oscillating current and voltage will oscillate back and forth or ring (resonate) forever—that is, if there was no resistance. This perfect condition only occurs at cryogenic (supercold) temperatures, of little interest in most practical situations.

Since reactance consumes no power, we can probably ignore it in most lightning victims. Active power that dissipates in paths taken by lightning through the body is current squared times resistance. Those body parts, such as nerves, muscles, veins and arteries—contain greater water and electrolytes, so are good conductors. But fat, tendon and bone exhibit high resistance. They resist current, and thus heat first.

Skin resistance is intermediate, but sweating lowers it to 2.5 kil-ohms, and being immersed in water further lowers it to half that resistance. This permits more current to flow over the skin, creating a greater chance of cardiac arrest but no surface burns. Luckily, the short duration of most lighting strokes translates into less internal current flow and flashover current going around and outside the body. More dangerous, "hot lighting" is no hotter than "cold lightning," but only lasts longer and thus the longer exposure is more likely to set buildings afire and kill its victims.

When a victim receives two mA from a 120-volt rms—root-mean-square or ac sinusoidal equivalent to that of dc—current, he feels a tingle. But above 15 mA, he will freeze to the circuit. At double that current, he will stop breathing. But in the 50-to-100 mA range, his heart will go into ventricular fibrillation. For "hot lightning" the victim is exposed to currents longer, and tissue breaks down, changing its resistance so unpredictably that it is impossible to predict or determine amperages.

High voltages above one thousand volts destroy more tissue and cause worse injury, including major amputations. There are no known ball lightning survivors with amputations, but there are some who experienced minor burns and shocks. A very few were burned seriously and killed, and several were permanently blinded. Old-timers believe that four-legged animals such as oxen and livestock are more susceptible to injury and death from ball lightning than are people.

The pathway taken by current determines risk. If the pathway is through the heart, it will end in cardiac arrhythmia; through the brain, respiratory arrest or paralysis; through the eyes, cataracts. In a few cases, pinholes in the eye leak out fluid and dangerously lower interoccular eye pressure, and if not immediately treated will lead to blindness.

Also, in America, between 900 to 1,000 persons are injured by lightning each year, but it may be higher because of sloppy reporting. And no one has an estimate how many of these injuries came from ball lightning. Best "guesstimates" might put it at five deaths per year from ball lightning and 20 injuries, mostly minor. Ball lightning has a wide variation in lethality. Some whom are touched only feel a tingle; others are burned; a rare few are totally unlucky.

Injury and Death

Fireballs rarely cause permanent injury or death. Although most exploding fireballs are harmless, some do great damage. Some experts believe that, under similar circumstances, animals are more often killed than people. Fireballs can create unpleasant death. Gruesome burns and streaks of charred flesh may mar the victim, and generally the corpse decays quickly—a phenomenon associated with high voltage electrocutions.

A fireball's explosion may cause injury and damage, intense heat, and various electrical effects that include massive voltages and current surges, electrical and magnetic induction, and possibly other effects not yet known.

For example, a fireball or some underlying phenomenon associating with the balls will burn out light bulbs that are not nearby—and the filaments never glow. Certainly, nuclear electromagnetic pulses or NEMPs from neutron and nuclear bomb electromagnetic pulses can do some similar "magic," but not so selectively. Furthermore, no such intense pulses have ever been detected. But it is the incidents of injury and death that captivate the imagination.

Chandelier ball singes mom, kills woman, burns her black. [527 and 584 Swank] My mother, when a young woman, was at a lodge meeting on the second floor of a two-story building, in a small town when a green fireball descended down a chandelier. It killed one woman, set fire to the clothing of another, singed my mother's eyebrows and scorched the turquoise blue velvet lining of her hat. I remember seeing the brownish scorch color on the blue. The woman killed was burned black. Three women were burned, our mother, least of all. The woman who was killed was presiding and our mother was an officer. A ceremonial was in progress and all the injured women were grouped around a podium. The fireball lasted ten seconds and there was an odor. I was five at the time and my brother and sister, several years older.

Fences shoot balls, exits stoves. [interview, 1974] Many of us older immigrants, particularly peasants and country folk, have innumerable accounts of fireballs from our native homelands. Some fireballs roll along fences, shoot off. People still see fireballs today, but not like before because more people live indoors and in cities, and are not as close to the outdoors as we were. We observed more carefully, we saw more. One time the thunderstorm hit and it began raining. Suddenly the stove cover began rattling, and out popped a ball that glowed like a blacksmith working a horseshoe. It went out an open window, exploded in a bright flash, and the house shook like thunder.

Falling blue ball explodes, kills two. On May 29 of 1938, a blinding blue fireball fell on the Bold Colliery at St. Helens in England. The explosion injured two men and killed another. Two weeks after this, a blue fireball fell upon a house on Brook Street, not more than a mile away, and soot and fire debris showered down into the street.

Chases and electrocutes girl, 21 deaf. When a violent storm broke out in Ouralsk in France on May 22nd of 1901, a festival crowd filled the streets. By five, twenty-two youngsters gathered inside a house, waiting out the storm. A seventeen-year-old girl sat down on a step with her back to the street. A fireball floated in, attached to her neck—her head slumped foreword—and rolled down her body, floated into the house and exploded, doing considerable damage. A streak of charred flesh marred her corpse. The others were deaf for life.

Killer ball burns boy. During a spring thunderstorm that lashed western lower Michigan, flying at three feet above the ground, a fireball that flew straight into the chest of a boy, blew his shirt off and burned his internal organs, killing him.

Shorted equipment emits injurious fireball. When a glowing machine short-circuited, from the socket (which burned out) came an 18-inch fireball that flew horizontal into a man standing six feet away. Hit, he fell to the floor with his clothes around the impact area on fire—later leaving a burned-out area of eight inches across. The victim was carried to the emergency medical room.

Gnome's blueball kills Richmann. On August 6th of 1753, a sudden and intense lightning storm raked St. Petersburg in Russia. Swedish scientist Professor Georg Wilhelm Richmann stood several feet from the gnome or pointer, with his head inclined toward it, when a fist-sized blue fireball moved from the rod of the gnomon or vertical column toward his head and exploded.

Solokow, an observer from the Royal Society who witnessed the event, was knocked unconscious. When he awoke, the door casing was split halfway though, and the door itself was torn off. Richmann lay dead. A red spot on his forehead spurted drops of blood. His left shoe burst open, exposing on his foot a blue mark. Richmann's clothes were unburned but many burn marks covered his body, which corrupted abnormally fast. It was only with great difficulty that the mortician was able to insert the bloating corpse into its coffin.

Benjamin Franklin also made this same observation, experimenting with geese and ducks that he electrocuted from charged Leyden jars. In addition, Franklin discovered that when the electrocuted fowl were immediately cooked that their meat was made tenderer. It is not surprising that the corpses of those electrocuted by fireballs should corrupt so abnormally fast. But why exactly do victims of fireballs decay so quickly? To answer this, we must first medically examine normal departures.

Death is natural. When the heart stops the body temperature drops (*algor mortis*) until it reaches the same as the surrounding air or ambient, in under three hours. Within two days, temperature rises due to activity from protozoans and bacteria, which first attack the blood and intestines. Other organs decay at different rates. However, death from fireball electrocution accelerates the decay process.

Decomposition follows six stages. At first the flesh seems normal, although the skin pales as gravity pulls blood downward. In seated victims, witnesses describe an ashen color that begins at the head and rapidly moves down. Skin near lower blood vessels turns purple (*livor mortis*). The body then distends or bloats from internal gases and the abdominal skin turns a green color that travels to the thighs and chest. The mouth and nose ooze blood, tongue and eyes protrude, and sulfurous odors waft free. Next, the flesh turns liquid and slips off, the abdomen collapses, and organs explode and release liquids, and intestines are forced out openings. In the butyric stage, fermentation removes flesh; and in the final stage of dry decay, the skeleton remains. Then, the skeleton dissolves, especially in acidic wet soil.

At death, the muscles briefly relax (*primary flaccidity*) but rigor mortis soon begins in the eyelids, jaw and neck, and in four to six hours extends to the other muscles and organs. This continues for one to two days, when decomposition is under way, at which time the muscles relax again (*secondary flaccidity*). Before death, Richmann's body operated both aerobically and anaerobically (with and without oxygen). Anaerobic activity generates lactic acid. At death, oxygen intake stops and lactic acid increases and turns flesh gelatinous-like and continues to *rigor mortis*. Upon execution by electrocution, cell walls burst, body fluids vaporize and seek exits. Richman's electrocution by ball lightning was similar.

Shocks People

When victims are thrown to the ground by a fireball explosion, it may be from the blast or from the electrical shock. Some who arise are unable to talk. When they regain their power of speech, they say their legs suddenly ceased to support them and they fell to the ground with flexed limbs. Voltages or potential differences because of temporary unequal charge distributions and current flows, usually invisible, sometime travel evenly outward but may concentrate in certain directions, and may prefer conductors or seek capacitance, such as a human body. Sometimes it is impossible to determine if the shock came from the fireball or a nearby lightning stroke.

When fireballs touch a metal conductor, such as an iron gate or cable, they may ignore it, glide or roll along it, or disappear into it; converting into a large surge of current that briefly flows in the gate or line. The mere presence of a fireball nearby is often associated with charges, voltages and currents in nearby conductors.

People differ greatly in their susceptibility to electrical shocks. For example, according to nationally renowned court witness, investigator and criminal authority, police Captain Massad Ayoob of Concord in New Hampshire, Los Angles' infamous and prison-trained, hulking bodybuilder Rodney King in 1991 took shocks—after a wild attempt to out-drive police cruisers—from the arresting LA police officers' taser that might stun a velociraptor, but with no more apparent effect than mosquito bites, which alarmed the officers.

Ayoob reports that some ex-convicts and felons zap themselves with tear gas and stun guns to build increasing tolerances so they can resist police; and, like evolving cockroaches, some develop strong tolerances. Ball lightning victims, animal and human, likewise differ in their tolerances.

Shocks hand, squeezing effect on body. [150 Reinert] During a 1934 lightning storm, a one-foot fireball appeared between two houses in Joplin, Missouri, and bounced on the ground. My mother and I were in the kitchen. She felt a shock in her hand, which held a paring knife, and felt a strong squeezing effect across her whole body.

Pole-decender rolls and shocks. [298 Ledford] Just after the worst part of a lightning storm in Tarrant City, Alabama, I walked out on the sloping sidewalk (an electrical ground) where rain still ran off. Telephone poles and wires ran overhead, over the sidewalk and street. Suddenly a yellowish-orange, ten-inch fireball ran down the side of a nearby pole and along the ground for twenty-five feet, and after a couple seconds, disappeared with a loud crackling. I had a table knife in my hand. Simultaneous with the fireball's departure, I felt a slight shock. I stood ten feet from the ball.

Ball flies off oak, electrifies observer, then explodes. [42 Strain] It was a late April afternoon in 1946, about five, with a low dark cloud cover and starting to rain large drops, with much thunder and lightning, and a gentle ten mph breeze from the southwest. I was looking out of a second-story Northwest corner window in the dormitory of the Columbia Military Academy of Tennessee. I pushed the screen out to shout to a friend who was running towards the barracks. He was forty yards from them when lightning struck a large Oak tree thirty-five yards to my left, with an ear-splitting crack.

After two seconds, a large orange-to-light-yellow sphere rolled off a large limb into space toward the northeast. It rose at five degrees, rolling rapidly, increasing in size and slowing its speed as it moved from the tree. When it flew thirty yards and grew to five feet, it exploded with a terrific roar, similar to a high explosive detonation. My hand—it was in contact with the screen—tingled throughout the entire flight. At the explosion, it received a shock.

After the explosion, a strong odor of ozone lingered. The oak was afire, with one eight-inch-diameter limb broken off. The young man was unharmed, despite the fireball exploding only twenty yards from him, almost overhead. He had time to stop running, look at the ball, and dive to the ground before it exploded.

Fence roller shocks guard in portal. [59 Brown] During an electrical storm one Saturday morning about ten-thirty, at the Central Portal of the [deleted] nuclear plant, I sought refuge from the rain. A tremendous bolt hit a hundred feet from the portal. An 18-to-24-inch orange fireball, making a frying sound, rolled from the top of the perimeter fence toward the portal at forty mph. At twenty-five feet from the portal (after fifteen seconds), the ball extinguished itself at a point were a ground strap attached to the fence. Heavy ozone lingered. The guard on the gate, with whom I was conversing, had one foot on the radiator in the guard shack. The instant the fireball disappeared, he received a shock from the radiator. He required no medical treatment and I experienced no shock.

Two fireballs attack and cripple victim. [583 Spears and brother] In 1957, I was hanging out some bleach blankets in Miami. Although we had an earlier storm, it became a beautiful July Sunday. As I took the blankets off the line, I noticed an orange-and-red basketball-sized fireball rolling (airborne) at me down the metal clothesline. When it reached the blanket I was hanging out, it went under it, and I felt a searing pain in my left leg. My pants leg was burned, and the hair was burned off my leg from the knee down. The noise was like high tension on a TV set.

My brother took me to Jackson Memorial Hospital in Miami, where they took x-rays and told me to walk on crutches. I was crippled for about three weeks. I stayed there overnight. When I returned, they wanted me to go to the University of Miami for further tests, but I refused. They said the bone in my leg was burned, and I believe it, for it still pains me so bad I cannot sleep on some nights. If you would like the hospital records, I will help you get them.

Ball electrifies fence, paralyzes boy. Former Deputy Director of the British Meteorological Office, James Durward and his twelve-year-old son were driving through a back road in Lock Tummel, near Pitlochry in Scotland. The sky turned dark and the wind picked up. Lightning flashed as a storm descended. When they came to an iron fence and gate across the narrow road, Durward stopped his car and his son opened the rusty gate. At this instant a twelve-inch white fireball floated from between some pine trees. It struck the fence and vanished silently. The boy jerked back from the gate with a shock so severe that his arm was temporarily paralyzed.

Three balls melt cable, damage equipment, shock workers. On April 26th in 1939 at three-thirty in Roche-fort-sur-Mer in France, lightning broke into three branches before hitting ground at a construction site. Near the first impact point, a worker installing a gas pipe had just climbed out of a ditch. Though shocked, he was uninjured. An eight-inch fireball appeared several feet above ground level, climbed up an iron cable, which then melted and smoked, and disintegrated. Several workers at the second impact-point saw an eight-inch fireball slowly float towards the top of a metal crane. It touched the crane and vanished with a great explosion. At this, a thick spark shot out 130 feet and hit a dockworker in his forehead, knocking him unconscious.

Twelve shovelers working between 30 and 160 feet from the crane were knocked down and one was thrown two feet into the air. Their shovels were hurled a dozen feet. A nearby electrician handling a wire received a shock and his hands froze to the wire. Intense current surged in the crane's electric cable, vaporized the circuit breaker and fused motor windings. At the third location, a grapefruit-sized fireball hit a lightning rod, descended, and reached ground in twenty seconds, then rolled rapidly down the road, passing houses. Inside each house, sparks jumped from various objects and several circuits blew out.

Author—Your author could not obtain a detailed map and description from that time period. And he cannot determine if the shocks were caused by the fireballs or by lightning. Probably both contributed. Or, could some common but hidden underlying phenomenon, a "Hidden Progenitor" or parent cause, about which we can only speculate, provide the energy and charge required in this and other cases? It is impossible to interview the witnesses. If this account is even half accurate, then the amount of charge contained in these fireballs is astounding.

Exploder shocks workers and punches lady. Seen by a forklift driver outside a British print factory, a dazzling bluish-white, tennis ball-sized fireball bounced along the roof and passed through netting to enter the building, then flew along the ceiling girders. It shocked three workers and forcibly hit one woman in her shoulder, and created such a ruckus. The fireball then hit a printing machine, showering sparks. Finally, after hitting a window, it exploded in an orange shower.

Fireball electrocutes woman holding flatiron. On a day windless, stuffy and hot, just after a rain, a woman outdoors put a flatiron behind her back. She failed to notice that a fist-sized glowing fuzzy fireball materialized nine feet behind her. The witness, a physicist, said that the ball flew into the iron and electrocuted her. Still breathing after a half hour, the woman's face turned blue and she died.

TV emits fireball that hits man. During a winter southern lightning storm, a basketball-sized, whitish blue fireball emerged from the television and landed on the lap of an awakened man, lying in bed. At this, lightning struck nearby. He began flopping about, but could not get out of bed and could not swing his legs over the side. Finally he succeeded. He was covered in goose bumps, his right side tingled, and the muscles on that side kept flexing uncontrollably.

Chicken wire fireball injures farmer. When a chicken farmer leaned over a gate made of chicken wire but had not yet touched the wire, he felt a tingling on the nape of his neck, and he stood up. But when he touched the wooden post, a large fireball rose from the chicken wire and hit his arm. Severely injured, he went to the hospital and took three weeks to fully recover.

Minor Injuries

From the earliest reports, injury from ball lightning embedded itself in our folklore. More recent reports confirm the early stories. Most intimate contacts with fireballs only result in minor injuries at worst and close encounters at best. Oak Ridge witnesses were no exception.

Blinded from fireball's glare. [464 Horton] An airborne, blue-white fireball changed to orange-red, followed a conductor for several seconds, exhibiting heat and scorching before it burst, temporarily blinding me.

Attic floater safely grazes lady's ankle. [541 Larsen] I was at the head of the attic stairs. A four-inch fireball appeared at the corner of the attic near a small window, where electrical wires attached to the house. The bright surface was spherical but not smooth, the color of a candle flame, and with the same range of color variant. It moved in a straight line, somewhat diagonal across the floor, starting two feet high and lowering to six inches at the stairwell. At that point, it passed by me, and appeared to touch my ankle, with no electrical or physical sensation. It then turned and floated parallel to the stairs to the back of an outside door in a hall at the bottom of the stairs. At this point it rattled a key that was left in the lock and passed from view. It covered twenty feet in three to four seconds.

Scorches trousers of seated man. [505 Smith] During a thunderstorm in 1929, while sitting in my living room, a two-foot fireball of a light color, floating a foot above the floor, entered the room through an open door, then traveled in a circular path. It passed close to me and scorched the leg of my trousers, numbing me, then left the room through the door. Six people were in this front room, but no one was hurt, although all saw the fireball.

Line ball bursts pole, hurls man. [386 White] I was walking on the old Scarboro Road in Oak Ridge at the railroad crossing. This was a rainy early evening at five with dark clouds hanging in the east. There was a flash. Within seconds I heard a hissing noise. A fireball hit a utility pole and exploded. Traveling the electric line to the traffic light, sixty feet distant, it burst into flame. I found myself getting up from the gravel shoulder of the road.

Ball descends pole, topples man, crosses road. [45 Baker] Over the island of Kauia, the early summer sky in 1945 was overcast. Rain fell. I ran along a blacktop road, following another man, past a power transformer pole. Lightning struck the pole. A one-foot, whitish-blue fireball traveled down the pole, across the road above ground level, and into some bushy undergrowth about 25 feet away, causing the man to fall down. He was uninjured.

Bursts over motorist. [210 Neal] After a rain, an 18-to-20-inch fireball came straight down and burst 15 feet overhead. On another occasion, I saw a two-foot fireball three feet above the ground during a rain. It came straight toward me and traveled 400 feet. It disappeared after several seconds. My son once saw a "fall of fire."

Midair Violet Globe Disintegrator Topples Porch Furniture. [558 Prescott] On a screened porch, after a flash of lightning and followed by thunder, I noticed the blue-violet eight-inch globe hung in midair. After four seconds, it suddenly disintegrated, knocking over a wooden table and toppled two wooden chairs. I felt no physical effects, just fear.

Fence-Roller's Explosion Hurls Man And Hammer Ten Feet. [115 Walsh] Another man and I were fixing a fence where the cattle had gotten out. I was pulling or stretching the wire while he nailed it onto a post. This was in 1940 just after a bad electrical storm. I looked up and saw a large orange-yellowish, ten-inch fireball coming down the barbwire so quickly that we had no time to escape. He was knocked about ten feet away, and I was about two feet closer to where the ball had come in contact with us. We never found his hammer. The ball traveled on the top wire, about three feet above the ground. At first it didn't travel fast. The fireball's color was that of a moon on a clear bright night.

Bounces off car antenna and hood, blinds man, shrinks. [530 Attrill] Lightning struck my father's car in 1949. A baseball of fire burned off the antenna, rolled down the hood and bounced along the road for 20 feet and gradually got smaller before disappearing. There was a sharp odor. I was temporarily blinded.

Fireballs play all over floor. [477 Presley] One night during an electrical storm my radio and electrical stove burned out. I was closing windows when several fireballs of grapefruit-sized fire played all over the floor. I experienced some blurring and loss of my vision.

Hits hand, goes numb. While at a video game, a player reported lightning struck the first neighbor's antenna and a faintly hissing fireball appeared and hit his hand and then vanished with a loud snap. Unhurt, he felt his hand numb for several minutes. Although there were no burns in his house, a microwave in the second neighbor's house shorted out.

Floor roller badly shocks boarding house waitress. [538 De Laguna whose 80-year-old mom wrote this case up on July 10, 1961.] I was a child of eight and did not see the fireball myself. My mother and I were staying at a country boarding house in Connecticut. We had been to church and were waiting for our dinner with the other boarders in the sitting room when the storm broke, and the house was struck. There was a flash and crash. Bricks and shingles fell from the roof, and wood loudly splintered. We later found a heavy attic rafter splintered.

The shiny patent leaver was peeled off and hanging in strips from the dashboard of the carriage out in the yard.

The waitress saw and felt the fireball. She was setting the table in the dining room and saw the ball come in through the window and roll across the floor toward her feet. It may have touched her feet because she was badly shocked. The house was not wired for electricity, but there may have been a telephone. I do not know if it had a lightning rod. The old large farmhouse had a stable and carriage yard. My new pink dress that I had worn to church lay on the bed. Above the bed was the stovepipe, with a hole stopped with a tin cover because there was no stove in the bedroom. The shock had forced or blown the tin cover out and covered my dress with soot.

Bright fireball hits trucker's face. A trucker and his wife, Eddie and Velma Webb, from Greenvillein in Missouri were driving on an early October night through southeastern Missouri. There was little traffic, so when he looked in the mirror and saw a bright light coming up real fast, he yelled for his wife to wake up. When he stuck his head out the window, the fireball struck his face, knocking off his glasses. He stopped the truck. Highway patrol Sergeant Ed Wright investigated and sent the glasses to Dr. Harley Rutledge, the head of the state Southwest State University physics department. Warped with one lens missing, and examined under a microscope, the glasses seemed internally heated, with some residue on the surface.

Blueball safely grazes four, explodes. A classic 1900 case was reported by the Baron of France. Some lightning balls touch people without harming them. Eleven people gathered in this French aristocrat's salon during an electrical storm. A blue fireball appeared and slowly crossed the room. It grazed four people without hurting them. It floated out the open door in front of the entrance to the grand stairway and exploded.

Ball smacks woman on cheek. A bright streaking fireball hit a woman on the side of her left cheek, and a large red blotch appeared later.

Throws tobacco farmer onto ground. When a tobacco farmer bent over to take a drink from a wooden barrel, lightning hit the wooden pole holding up the shade netting. A yellow basketball-sized fireball traveled down the wire and jumped over the barrel. The farmer, thrown back and onto the ground, although shocked and stunned, soon recovered.

Bounces off metal table, sticks to finger. A one-inch fireball slipped through the glass window, bounced off the metal washer, very quickly bounced off a metal table, back to the washing machine, bounced off that onto the mother's small finger, where it painfully stung. She quickly shook her finger and the glowing ball flew into a corner and silently vanished.

Lady's finger and flyswatter attract fireballs from floor. With her hand on her waist, a lady stood on a rug during a thunderstorm. The air between her finger, which was slightly extended, and the floor wavered as if warm. Something from the floor slowly rose up to her finger. Standing five feet away, she saw a pecan-sized oblong fireball that glowed through a haze briefly attach itself to her finger. Lightning flashed outside, the ball vanished. In another nearly identical occurrence, the fireball arose and attached to a flyswatter.

Glowing bubble harmlessly bounces on girl's head. In early evening, after an electrical storm passed, the air turned cool and windless. A 30-inch whitish-gray, glowing "electric soap bubble" that was bright (the core was more intense) appeared two feet above the head of a pre-school teacher walking past a railroad track. It circled her, moving up and down, before it suddenly floated above her and rapidly jumped up and down harmlessly on her head. It quickly ascended and vanished.

Exploder hurls two climbers to ground. On a clear day on a Colorado mountain, a bright large fireball exploded 12 feet from two eyewitnesses and in a loud explosion hurled them both to the rocky ground.

Chases and burns woman. During a thunderstorm, a woman was deliberately chased by a four-inch blue ball of fire surrounded by a halo. It burned a four-by-three-inch hole in her skirt and then exploded. She had first-degree burns on her left hand from trying to remove or ward off the fireball.

Kills Animals

Popular folklore held that fireballs spared humans but killed animals. True or not, no one seems to know for certain and the statistics prove nothing. Should horses and cows withstand lightning better than humans? Conventional wisdom says otherwise. Not many of the Oak Ridge witnesses knew of deaths. Although many times the concurrent lightning might have caused death, sometimes we know there was no lightning, and a fireball was the killer. Fortunately, these cases are rare.

Kills livestock. [233 Huffine] After a flash, a twelve-inch red fireball was airborne for sixty seconds and then killed several head of livestock.

Fireball inspects, then kills horses and mules. [551 Collins] Three adults and I took shelter in 1923 from a sudden morning rainstorm in a small split-rail cotton shed, square, with a dirt floor. It was nine or ten and had only begun raining. A saddle horse and four work animals, with gears on, were standing in a close group with their heads at the door and their bridle reins in the doorway. A yellow-orange fireball suddenly appeared on the animals and moved from one to another across their backs and down a leg of one to ground and back up a leg of another. After a few seconds it exploded in a blinding flash that created significant heat but not enough to burn. We were eight or ten feet from the animals and the flash temporarily blinded us and caused us to fall back onto the floor.

We were uninjured, and as we regained our composure in a minute or less, we saw that four of the five animals were on the ground. The fifth was running in circles. Two of those on the ground were already dead. One got up a few minutes later and survived. The other one died very quickly. We saw no evidence of burns on any other animal or the metallic portions of their gear or saddles. There was no other lightning or thunder in this rainstorm. The noise from this was very loud and heard by others in the distance. An unusual odor of ozone was present for several minutes.

Bouncing fireball kills Iowa horse. Martin Uman described a yellow-white "washing tub sized" fireball that a witness saw bounce down an Iowa dirt road, moving faster than a runner. It went partly around a block and then hit a small shed and blew it apart, killing the horse inside. The witness looked about when he first heard a loud rushing sound like a powerful wind, so did not see where the ball came from.

Fries squirrels. When lightning struck a utility pole, a white fireball rolled down the pole, around the tree and then ascended the pole. When it reached the top, it vanished with a loud crack. At the bottom of the damaged pole lay two defunct squirrels, both fried.

Spontaneous Human Confabulation

Creepy are the stories of victims consumed by a heat so intense that it eats their entire body. It consumes bone. It leaves greasy black soot around the room. Many victims leave behind just a leg or arm—a grizzly memento of what had once been a human.

Can a fireball materialize inside a person? If so, do such fireballs incinerate them? To incinerate a human body as allegedly occurs in some spontaneous human combustion (SHC) cases requires a fire at a temperature of 16,600 degrees Celsius or 30,000 Fahrenheit, which skeptics correctly claim would do considerable damage to everything near them. But this does not happen. In fact, the ceilings, beds and floors—although coated with black soot or grease—often go otherwise undamaged. Case closed? Not quite.

In many articles, the *Skeptical Inquirer* journal and CSICOP researchers exposed the sloppy investigations of researchers and writers. This journal describes how some body parts do not burn because of temperature gradients, or how body parts were removed before the photos were taken.

Popular articles claiming temperatures exceeding 1,500 degrees Centigrade are needed for SHC are incorrect, since tests proved that 500 degrees would almost totally destroy children and 800 to 1,000 degrees will do so for adults. At 680 degrees, arms are charred in 10 minutes; legs, 14 minutes; skull and bones, visible in 15 minutes.

When there is no external heat, fat under the skin promotes burning. An average adult has 15.4 kg of fat. If a person wears flammable fabrics, the temperature will reach 500 to 600 degrees. Mark Benecke, a qualified forensics biologist that wrote in the March/April 1998 issue of the *Skeptical Inquirer* that there are no known cases where internal organs are damaged more than the outer parts. Combustion never starts from inside the body. There are even cases where criminals were covering up evidence of their crimes.

The very skeptical University of Colorado team, headed by the late Dr. Robert Condon and financed by the USAF, investigated the following case in 1968 and concluded that it is reliable and accurate. The Condon Committee reported roughly as follows.

On the morning of Monday, July 2 in 1951, Mrs. F.M. Carpenter of 1200 Cherry Street, Northeast, in St. Petersburg, Florida, went to the door of a tenant, Mrs. Mary H. Reeser, age 67, whom she had just seen last night. She did not answer when a Western Union boy tried to deliver a telegram. When she touched the hot doorknob, she quickly pulled her hand back. Fearing the worst, she ran outside to get several house painters. They broke the door open.

Though the windows were open, the apartment was insufferably hot. Near the front window stood the remains of a once-big armchair. Among the springs of the chair were charred remains. A 187-pound body became ten pounds of charred carbon. The fire consumed her head like no normal flame. All that remained was her left foot, her skull, and a few backbone vertebrae. The police and fire department were called in to investigate.

Mrs. Reeser was last seen alive at nine the preceding evening when her son, Dr. Richard Reeser, her landlady and another friend said goodnight to her. At the time she sat in the chair she would die in, and wore a coat over that, and cloth bedroom slippers.

Her room showed little effects of the inferno, and visitors of the prior evening could provide little information. The walls above four feet were covered with a sooty black—the remains of Mrs. Reeser—and also heavily coated the drapes. The intense heat radiated from the chair had cracked a mirror ten feet away. Across the top of her dressing table two tall candles melted into pink puddles.

There were few signs below the four-foot level, and the fire left only two marks. A small spot burned in the rug beneath the chair and the plastic electric wall outlet near the chair melted, causing a short circuit that below a fuse and stopped the victim's clock at twenty minutes past four that morning. Edward Davies, arson specialist for the National Board of Underwriters, concluded the heat was so sudden and intense that he could not begin to guess what happened. Dr. Wilton Korgman, famed pathologist from the University of Pennsylvania, wrote in his report: "Never have I seen a skill so shrunken nor a body so completely consumed by heat. This is contrary to normal experience and I regard it as the most amazing thing I have ever seen."

Suppose, by a rare coincidence, a person sits or stands at the very spot where a lightning ball is trying to materialize. Imagine the millions of joules of heat energy inside the victim, all occurring within an instant—burning for less than a minute. If so, then the temperature was far above the 2,500 degrees that Dr. Korgman reported necessary to incinerate the corpse.

If true, then the heat that ate Reeser was so brief that nothing in the room below four feet had soot, and the room appeared so free from fire damage that it was easily cleaned up. Yet, so incredibly hot was the object—ostensibly so—that after it vaporized flesh and bone, the heat ostensibly burned the chair so quickly, it suddenly went out without damaging the rug. Despite this innovative theory, your author remains a rational skeptic. Exhaust ordinary explanations before you consider the extraordinary. There are ordinary explanations.

A seated body can burn and then stop before the muted fire reaches the legs. And certain fabrics will continue to burn, wicking out subcutaneous fat that liquefies, while the unclothed body parts are left intact. Easy victims, the elderly, infirm and intoxicated are unable to move quickly or awaken.

Most spontaneous human combustion cases are straightforward. The final solution to this macabre mystery came recently when British forensic experiments duplicated these cases. By placing dead pigs in clothes on beds in simulated bedrooms, these experts confirmed that a very-hot-but-slow-burning fire—with the clothing acting on melting body fat like a wick—would slowly incinerate flesh and bone of a carcass with a very hot but small flame, yet leave the surrounding room and furniture unscathed. There was very little smoke or soot. Your author saw these films and, yes, they do scientifically solve this mystery. Spontaneous human combustion occurs only under unique conditions. And, although it does occur, it is not spontaneous.

The researchers stated that bodies contain over 70 percent water. The body is too wet to catch fire. Professor Dr. Douglas Drysdale explains "the wick effect." On rare occasions the body can burn like a candle. The clothing will wick out the fat. This was only a theory and had never before tested—until recently.

In the British experiments, dead pigs were wrapped in clothes and placed on a bed with a blanket in a bedroom. After three hours, the tiny flame was 812 degrees Celsius; body fat was being wicked out; and after five hours, bones were reduced to a state where they would crumble. Nearby objects in the bedroom escaped. After seven hours, the fire extinguished itself. Looking at the film of the tests, your author saw little damage to the bed and room.

But are fireballs capable of burning flesh and bone? Maybe, maybe not. If so, it seems rare. The following Russian case is not the only one like it.

Fireball burns man's hand to bone. A dazzling eight-inch, violet-tinged fireball burped out above a bed on a light switch that began to burn with a yellow fire. Alarmed, the man hit the flame and fireball with his palm. The shinny ball broke apart into pieces that fell onto the woolen blanket that started to smell as if burning. It lasted three seconds. Thunder struck a second time and a second fist-sized fireball appeared on what remained of the damaged switch. The cord broke and the ball disappeared after a second. Only then did the observer notice with terror that the odd flame had burned his hand right to the bone and charred the sides black. Burned, his fingers were as black as coal.

Although this case was not spontaneous human combustion, it confirms that some fireballs possess flesh-burning capabilities. It suggests that some SHC might be ignited by ball lightning.

Eminent science writer Arthur C. Clarke once stated, perhaps in ironic jest, "There's one mystery I'm asked about more than any other—spontaneous human combustion. Some cases still seem to defy explanation, and leave me with a creepy feeling and very unscientific feeling. If there's anything more to it, I don't want to know." I agree with Clarke. When I first read these stories, they creeped me out, too.

Not to be a total skeptic, your author asks, if spontaneous human combustion is true, then why does it fail to materialize inside animals—like barnyard cows, pigs, zoo elephants, and wildlife? Or form inside trees? Or inside rocks? They are far more plentiful than people.

9

Mechanical Damage

Damage is usually mechanical, thermal, or electrical, and some cases combine the different types of damage. Sometimes there is electromagnetic or chemical damage. Mechanical damage is physical damage and involves kinetic energy. But what is serious and minor damage is subject to personal subjectivity. Minor property damage is when another farmer's cow is killed. Serious damage is when it is your cow.

Serious Property Damage

Although the current trend among some experts is to deny that ball lightning can do any damage—because there are difficulties for traditional physics in this—but the cases prove otherwise. In fact, some of the damage is serious.

Enters door, knocks artillery shell to floor. [86. Greshy] Lightning struck outside. A crackling, eight-inch white-to-orange fireball entered through an open door, and in a rolling motion moved horizontally five feet above the floor, in a direct line toward the mantle. It struck a four-inch diameter artillery shell casing that fell to the floor. After several seconds the ball burst.

Author—If this had been a large live artillery round with primer and propellant, or just a box of live grenades, military explosives, solid rocket fuel propellant (as with the Minuteman motors at the Hill Air Force Base Liniac), plastic explosives like C-4 and Sematek, jet fuel or volatile chemicals, then detonation by ball lightning is possible—leaving no witnesses. How many times have accidents occurred for which other things were blamed, but that fireballs caused them? No one knows. Probably not often, but the accidental formation of fireballs may have cost loss of life; and they remain possible sinister culprits for several unexplained rocket and airline disasters. The possibility exists—as stated to me in a letter

(Evans) from the Central Intelligence Agency—that there are military and terrorist applications.

Ball falls into barn, explodes, instant inferno. [573 Duncan] A ball of fire fell into a barn and exploded, causing widespread ignition throughout the barn with such rapid burning that it was impossible to remove any of the contents.

Slow descender sets Sears warehouse afire. [496 Freeman] During a July storm of 1946 in South Bend, Indiana, I saw a two-foot fireball, falling at a 45-degree angle, hit a Sears warehouse across the street from our house. There was a loud crash and an odor. It set a small fire on the warehouse dock, which did little damage. The flight lasted ten seconds.

Ball ignites barn, burns immediately. [571 Martin] My father in Kansas saw an 18-inch ball of fire bounce along the ground of our farmyard and fly up into the large barn loft, piled with dry loose hay. Yes, the barn burst into flames and immediately burned.

Orange fireball instantly ignites wet wheat field. [554 Colborn] I was riding horseback in a heavy rainstorm in a wheat field in southern Nebraska. Lightning hit behind me and the loud crack caused me to turn my head in time to see a yellow-orange three-foot fireball. Whether from radiated heat or from discharge to the ground, it set the wheat stubble on fire and burned eight acres. The heat must have been intense because much of the area burst into flames immediately. It had rained and continued to rain hard for some time, soaking the stubble and wheat. This confined the fire, since the whole field was forty acres. I am with the TVA Hydraulic Laboratory.

Author—As with many other cases, this case illustrates the enigmatic switchable behavior of fireballs. For example, sometimes fireballs exhibit little heat or energy, but then at other times they manifest intense heat and great power. Also, to complicate any explanation, at times, a single fireball can switch from nice to nasty or the reverse.

Yellow carport-wrecker dumbfounds six lunchers. [204 Shelton] This fireball or "fall of fire" broke two twelve-inch boards, shot out the end of a carport, went into the wiring inside the carport, burned all the insulation off the wiring, came into the kitchen wall switch, blew it out of the wall, and melted the switch parts. There was a loud report much louder than a high-powered rifle, and the six-inch fireball went over the electric range. This pale yellow ball disintegrated and left six sitting people, eating lunch, dumbfounded.

Fireball explosion demolishes chimney, hurls bricks. [172 Lundell] I drove home from Dandridge to Sevierville in Tennessee during a light steady 1942 August rain, but no lightning. I arrived at five p.m. and was washing in the bathroom. Suddenly a terrific explosion shook the house. I ran into the living room, as did my niece and wife. At the same time, a crashing sound like a house collapsing rent the air. Since our house was undamaged, I looked out the window. Although my garage and my neighbor's house were intact, his chimney—which projected eight feet above the roof—was disintegrated completely to slightly below roof level. Most bricks lay in a ten-square-foot area in the garage driveway in front of its door. Some of the bricks lay in the next yard, thirty-five feet from the chimney.

Author—When a chimney is damaged, it is almost always by hot (longer lasting) lightning, not by a lightning ball, even if one is seen nearby

His furnace door was blown open. My niece, who was in her bedroom listening to the radio, was looking out the window. Suddenly a six-inch fireball flew horizontally from the southwest across the next-door house, at thirty feet above the ground, heading right into the garage chimney.

Author—Although fireballs do hit chimneys, as in the above case, your author remains skeptical of most other reports that blame balls for exploding chimneys. Lightning may hit the chimney—the fireball may be a byproduct, not the cause. Lightning can totally shatter even a large chimney like a bomb hit it, and blow a large hole in the roof. If it is cold lightning (with its shorter flash), they will do damage but no fire. In this case there was an eyewitness. But fireballs that depart up through a chimney often explode like a bomb, damaging chimney tops.

Two-minute roller shreds tree. [423 Buchanan] During one storm, a pale yellow twelve-inch fireball hit a large tree and tore it up like an explosion had done it, and rolled on the ground, then gradually disappeared after two minutes.

Repeat visitor torches dishwasher. [328 Henley] Blue-green fireballs appeared several times inside my kitchen during lightning storms. At all times they moved close to the floor; and, with each appearance, something occurred, such as ringing the phone or setting fire to the washing machine.

Barn roof-roller bursts silently. [236] My father and I were driving to his farm to check his cattle. It was May in 1924 and dark clouds raced across the sky. Since our car was parked too far away, my father suggested we walk to a nearby farmhouse and seek shelter. We sat on a long porch facing a large barn fifty yards away. A hot moist wind blew and lightning flashed. I gazed intently at the old barn. A concentrated flash of lightning appeared at the crest of the barn's galvanized roof. Quickly this flash developed into a pinwheel of fire, first yellow and then blue, rolling along the full length of the roof along its apex. The one-foot fireball moved off the barn into the air and burst silently. Rain began and we went into the house.

Red-yellow-green fireball destroys forest. [395 Parks] Lasting a couple minutes, a four-foot red, yellow and green fireball followed the surface and gradually vanished, but resulted in the destruction of a forest.

Author—Your author is skeptical. This case is frustrating. If you decide to investigate cases, never let respondents do this. Demand details. Although this report is tantalizing, it is useless. And it could have been a great case, but the details are sketchy and there was no follow-up by the investigator, who was overwhelmed by hundreds of cases.

This alleged fireball was unusual and it lasted a long time and set a forest on fire and vanished slowly. What surface it followed we will never know. Nor will we ever know why it was so destructive. But there are other similar cases where fires where set, with fields torched (even during intense rain), and even large buildings burned. This suggests that perhaps fireballs are sometimes far different in some unknown way from typical incendiaries.

Pinwheel rug-scorcher melts two screens. [408 Hampton] A basketball-sized "pinwheel" fireball bounced along, without diminishing in height with each bounce, scorching the rug. It melted screens in two doors, leaving an odor.

Author—Scorching rugs is atypical. Why do most—but paradoxically not all—rollers fail to scorch or leave any sign on rugs, grass or linoleum? Some fireballs that touch or roll on soaked grass during a drenching rainstorm, or along a warehouse roof in one case, will set it ablaze. Paradoxically, many bright slow rollers may not scorch, but some dull fast rollers will scorch and even set fires—but there is no clearly discernible correlation.

Minor Property Damage

Not surprisingly, during appearances, minor damage occurs frequently. Fireballs exhibit mechanical, electrical, magnetic, and thermal effects. To say that an effect was minor does not mean that the ball had less energy or less potential to do serious damage. Some minor effects could have become serious or deadly if given the chance.

Disturbs magnetic objects in nuclear bomb facility. [319 Orange] I saw a cherry-red five-to-six-inch fireball strike a vent pipe in the [deleted area] inside the [identification deleted: nuclear bomb building]. It disturbed magnetic objects.

Fuzzy fireball hurls telephone off wall. Lightning struck near a house. A fuzzy six-inch fireball sprouting a coat of sparklers followed the phone, blown off the wall and hurled across the room. The ball made a right angle turn down the inside wall. Exiting and leaving a desk, it blew papers all over the place. Following the wall to the television, it knocked pictures off the wall, and hit the set. The cable repair technicians discovered the cable at the pole was melted. The television itself was unaffected. Outside, where the phone was on the inside wall, the vinyl siding had a round bent and distorted area, but lacked signs of burns and melting.

Can fireballs debark trees? Lightning struck a large tree, the top exploded, a basketball of fire spiraled down the tree—removing bark—and reached the ground after ten seconds. The fireball then shot against the wall of a building. Did the fireball cause the spiral scar or did lightning do it and then the fireball merely follows the downward path made by a spiral lightning scar or groove in the bark?

Author—Most tree scars created by lightning are spiral, and the average scar from lightning strikes on trees will extend along 80 percent of that tree's height. Although lightning scars never reach near the top, most reach close to the ground. The spiral pitch-rate varies. The slab blown off can be from just a single strip of the bark. Or the slab can come from both bark and wood, which usually comes out in two vertical parallel strips.

Although whether or not ball lightning can debark trees is still unanswered, in either case, it is not likely that the tree was a deciduous tree with its smooth bark, such as a birch, because in lightning strikes, their barks are torn off in large ugly patches that fall off. If ball lighting can do this, then it raises more difficulties needing explanation, in part because of the slowness of the ball and how it could physically reach under the bark.

Red ball crackles, explodes, rips bricks. [Stalzer] Today [1976], I am an American engineer and patent attorney. But this event took place in Eastern Europe. The temperature was 70 degrees and the humidity, 70 percent. Dark clouds covered three-quarters of the sky, and it looked like a summer rainstorm developing. But there was no lightning, thunder or rain, and an hour later the sky cleared again. The house I was in sits in open countryside, with only three buildings in a half square kilometer, and there is no nearby radio or radar station.

I sat lazily in a chair facing north, when a 12-inch ball of red-orange fire materialized instantly right in front of my eyes. It remained stationary, three feet above the floor. Electrical discharges—an eighth of an inch high—totally covered the ball, which sparkled, hissed and crackled. Then it came toward me—within 18 inches from my face—and then turned left, passed through a door, all the time floating three feet above the floor and moving at the speed of a man walking. It then turned right (north). As the ball changed speed, it remained undistorted. I felt no heat and smelled no odor. By now it had made two right angle turns and was in the kitchen. When it hit the water faucet, it exploded and ripped part of a brick from the wall.

Makes hole in pier. After a thunderstorm passed, a fireball descended from the sky and at 300 feet from the three observers, it hit a pier and made a hole in it.

Farm trio wanders, hits objects and explodes. Before rain fell, lightning struck a large walnut tree and several glowing fireballs emerged from under the tree to float rather randomly: Three soccer-ball-sized blue-white fireballs came at half-a-foot above the earth straight towards the farmhouse. Hitting the wellhead, the first ball exploded, and the second crackled out when it hit the base of the windmill tower. Neither left any marks. The third wandered within 15 feet from the house, drifted left, right, and back around the yard. It then expanded, fizzed, and vanished. The event lasted almost two minutes. In 20 minutes, a tornado passed nearby.

Explodes and topples power lines. When an orange fireball flew horizontally down the street, it hit the telephone and power lines. The explosion took down the power lines.

Through door, heats pail. On the South African highveld, when an intense lightning storm raked the countryside, a fireball came in through the closed front door, sped around and down a hall, hitting a pail with a loud clang. With its paint blistered from the encounter, the metal was too hot to touch.

Zigzager hits faucet, explodes, flies outdoors. A mild thunderstorm had passed earlier, but it was now sunshine peeking through the clouds. On this spring afternoon, all was calm. After sitting on the sofa, watching television, the three boys were about to get up and walk to the kitchen, when a soccer-sized, loudly hissing, yellow-orange ball of fire floated in through the open kitchen window. It zigzagged across the room, hit the kitchen faucet, and banged loudly. The brick wall's plaster fell. After two seconds it exited the front door, leaving behind ozone, a burned television, charred outlets, a smoking and burned iron, and a large wall crack.

Noisy roller burns floor. Lightning struck near the house and a softball-sized fireball came though the window, bounced onto the floor, then rolled into another room and loudly vanished and left burn marks where it rolled.

Hits metal bedpost and rings. After slowly passing though a windowpane—without changing shape—a two-inch fireball hit a metal ball atop a bedpost, made a melodious ring like a tuning fork, then bounced back to the window and passed through the glass as if it was not there, having lasted eight seconds. Which one was it? Did the fireball or bedpost ring?

As serious as mechanical effects sometimes are, the damage from induced voltages, electrical currents and heat as evidence from the cases in the next chapter can be more dramatic—and more dangerous.

10

Electrical and Heat Damage

Fireballs pass through glass, sometimes vaporizing holes but in other appearances, leave no evidence, and occasionally instantly melt holes through metal. A few ostensibly pass through walls, without damage. Because fireballs exhibit electrical effects, it is not unusual to expect them to manifest this in terms of damage to metals and semiconductors. How fireballs sometimes cause electrical transients and melting—even vaporization—in some adjacent objects is a mystery. It is not by induction.

Melting and Vaporization

Voltage is electrical pressure arising from the separation of charges, usually electrons, but sometimes ions (atoms that lost an electron). Under pressure, these charged particles move to lower electrical pressure through a medium. If that medium is a conductor, and if the energy is sufficient, then it may melt and even vaporize, as occurs with a fuse.

Melts outlet, blows fuses. [191 Andrews] During a storm, a fireball came through the door, across the floor, to the opposite side of the room. It struck an electrical outlet near the floor, which melted the outlet and then blew out the circuit fuses.

Utility ball sparks, house bulbs go dim. [488] Lightning struck a power pole across the tree and three poles up away from my house. I did not see the actual lightning but saw the brilliant blue one-foot fireball atop the pole. Sparks or streamers of fire fell as if something was burning. But afterwards I could see no indication of burning on the power wires. During this event the incandescent lights in my house—the ones that were on—glowed dim. Others that were turned off did not glow or seem to be affected, except two that were burned out by this. There were several guy wires on the pole, one within a foot of the top of the pole, and they were well grounded. The ball seemed to make no effort to go to or down these wires.

Making a sizzling hum, it lasted ninety seconds and decayed suddenly but not instantly. It left me with spots in my eyes such as an arc welder would cause. Other witnesses were closer that I was—Crawford, a pipe fitter at Oak Ridge Labs and a neighbor, Mrs. Powell. This incident occurred Saturday afternoon of September 10, 1960 on Faust-Carney Road in Powell, Tennessee, about half a mile west of the Clinton-Knoxville Highway.

Quickly heats cable. [468 Hickey] For several seconds, a basketball-sized fireball played on a cable. It happened very quickly. When I felt it, the cable was still hot. Did the fireball heat the metal cable? Or was there some other underlying causative phenomenon?

Orange fireball does not affect crane. [428 Hames] I saw this eighteen-inch orange fireball that lasted for several seconds in a machine shop several years ago during an electrical storm. It was on or near a small overhead electrical crane but it had no apparent mechanical or electrical effect on the crane.

Burns electric motor and melts armature wires. There was lightning but no rain. A blue-yellow, six-inch pear-shaped fireball with a vibrating surface as bright as a 25-watt bulb flew and hopped over concrete slabs. After three jumps it instantly entered the electric box of an electric motor that powered a water pump, and disappeared. It burned out everything inside the box and melted part of the armature. The witness, a professor of chemistry, could not calculate how much energy was dissipated. For the fireball to do this so quickly must require an enormous and sudden dissipation of great work, which means that the power or dW/dt was incredible.

Skinny fire-snake jingles phone. A thin snakelike fireball "crawled" into the machine hall room of a high-voltage substation through a slightly open window Shining like a kerosene lamp, the yellow-blue fire-snake was 20-inches long by 0.75-inch wide. Woven from distinct "threading," the fire-snake was not uniform in shape, and though the upper edge was smooth, the lower was saw-shaped. After it flew alongside a telephone wire to an adjacent desk, it quickly entered the phone, which made a sound. The phone ceased to operate and its coil was burned out, despite all equipment being grounded. It lasted several seconds.

Burns wood, melts TV, blinds man. A basketball-sized orange fireball came from the sky, went through the tree limbs of a big elm, and then broke a living room window pane, hit the floor and burned a one-inch hole in the hardwood, and then touched the unplugged power cord to the TV set. This melted the insides of the old vacuum tube set. The ball rolled and exploded, blinding one witness for 15 minutes.

Exploder activates light sensor. When a three-foot fireball exploded, the outdoor light—which had a photo sensor and was on because it was darkly overcast—briefly turned off because of the flash, but it shortly came back on.

Blue ball enters TV. When lightning struck nearby, a silent eight-inch blue fireball moved down a metal railing, floated from it for five feet and harmlessly passed through a window. Inside, it silently floated across the room at four feet above the floor to a television, which it entered through the set's glass screen. With a loud pop, the screen went black and the set would never work again.

Exploding basketball magnetizes computer monitor. Just past five on a mid-July afternoon, a thunderstorm blasted a Midwestern city. In a kitchen, two friends looked out a window. Suddenly, their hair stood up from static electricity. A bright white basketball-sized fireball appeared near the ceiling and three feet from the inside wall. After several seconds the ball vanished with a loud bang. Smoke and ozone wafted for minutes. The explosion set off the burglar and fire alarms. A television in the family room and two computer monitors, on opposite ends of the downstairs, and three modems on as many floor levels, and one CPU were blown. Badly magnetized, one monitor was later de-gaussed, but this only partly restored it. Surprisingly, the three phone lines continued to operate and circuit breakers were not tripped.

Fireball vaporizes TV's traces. From the rear of a television came a fireball that hovered motionless for several seconds, then exploded. Later that night, the air conditioner emitted a softball-sized blue fireball with an orange-red fire burning or sizzling covering its surface, above the living room. The television, which went white just before the ball came out, then went dead. The fireball exploded into bits that hit things and instantly disappeared. The set's turret-style tuner and gold-plated contacts were welded together. Although the PC board remained, its traces were vaporized.

Powers fridge into overdrive and flickers lights. When lightning struck an electric pole, a brilliant, white grapefruit-sized fireball, coated with a half-inch glowing fuzz, materialized and slowly rolled atop a power line to the house, floated into the kitchen window through steel security bars into the dining room. As the sizzling fireball floated through the kitchen, the refrigerator went into overdrive. Several house lights that were off, flickered. One lamp that was turned on burned out. The fireball then floated back out the window the same way and rose back onto the cable, and then exploded. Did the lightning cause the electrical effects, or as the reports seem to suggest, did the fireball or some underlying cause create it?

Fries car radio. Walking from a stadium after a game in mild rain, a couple saw an intense beach-ball-sized fireball near their car. In a bright flash and loud crash, it disappeared. The car lights were dimmer and the ruined radio was "fried."

Fries air conditioner. After a late thunderstorm, when a slow-moving silent fireball gently bounced up to a house and entered the air conditioner and ruined it, frying a one-foot circle in the exterior paint of the metal casing.

Pulsating ellipsoid melts power cable. After lightning hit a telephone pole and struck two of its wires, an ellipsoidal fireball formed on two lines with its major axis parallel to the three-phase 13.2-kV main feeders. The minor axis was vertical. The ellipsoid alternately shrunk and grew, until it shrank to nothing and melted in half the two wires, which fell to the ground, sparking.

Through Screens, Glass, Keyholes

Not only do properties conflict with one another, but fireballs themselves behave differently. Some pass through glass or even metal screens without any visible effect, but others instantly melt holes. Why? Some pass through both a screen and windowpane without any evidence, but others melt holes through one but not the other.

Fireball slips through hospital's glass window. [Eddy K] I drove from Taunton in Massachusetts south to visit a friend's aunt at the old Truesdale Hospital (now a clinic) in Fall River. Since the hospital was short on beds, she lay in a bed near the end of a second-floor corridor. Six feet above the bed was a small window. Three visitors and I stood close to her bed. It was after nine on a stormy night. When lightning hit nearby, a reddish orange fireball the size of a human head slipped through the windowpane, and floated down toward her. But I don't remember if the edges were slightly fuzzy or clear. The thing looked like a basketball made of fire, but it wasn't dim, about like a hundred-watt bulb. After four seconds, it vanished a foot above her without a sound, just winked out. I yelled, "Did you see that!"

Switch box emits blue fireball, descends stairs, penetrates screen. [459A Evans] I visited relatives in Greensboro, North Carolina. The main power panel to their two-story home was mounted in the wall at a landing halfway up the stairs. After a tremendous explosion, a blue-white twelve-inch fireball floated from the switch box, rolled down the stairs, and came out the front door—passing through the screen on the lower part of the screen door. We were on the front porch. The fireball continued across the porch and disappeared into the ground. Although some fireballs that pass through metal screens burn or vaporize a hole, most do not, leaving no trace.

Through screen, pops, scatters sparks over living room. [390 Greene] During an electrical storm, I was near an open window and a five-inch fireball came through a screened open window. It flew through a doorway into the center of the living room. After a second it exploded with a loud pop directly under the electric light fixture. It scattered sparks all over the room.

Penetrates window screen. [136 Hurst] During a later afternoon electrical storm, a two-foot bright-yellow or yellow-white fireball came in through an open, screened window at the Union Carbide Nuclear plant, where I work, in Oak Ridge. It moved rapidly across and in contact with the floor of the large room, until it reached the wall. Here it dwelt momentarily, then disappeared.

Though two screen doors, splits plum tree. [114 Ross] An airborne eight-inch, yellow fireball harmlessly passed through two screen doors into and out of the house, traveled a hundred feet and split a small plum tree in two. There was no lightning.

Enters screen, ascends stairs. [ORNL] Lightning struck a large tree, from which came a reddish eighteen-inch fireball that rapidly traveled to the storm door and through the screen, and up the stairs and disappeared.

Blue tree-descender passes through screens. [278 Southard] I lived in a farmhouse in the small town of Himyar, Kentucky, seventy miles north of Oak Ridge. One June day in 1949, my children played on the front walk. Thunder and lightning began. I started to the front door to call them in, when lightning struck a sycamore tree growing three feet from the front walk. A greenish-blue, ten-inch fireball rolled down its trunk and between my daughter's legs, then up the walk around four more children. The ball went up one step across the porch, right through the metal screen door, and skipped across the floor. It left no burn marks on the floor and the children were unhurt. An odor lingered, like the smell of helium around Heli-Arc welding.

Passes through a wall. [213 Gregory] I was working one night during a storm, when lightning hit one end of the building we were in. A bright-blue six-inch fireball came from that end. It floated at the same height, flying in a straight line at constant speed. After it passed fifty to seventy feet from me, it went for about another two hundred feet. It went straight through the wall. The lights went out. One eyewitness got a good look at it.

Strips line insulators, penetrates walls. [154 Stoddard] I am the Industrial Hygienist with Union Carbide Nuclear Company. I once lived at the end of an electric power distribution system amidst tall pine trees. Frequently during storms, lightning hit the pines. Fireballs often materialized on the lines leading to the house—from the distribution pole in the middle of the street. These three-inch fireballs, white-to-light-yellow in color, moved swiftly from the pole to the house, sometimes along the top of the wire, but at other times like a bead on a string.

On one appearance, the ball appeared as a bead and moved more slowly. As it moved toward the house, the insulating material stripped off in strings and pieces until half the length from the house to the pole had sections of insulation hanging from bare wire. At other times the fireballs penetrated the house walls and followed cracks in the floor before jumping to some metal object.

Fireball strips insulation off pole wire. [238 Johnson] A white and slightly red three-to-four-foot fireball rolled along an electric wire—stripping off insulation—and after six seconds, exploded at the pole.

Ball with fingers enters screen, seeks faucet. [80 Brown] My father, mother and I were in the kitchen shortly after breakfast just as a June electrical storm descended, but no rain yet. The kitchen range was still hot. I looked out the back door (across the rear porch) when there was a crack of lightning and I saw a nine-inch bright orange fireball coming towards the house. Sharp fingers or sparks extended from it on all sides and it made a high sizzling sound.

The fireball came right through the porch screen door without changing shape or damaging the metal screen. It went straight through the porch and entered the kitchen (through its open door) and made an abrupt left-angle turn, floating to the water taps on the kitchen sink, where it exploded with a loud bang and a brilliant flash. All three of us were momentarily deaf (several seconds) and blinded (thirty seconds). We were not shocked or burned. The plumbing system was undamaged and there were no marks on the taps. The ball lasted thirty seconds.

Enters screened window, hits metal table, exits screened door. [71 Krohn] One summer while I worked on a metal table (the type with a hole in the center for an umbrella) a storm arose. I moved the table inside and put it in front of a window to continue my work. Soon after, a baseball-sized fireball entered through the metal screen of an open window, hit the table, bounced randomly around the room, then exited through the screen of a door. It did not damage the table or room.

Yellow fireball passes through door. [Hess] I once lived in a house located in open country on Main Street in Quakertown, Pennsylvania. It was just after one during a July afternoon. Water ran in the second-floor bathroom and in the first-floor kitchen, with a vacuum working on the second floor. There was no sign of a storm. It came through the bedroom, which was at an angle to the hall. It entered through an open window and didn't stop, moving as fast as a person walking. I was vacuuming the second floor hallway and my sister Betty of Bethlehem was working on the first floor and saw it. My daughter Sue was cleaning the bathroom on the second floor and also saw it.

I was looking east when I first saw it—an eight-inch yellow-white fireball that was partly blue. It did not change color, but it hummed as it came out of the bedroom. It passed a foot from me when it went down the hall, turned right a bit, and went down the stairs. It passed the kitchen door, turned right and went through the living room and out right through the closed French doors! There was no smoke, but I was too frightened to remember if there was any heat. It smelled an electrical smell. The home had many lightning arrestors on the roof.

I live in Northfield, New Jersey, only fifteen miles west of Atlantic City, and worked as a Federal Aviation Agency Supervisory experimental simulator operator at NAFEC Air Force base in Pomona, which now is also used to fly in casino patrons.

Passes harmlessly through steel wire-screen but melts enamel. [49 Woods] Before coming to Oak Ridge, I lived in Palestine in southern Illinois near the Indiana border. One late summer day in 1939, a thunderstorm was approaching. My little sister walked into a bedroom to make a phone call. She screamed for me when she saw the eight-inch, white-yellow fireball descend rapidly outside the window. It came down and slipped through the steel wire screen and bounced on the windowsill. I ran into the room.

Constantly making a loud spark discharge sound, it settled to the floor and rebounded slightly, much like a balloon but with a rotary motion. It deformed, making a brief flat spot, with each bounce. Glowing like a light bulb, it bounced (damped bounce) across the floor toward the leg of an iron bed. With a blue flash and a rather loud spark discharge, it vanished, leaving a little vapor or smoke. Total life span was ten to twenty seconds. It had burned some of the white enamel on the bed leg. The phone, on a stand near the other end of the bed, malfunctioned.

Author—The ball moved slowly and lasted over ten seconds, long enough so that the two witnesses could observe it. In general, a longer-lasting ball is more likely to be more accurately witnessed. Notice the apparent contradictions. The ball slipped through and left the metal screen undamaged, but made sparking sounds.

Yet upon touching the iron leg of the bed, it exploded in a blue flash, burning off white enamel. It also caused the phone to fail—yet never touched it or any wires. Is it possible that somehow the fireball, which slipped harmlessly through the metal screen, emitted an electromagnetic pulse so intense that it caused the phone to fail? Not likely. Researchers never detected such electromagnetic pulses of the necessary intensity. There are other cases where this mysterious phenomenon occurs.

Sunny day "basketball" exits harmlessly through metal screen and two windows. On a sunny afternoon, a boy sat facing south on a sofa with his grandmother. On their left in a chair sat the aunt, next to a French door that led out to the patio. Floating four feet above the floor, a basketball-sized sphere of intense yellow-white entered into the east patio door and floated to the right side of them, then made a one-foot motion to the right. The ball sensed the wall, avoided it, and floated into the bathroom and out through the closed windows. It passed through one metal screen of the patio and two house windows—with no damage.

Slips through screen, rolls and hops, melts metal chain. Pastor John Lehn of St. Mark's Church in Jim Thorpe, Pennsylvania, in 1958 encountered a metal-melting neurotic fireball during a summer lightning storm that lashed York, Pennsylvania. A grapefruit-sized yellow fireball appeared and slipped through the screen of an open window, then descended to the floor. It swiftly rolled around the witnesses' feet. Silent throughout, it hopped into the washbasin and melted in two the steel chain holding the rubber stopper. Several weeks later during a storm, the phenomena repeated, except that after circling his feet the fireball moved away and vanished.

Author—In this classic case reported elsewhere in the literature—and there are others like it—it is not uncommon—the ball slipped through the screen (apparently metal, as most were back then) and rolled around the minister's feet. How many times it circled his feet, or if only a semicircle, was not reported. In some other cases—a nearly identical case occurred earlier to a nuclear physicist at Oak Ridge—the fireball does a complete circle, and other rare cases where it does six or more circles as it rises up to the eye level. But a silent rolling ball is capable of rising rapidly, as the witness's use of the word "hopped" infers.

Despite the fact that it did not damage the screen or floor, and that he reported no heat, this ball melted the metal chain. Could a brief transient of heavy electrical current melt the chain? If so, why did it not earlier flow in the metal screen? On the other hand, non-conductors have been in many cases scorched or burned, but in those cases the cause was often probably direct heat, but that would be an assumption. On the other hand, it could be due to local but intense currents. Or something else? Assume nothing, question everything.

Punctures screen, slips through glass. [Reported to Uman] During a lightning storm in Omaha, Nebraska, a woman in her kitchen heard a sharp crackling from her left and at the window. A crackling, baseball-sized blue fireball punctured the metal screen and flew toward her, curving above her head. She felt a brief electrical tingle. It moved to the oven and slipped through the door's mica glass and hit the rear of the oven, then spattered into bright streamers. It left a small hole with scorched marks on the back.

Although the above account fails to state dimensions, these mica glass windows were usually smaller than a postcard. When a witness reports an electrical tingle, it indicates that a great potential gradient exists, that is, very high voltages lurk nearby.

Two spinning sparkling balls slip through glass pane. Materializing simultaneous with a lightning flash, a dazzling one-inch yellow fireball flew in through a window—but did not damage the glass—and struck the wall near the switch and bounced off it. The ball then stopped, spinning on its axis—with tiny sparks sprinkled on its surface—and then vanished with a soft explosion. Several minutes later, lightning flashed again, and a second white dazzling fireball slipped in at the bottom of the lower windowpane. After it flew at 15 feet-per-second down a corridor, it darted back and forth, and then after three seconds, it vanished. But before it disappeared, it lit the nearby area like an electric arc.

Multi-color ball penetrates glass. Tinged with red, a blue-and-violet fist-sized fireball slipped without effect one yard into a room, after passing though a glass window, then made a right angle turn and floated parallel to the inside wall. After three seconds it vanished with a loud explosion, leaving an ozone odor but doing no damage.

Glass rings, ball enters wall. The upper windowpane made a loud ringing noise of glass vibrating, and a nine-inch orange fireball appeared outside a hospital window and then harmlessly slipped in through the glass. It flew at one foot per second across the room and quietly penetrated the wall above the door! Although it flew past two hospital visitors at three feet above their heads, they felt no heat. Although the walls were thick and brick outside, whether the ball exited outside, as in some other cases, could not be determined. In a few cases, fireballs that penetrated a solid door or wall—without damage—were also seen by witnesses, who were on the other side, as emerging from the other side.

Two glass-penetrators harmlessly injure man. After a dazzling four-inch, light-yellow fireball flew at one yard-per-second passed through the window glass, with a crash, it exploded against a stool upon which sat a man. Aside from temporary foot pain, he escaped permanent injury. Ten seconds later, an identical fireball repeated the event, also disintegrating into flaming pieces, just like an arc welder scattering sparks.

Gray-blue ball slips outside through window. During a lull in the storm, lightning struck nearby and a nebulous, bluish gray fireball—it resembled a ball of smoke—slowly floated across the kitchen and through an open archway into the living room and then slipped outside through the glass pane of a closed window.

Puffs out of socket and flies through glass. Simultaneous with a lightning flash, a glowing orange fireball "puffed out" of an electric socket—just like an expanding bubble on a child's soap pipe—until it grew to the size of a basketball. At this, it flew from the socket and across the room, and without damage passed through the glass window as if it was not there. Outside, it flew several yards and then exploded.

Crushes glass and stings finger. Glowing like a hundred-watt bulb with a blue tinge in its center, a four-inch yellow fireball flew through the glass window—crushing the glass and leaving a round hole with blackened edges. Flying at six feet-per-second, it flew to the finger on one witness. It hurt like a hot needle. It then detached and flew out through a small open hole in the door for the cat, out onto the porch and exploded after a five-second flight.

Melts two holes in two glass doors. A blue-tinged eight-inch fireball melted two holes though two glass doors of a hospital and moved towards a working machine. The ball emitted an electrical crackle and disappeared.

Melts hole in glass. On a cold autumn day, a three-inch red fireball flew out of from between cooking rings of the coal stove and slowly flew back and forth in the room. It would never come closer than six inches from the walls. It floated to outside through the glass window, without any sound, and without changing itself. The only damage was a quarter-inch hole in the glass with a cindered edge and melted outward. Old timers claim that those fireballs that melt holes in glass are more likely to explode and also cause greater damage when they do.

Fire-snake cuts out glass cone. With a sharp blast, a glowing "little snake" flew through the balcony glass door into a room and "melted" to form a bright puddle on the floor, and then quickly vanished. The ball knocked out a truncated glass cone—from a two-millimeter inlet hole and six-mm base outlet hole—was knocked out. The glass cone and the hole both had sharp edges without melting. Why do some fireballs freely pass through glass without damage and others cut holes without melting the edges? But some do melt the edges. And a few crack the whole pane. Why are some holes conical, albeit usually oval, and others cylindrical?

Through glass pane, burns man. A large orange fireball entered through a window and hit a man taking a bath in an old metal washtub on a rural Midwest farm. Stunned, he ran into another room. Where his leg and opposite arm touched the tub, he saw red marks.

Melts oval hole in glass and ruins rectifier. Tinged with blue and red, a football-sized fireball that seemed somewhat transparent, and not too intense, flew through the top glass pane—leaving an oval hole with round edges that measured seven-by-ten centimeters through the glass. Silent and slowly flying near the ceiling, it was covered with blue sparks. A stuffy blue gas filled the room. The ball approached a battery of a vacuum tube radio and the power supply burned out—even though he ball never touched it. The ball then flew outside though the lower pane, breaking it.

Cuts glass sharply. Lightning flashed near a fifth floor apartment and an apple-sized golden fireball tinged with red formed on the glass window—cutting it as if by a diamond—separated from it, and flew above a table and hit a cupboard—cracking it from top to bottom—only to bounce back. It then flew near an electric switch, which was undamaged, and after half a minute, broke apart in sparks.

Turns glass blue white, chars arm black. An Indian physicist at the Raman Research Institute saw a 15-inch blue-white fireball, as bright as a 100-watt bulb that passed in through a glass pane and left a 10-inch-diameter opaqueness in the glass. It then touched some wooden object, charred it, and then touched the arm of an unlucky person, whose arm was not burned but charred and blackened. Alarmed, another man opened a window and the ball floated out.

Cold-day kaleidoscope ball penetrates glass. On a day cloudy and beset by wind and snow, and with the thermometer hovering near freezing, a woman saw a ten-inch fireball float through a double-glass window. Within this ball played several colors—orange, dark purple and bright red—much like firewood in a Russian furnace or a large bonfire; and the glow illuminated the room. The ball then floated six feet into another room, stopped in the center of that room and changed its appearance to resemble a white-colored gas (or like a cloud in the blue sky) before the ball disappeared, leaving the odor of ozone lingering for two hours. It lasted four seconds.

Bubble passes through glass and enters furnace. Just before a storm, on the glass window there formed a glowing six-inch pale-yellow and pink "soap bubble"—not completely spherical—that passed through the glass and "vibrated" for a second. It silently maneuvered between two witnesses and flew into a furnace to disappear in the fire. It lasted four seconds. The "bubble fireball" may have slipped in between the frame and the glass but there was no evidence.

Morphing gray soap bubble penetrates glass. Twenty seconds after thunder, a faintly glowing light-gray six-inch fireball slowly flew in through a closed window. Like a large soap bubble, it constantly changed shape. After four seconds, with a loud crackle, it exploded into a spark that flew into a socket, burning it black and the wall around it. It burned out all the underground wiring. Until ten seconds later, the witness could not move.

Splatters on glass window. A bright white six-inch ball appeared outside a window and flew into the glass—toward the face of the witness!—but with a sharp clap broke into small sparks.

Girl chases red-blue fireball through glass door. During a storm, a young girl chased a two-inch red-blue fireball in her house. It passed harmlessly through a glass door—the girl did not, and though she broke it, she escaped harm. The fireball flew into a tree.

Rod of fire enters window glass. During a storm, a one-foot-long rod of yellowish-white fire came inside through a glass windowpane, passed silently one foot from the observer's head, and floated down a hallway. It left no odor.

Big ball larger than window enters through it. On a clear sunny day in early June, two medical doctors stood in a room with an x-ray machine and other electrical devices. Then through an open window, measuring about 60-inches-by-100-inches, there entered a giant golden orange fireball—about as bright as a 150-watt lamp—and it was larger than the window opening—with a more intense 30-inch core. Near the core's indistinct boundary the color was not as bright and on this boundary were many tiny "snowflakes."

The giant ball floated evenly and slowly straight from the window through an open door, entered the corridor and broke apart with a thunderous noise, into yellowish orange "snowflakes" that then vanished when they were 16 feet from the doctors. It traveled 38 feet and left no traces of its passage, and did not affect any equipment.

No damage to glass pane. When lightning struck the house of an electrical engineer, he saw a six-inch fireball come from a wall plug at four feet above floor level. It floated clockwise around the room. After not touching anything, it passed through one glass pane of a French door, leaving no evidence in the glass, but leaving the conductors to the plug and light switch charred up to a junction box and were replaced.

Burns hole in screen. After lightning struck near an Oregon house, a large white fireball passed in through the screen, instantly burning a basketball-sized hole, passed by an observer two feet behind him—he felt no heat—and floated down the cellar steps, bouncing harmlessly on several of them, then did a left angle turn, floated down ten steps, turned left again and flew into a gas mangle, used to iron sheets, ruining it.

Windowpane glows and emits fireball. During a lightning storm, a windowpane started to glow and a 3-inch white fireball that was difficult to look at formed and passed in, cracking the glass. It floated into the corridor in a non-smooth flight that lasted for 20 seconds, but the biologist did not witness its silent disappearance.

Harmlessly penetrates screen. On a dry summer day, a 30-inch, whitish-yellow dim fireball without a fuzzy boundary flew into a fourth-floor through a screen window without damage and floated 6 feet above the floor and into the hallway where it crackled and disappeared. It left a sulfurous smell. A man with a doctorate in engineering stood 18 inches from the ball but felt no heat. After covering 18 feet in two seconds, the ball vanished.

No damage to glass or plastic. Lightning struck nearby. A three-inch orange-and-blue fireball passed through a plastic window and glass pane into a den room, moved across the room and then through the door into the living room. It then left, passing outside through the front storm door. After 20 seconds, like a ghost, it passed through the glass, leaving no damage or indication it had down so. The witness sat on a couch with two cats that both stood up, hissing at the fireball. The fireball passed through not just glass, but also plastic, without damage to either.

Glass-passing fireball inspects pair. When a two-inch fireball came through the glass of their kitchen window, it moved right in front of the two women's faces, floated towards the right to some cabinets, then back before them, then left to the refrigerator, then back before them, and then back out through the glass without damage. Ozone odor lingered.

Passes through curtain. While shaving during a storm, a man saw a five-inch fireball appear behind the shower curtain, pass through it (leaving no mark), circle his head, then go down the sink's plug hole, melting the chin. After several minutes, another fireball, this time faintly green, repeated the same motions, but instead went to the bathtub drain hole and melted that plug's chain. The witness did not say if the curtain was plastic or cloth.

Vandal melts screen. A ten-inch fireball entered into a window screen, burning it, and then darted rapidly around and vandalizing the kitchen. It hit a cast iron light fixture above the table, leaving it swinging and smoking with its bulbs' glass shards scattered.

Harmless to glass, lethal to switch. Softball-sized, it came through the picture window without breaking the glass and floated into the living room light switch/dimmer, which ceased to work.

Through melted screen and glass. Seconds after lightning hit a large pine tree in the backyard, a one-foot blue fireball flew from it, directly through the porch metal screen, through a closed glass sliding door, up to the metal chandelier above a table, under the witness' head, then back out through the front metal door knob.

Glass hole lacks stress. A two-inch orange fireball bore a hole in a glass window and then disappeared. Analyzed in a Russian laboratory later, the hole was like that created by an infrared laser. There was no track of mechanical tension in the glass.

Glass hole grows, has sharp edges. A hairy red two-inch fireball approached a second-floor double-pane glass window, and formed a small round hole with a glowing red contour in the glass. The hole's diameter grew to exceed one inch, and after five seconds the ball vanished with a blast and flash, leaving a hole and sharp edges in the outer pane. At this, a teacher holding a epidiascope projector received an electrical shock. Puzzled researchers determined that although the glass hole grew in size, it was cut—not melted—and had experienced no mechanical stress. Attempts were made by Russian scientists to duplicate the phenomena with the slow lasers then in use, but with unsatisfactory results.

Today, ultrafast lasers evaporate transparent materials, including glass. By the mid-1990s, physicists created ultrafast lasers that generated ultrashort femtosecond-light pulses of immense power that totally annihilate matter. But unlike slower lasers, they do so without affecting surrounding material. So fast are femtosecond lasers that they cut crisp clean holes and channels in metals like aluminum, very unlike longer pulsed lasers that melt surrounding metal and leave sloppy irregular solidification along the wall. Preternaturally round, these ultrafast holes exhibit untouched edges.

Without microscopic pits, burrs, slag or cracks on the edges of these ultrafast holes, we cannot even detect that atoms only nanometers away were smoothly blasted out. The lasers blast clean micro-holes several nanometers wide. Used for super-accurate scalpels in eye surgery, these lasers also perform pinpoint surgery on separate cells.

With pulses lasting ten picoseconds, a rapid thermal expansion of irradiated matter transmits a damaging shock wave into surrounding matter. Heat surging from this irradiated matter alters surrounding matter by melting it. Since power is the rate of work, ultrafast lasers may not transfer much energy but do it so quickly that their power exceeds that from the combined earth's generators in that ultrabrief pulse.

Cascading outward, liberated electrons explode, pulling along a plume of positively charged atoms so rapidly that there is no time to transfer energy and heat to the walls. Because the photons are so densely packed together, the bound electrons of transparent substances like glass are hit by at least two photons, sparking an electron avalanche. Amazingly, by focusing laser beams to a depth inside the glass, operators select how deep inside the surface they want to machine, without affecting the surface. Although slower lasers could not duplicate the holes in glass made by fireballs, femptosecond lasers probably can.

Femptosecond lasers are not the end of the line. They emit several cycles of a light wave. Although they are routinely used to investigate ultra fast atomic processes and are standard bench top instruments, they cannot control the exact phase, so cannot tailor pulses. Even faster, sub-femptosecond lasers will do so. For compact electron accelerators and x-ray lasers, hundred-attosecond lasers will become the norm to probe processes taking place when an atom is excited, and snapshot chemical ions forming. How they will affect glass relative to fireball holes will one day be known.

Through Walls

Rarely do fireballs pass through walls, roofs, ceilings and floors. Usually they avoid touching walls, or if not, they bounce off them. But there are puzzling exceptions. Sometimes several witnesses see a fireball enter a wall, and some see it re-emerge from the other side! Since fireballs squeeze through tiny crevices and skeleton keyholes—and few serious researchers doubt this—then passing through walls should be no surprise. Despite this evidence, your skeptical author doubted that fireballs pass through walls. It violates both my "common sense" and my own prejudice. But if proven true, this property is important to any proposed theory.

Chases girl, exits through wall. On a December 1928 night cold and dreary, a girl alone (her parents were at church) listened to a crystal radio, from which came a white fireball, when lightning struck the house, and chased her "around the room" before it passed directly through the wall.

In through one wall, exits through the other. During a lightning storm, a basketball-sized fireball entered a wall through one wall, spun in the air for a minute, and then slowly moved directly into the opposite wall.

Came in through the wall. During a humid-but-stormless night, a silent and reddish-orange fireball came out of a wall. After the metallic-looking fireball drifted across the room, it merged into and passed through a second wall and emerged outside.

Into a house through a wall. A two-foot fireball seen at 20 feet away glided slowly at a shallow angle towards a house and entered the house after it silently merged into the wall.

Exits through wall, explodes outside. A white-blue fireball rolled across a living room and merged into and disappeared through the wall. Several seconds later, the one-foot-diameter fireball exploded outside.

Fireball passes through three walls. During a thunderstorm a silent one-foot fireball—floating at all times at three feet above the floor and ground—passed into a house, passing directly through the kitchen wall. Then for half a second, it hovered motionless above the wall rotary phone—which rang weakly—and then floated through another wall, then through another wall out to the front yard, where the neighbor across the street saw it when it floated into a tree and popped out, but did no damage.

In one wall, out other wall. On a clear-weather day, a golf-ball-sized, white fireball entered a bedroom directly through one wall, floated across the room, and then exited through the other wall.

Rod of fire penetrates door. As bright as a 75-watt bulb, an orange-and-red fire-rod—three centimeters wide and forty centimeters long—flew through a glass window as if there was no glass at all, and then passed directly through a solid door to get outside.

Penetrates two walls. On a sunny and windy day, a grapefruit-sized fireball came from a clump of bamboo, floated right through the front wall of a house, and then flew rapidly through the living room and directly through the rear wall.

Sinks into marble floor, then rises. Lightning struck close by and a pale yellow fireball materialized in a half-open, steel-frame window. Hissing loudly, this basketball-sized apparition floated motionless, then began slight Brownian-type jiggling movements, and slowly floated to the marble-tiled floor. The fireball sank completely into the floor, vanished, and then after one second came back up out of the floor and floated up to three feet above the floor. It bounced on two steps past a computer analyst, who saw that the ball contained both yellow and white half-inch sparks coating the surface.

In through roof and floor. A red-orange fireball entered through the roof, passed through the weak floor into the basement (where it appeared blue) and blew up a toy truck.

Digs Holes, Melts Sand

In physics, the product of the magnitudes of force times distance is work, a scalar or directionless magnitude measured in foot-pounds. And power is the rate of work, and in the metric (mks) system is measured in watts or joules per second. In the American system, it is measured in horsepower or 550 foot-pounds per second, which is about 745 watts.

The amount of energy needed to rapidly dig holes and melt sand is enormous, far above even the most energetic estimations. Frankly, for this reason, if these cases are accurate, they hint that something far more energetic is occasionally involved than simple storage of potential energy.

Red fireball hits house, follows stream, digs trench. [520 Peters] While watching a storm from my living room window, I saw an orange-red six-inch fireball approach the house in a near-vertical plane. It struck my house, followed a water stream across the yard, and dug a four-to-six-inch trench.

Author—Unfortunately, Peters left out important measurements. In other trench-digging cases, fireballs dug trenches and even traveled below the surface and re-emerged. Trench-diggers are rare. In such cases, are the soil surfaces glazed from thermal or ohmic heating? Can fireballs create fulgurites? This glazed sand is formed when lightning travels underground. If any field researcher obtains accurate measurements, then any physicist can use these measurements to calculate the rate of energy expenditure and also determine the ball's energy density. If so, they will be enormous and far exceed that of singly ionized air at STP.

Digs hole in ground. [465 Patterson] One hot evening about four, rain fell mixed with marble-sized hail. Lightning flashed. A basketball-sized red fireball lit on a post, danced and made a noise. It struck the post and knocked off a piece, and then went into the ground, leaving a hole.

Author—In this report, again the details are missing. How long did the ball dance and how high and how fast? How big was the piece it knocked off? Describe the post? Was the piece ready to come loose? How deep did it go into the ground? What was the ground like? From this missing data, we could calculate how much energy the ball expended.

Fireball makes ground hole. [371 Morton] A noisy, ten-inch, bright red fireball, not associated with any lightning, remained in contact with the ground. There was scorching. It left a hole in the ground.

Author—Your author must criticize this interesting but data-deprived report. How deep was the hole? The witness reported a hole, not a mere indentation. Describe it. Was earth tossed out? How far? Was there scorching? How long did it remain in contact with the ground? What was the ball's life span? If Morton had listed data, then we could make some calculations on energy dissipation.

Six-foot orange roller tears earth, scorches grass, explodes into rainbow. [314 Gouge] On a July afternoon in 1928, when five of us played baseball, a storm came—more wind than rain and lightning. A power line within a hundred feet of us had large copper or brass wire strung from one tower to another. These high power lines are just outside the Knoxville city limits on the old Ball Camp Pike, now called the Solway Highway.

We watched lightning. One came down like a tree limb, with one streak wider, to hit the tower 60 yards from us. At this, a basketball-sized very orange fireball briskly rolled from this tower. It rolled 300 yards, and just before it got to the lower tower, it dropped to the ground and exploded like dynamite into rainbow colors. A blue six-foot-diameter haze rose from the ground. After thirty minutes we examined the site. It had torn up and cracked the earth for a twelve-foot radius around where the ball came down. It burned the grass and weeds to stubble. A sulfur smell lingered for thirty seconds.

Author—There is a curious but common occurrence—a fireball balancing itself and rapidly rolling on a cable—a feat unduplicated by the best circus wirewalkers—and then deliberately dropping from it when the ball approaches the lower tower. Why did the orange ball explode like dynamite into rainbow colors? How powerful was the explosion?

Yellow ball digs trench, chips boulder. While eating in a Norwegian hotel in the summer of 1900, Englishman A.W. Crossley saw a yellow fireball dart for the window that he sat behind. Large fillet flames shot out and a crash shook the hotel. Next day, Crossley found a furrow track in the ground that began 65 feet from the window and proceeded for 144 feet. The furrow varied in width from three to five inches and meandered around a large granite boulder from which it broke off large pieces—one so heavy he could barely lift it, and while the other was only five inches across, he could throw it sixteen feet. In some places the track disappeared underground and reappeared several yards beyond. Whether the track left fulgurites—coating the tunnel with melted sand or glass—was not mentioned, nor how much energy was required, which must have been enormous.

Substation Transformer's Pumpkin-Balls Melt Sand Into Glass. [376 Hix] During severe electrical storms, lightning would strike power lines near a sub station and run into transformers, then spill over, and these balls would form on the ground. They moved in a wandering line in the general direction of a steel bridge near the transformer. These fireballs varied in size—from 15 inches to three feet—and varied somewhat in shape, which were near spherical, but closer to a pumpkin with a flat top and bottom. A very bright blue-white at first, these fireballs gradually changed to a reddish-orange just before vanishing—a change that took ten seconds. After the storm, I found burned traces of debris on the ground, which looked as if seared with a large arc—with melted crude glass.

The last two times (one in 1956) on Watts Bar Lake, close to my home, just above the steam plant where lightning struck a large tree and a 20-inch fireball appeared at the base. Afterwards, I picked up pieces of stone and sand that were fused together. The other fireball was in the spring of 1960 when lighting struck near the water tower at the X-10 [nuclear] area and the small ball appeared at the base of the water tower for a brief time. Several Oak Ridge scientists also saw it. All the balls sizzled. Every place the balls rolled on the ground, that area was scorched—and in some places the balls melted the sand and/or stone. Ozone odor lingered in each case. The usual colors ranged from bright yellow-orange to reddish-white. The after image was like when I closed my eyes after looking at an arc.

Author—The aforementioned case raises a question: Do fireball fulgurites exist? When lightning strikes and passes through sand, the grains melt and fuse, then solidify into meandering hollow long tubes of glass that are half an inch to two inches across and up to 60 feet long. However, although fireball-created fulgurites probably are occasionally created, as far as is known, there exists no solid evidence that fireballs do create fulgurites. These glassy tubes occur when a lightning channel travels through sand and causes it to melt and solidify into long hollow tubes corrugated with glassy walls. They can extend down 60 feet into sand and are one half to two inches in diameter.

Fulgurites are nothing new, and 260-million-year-old fossil fulgurites exist from the Permian, at the end of the age of amphibians of the Paleozoic Era. Fossil fulgurites certainly also must exist on other planets, including Mars, which once had running water with lightning storms. It is also certain that transient fireballs existed in Earth's early atmosphere—probably before oxygen. Could ball lightning form organic molecules and the precursors to early life?

Ruby red fireball comes from lightning on sand. Lightning struck 60 feet away on an earthen dam and a bright hissing three-inch fireball—pure red like a ruby laser—came from the strike site. It rose up to ten feet and flew 30 feet, then descended to the ground and exploded, lasting 15 seconds. Where the lightning struck there was a three-inch-crater with a corkscrew pattern.

Scorches Floors, Burns Vegetation

Although two scientists studying ball lightning derived thermodynamic equations and plotted data points on log-log graphs, they greatly under-calculated energy and made big mistakes. Fireballs fly against winds, which requires energy. Fire-balls scorch and ignite wood. Sometimes they burn grass instantly. But at other times they do not. There are too many contradictory puzzling aspects in their ability to burn objects so quickly.

When calculating energy content from cases, you must consider time of exposure, size of fireball and distance to the surface—assuming that only heat damage occurred, and that there was no explosive or mechanical damage, which will increase energy density even higher! Usually, time of exposure was too short to account for the burns. Thus it often requires amazingly higher estimates of heat and energy density.

Slow kitchen roller scorches floor. [60 Burns] Lasting ten seconds, an eighteen-inch blue fireball rolled on the floor through the house, scorching it and leaving an odor.

Author—We might ask what was the floor made from? Describe the scorch? How long was the path? If this data had been obtained, we could calculate the energy density. In some cases, physicists have calculated the balls' energy density and found it to exceed singly ionized air, sometimes by a considerable amount.

Like match flame, fireball rolls and scorches. [291 McKenney] I had earlier injured my leg and was unable to walk. During a storm I was sitting on the porch, when lightning struck some electric wires running into the house close to me. A basketball-sized yellow-orange fireball (it had clear edges, not fuzzy) bounced off the wires and rolled on the ground. Where it went I do not know, because in that instant I decided my feet were better and I ran into the house. It scorched and melted. [Witness failed to state what was scorched and melted.]

Rising tree hugger strips bark. [73 Beard] After a nearby lightning flash, a six-inch fireball that made no change in size hit the bottom of a tree and then went to its top, taking off a four-inch strip of bark.

Ball strips bark off elder tree. [547] I stood 200 feet from an 18-inch diameter, tall box elder tree. This bright 15-inch fireball appeared six feet above the ground and three feet from the tree. After a second it moved very quickly to the tree. The tree was struck and there was a crash. The fireball tore a long strip of bark off the elder.

Author—The box elder is a short—"tall" was probably 22 feet high—tree with leaves which are cleft to the stem, irregularly-toothed leaflets. Fast growing, hardy, and able to grow in treeless regions, it was planted by the early settlers moving west. Attractive when young, it grows up unsightly and tawdry. Its electrical impedance (resistance plus reactance) at higher frequencies normally encountered in many lightning strokes is lower than in most hardwood trees. Did the lightning strip off the bark or did the fireball do it? In either case, the question remains: Is it possible for a fireball to remove bark? If so, then by what mechanism?

Bouncer scorches vegetation. [96 Keilholtz] During a storm, a five-foot orange fireball descended onto an open field. It hit the ground and bounced. Vegetation was scorched and destroyed in a fifteen-foot circle.

Author—A five-foot ball is large. How fast was its decent? How did it hit? And what was the bounce like? What happened then? How did it destroy the circle? Was it from the emission of a tremendous amount of radiant heat? Or from electrical current surges? Or from something else? And were the roots affected? If so, perhaps an intense inductive field penetrated the soil, especially if wet.

High frequencies do not penetrate wet soil very deeply, and x-rays or high-energy particles could not do this damage so quickly. I once tried to duplicate such phenomenon with an oxy-acetylene torch to simulate such damage, but the experiment failed. Ohmic heating of the soil seems possible, but also fails.

Diver bounces, scorches grass. [40 Burden] A basketball-sized bluish-orange fireball came down quickly and hit the grass and bounced to hit again, four feet between the bounces. The grass was scorched brown in area four-by-eight feet. This occurred again a year later.

Author—The fireball instantly scorched the grass brown. Your author attempted to simulate this and scorch grass with a torch but could not duplicate the results reported in this and other reports. If grass was dry from a drought—which this report did not say—it would already be dry and catch fire.

Sphere burns house. [337 Postma] An eight-inch fireball slowly came from the sky in a straight line and hit the house, bursting on contact and setting it afire. It burned. There was a clap at the fireball's formation and at its contact.

Author—This superficially resembled a meteor, but the report describes the fireball's diameter and a clap at its formation. Most meteors are fairly cool by the time they strike earth. Most fireballs navigate to avoid damaging objects, but some do not.

Roller burns grass circle. [266 Brakerbill] During a visit with my mom and sister in St. Petersburg, Florida, during an electrical storm, I stood near a window. Lightning struck nearby. Through the window, I saw a baseball-sized fireball roll around in a circle. It grew smaller. We all saw it. I thought it was a wounded bird or animal on fire struck by lightning. Later we discovered a one-yard circle of burned grass where the fireball had rolled for fifteen seconds.

Window visitor scorches wallpaper. [162 Preston] A two-inch fireball came through the living room window, moved to the ceiling, then around the room at the upper corner, and back out the window. It scorched the wall and ceiling paper.

Radio Static and Damage

Fireballs have damaged radios and tubes. Vacuum tubes are tough. It takes a lot to destroy tubes, especially when turned off. And, unlike semiconductors—such as transistors, diodes, triacs, LEDs, and integrated circuits—all the old vacuum tube triodes, pentodes and rectifiers were rugged and difficult to destroy with heat or high voltages and excess current. Tubes survive "nuclear bomb electromagnetic pulse" shock waves or NEMPs that destroy hardened semiconductors. Tubes' heater wires glow orange-red. This boils electrons out of a tubular cathode. The electron cloud shoots through the coaxial or tubular grid screens that encircle the cathode, shooting outward to the positive plate.

The plate is at 70 to 250 volts dc and will kill. Most tubes are glass envelopes, and though they are electrically resilient, they are physically fragile and susceptible to shock and vibration. They consume far more energy than do semiconductors, and because of their large dimensions, operate at lower frequencies. (Large sheets of rugged microscopic tubes—thousands per sheet—were recently constructed and operate at high frequencies.) Aside from their mechanical fragility, tubes are rugged.

On the other extreme, some semiconductors (CMOS) are so sensitive that if unprotected, they are ruined by the slightest static electricity. Although some fireballs never entered radios, they still damaged tubes—but sometimes not their electrolytic capacitors.

It is impossible to tell how fireballs destroy tubes. As in many similar cases involving electronics, some did not touch the radio. The only known way to destroy such tubes is with significant current surges. Could lightning have come in through the line cords or antennas, assuming that there were external antennas? In several cases, the radios were unplugged.

Fireballs can be within ten feet from a conductor and sometimes—in some way not understood by physics—cause a huge current surge at that distance. Oddly, the phenomenon can be very selective. It is not electro-magnetic induction. We should not jump too hastily, and we must ask if things are not so directly causal as they seem. So, is it the fireballs (at least some of them) that cause this damage at a distance, through some inexplicable means? Or is there some other causative or associative phenomenon—not yet recognized—perhaps a "Hidden Progenitor"—that is involved in both causing and sustaining certain fireballs and doing damage?

Red dazzler rolls over radio, damages it. [344 McCloud] During a thunderstorm, a five-inch yellowish red dazzling fireball floated in through a window, making a popping sound. A radio was in the corner of the living room under this window. The fireball rolled over the top of the radio (plugged in but turned off) and rolled off it and disappeared. It damaged the radio and destroyed a tube.

Big red ball torches radio. [134 Warmley] During electrical storms, my mother always had us sit down and stay quiet. During one storm, a red basketball-sized fireball entered the room through the window or door and went past me and straight to the radio in the corner of the room. It struck the radio, which caught fire, and the ball went out. The fire was difficult to extinguish.

Floor roller melts radio. [156 Corn] I did not see where this one-foot fireball came from, but it seemed to roll across the floor, diminishing when it hit a radio, melting filaments and wires, with only a slight popping sound. It blew out the fuses.

Author—How did the ball melt tube filaments and wires? The radio probably was off and the switch open. The phenomenon was sufficient to create a momentary short-circuit and a resultant current-draw that blew more than one fuse. However, a short usually blows only one fuse on the circuit. Could the ball have electromagnetically induced a large current surge in the house wiring, which would once again only blow one fuse on that circuit? Such a pulse would be enormous.

Buzzing blue dazzler ruins radio. [39 Fuller] Seated in a darkened first-floor room during an early March 1950 thunderstorm, I listened to a small radio plugged into the wall, connected to an external antenna. I sat ten feet away, facing it, with a window to my right and another to my right front. The house was a wood frame structure. An outside roofed porch sheltered the windows. Lightning struck nearby. A blue-white, six-to-nine-inch fireball—bright but not blinding and making a high-voltage-discharge buzz—appeared on the electrical lead connecting to the radio. The radio failed but there was no fire, and the fuses did not blow. The outside eighty-foot antenna had glass insulators on its ends and was thirty feet above the ground, extending from the peak of our house to a large oak tree across the street. I believe the lightning-arrestor ground line was disconnected at that time.

Follows wire to radio, goes snap. [486] I twice saw fireballs, one inside a house, one outside. Of the same size, shape and color, both were eighteen-to-twenty-four inches across and orange with a yellow center. The one inside came in on a wire to a radio but failed to reach it—four feet away—and it moved off and vanished with a loud snap. The other one lasted 45 seconds and caused temporary blindness and ringing in my ears.

Author—These balls had a yellow core surrounded by an orange exo-sphere. They entered into a wire antenna. In those days, radios had long out-door antennas. To enter the house, most antenna wires went under the sash, and others in through a hole in the wall. So that the window could close tightly, the wire from the antenna connected to a flat and bendable metal ribbon, usually coated black that ran under the window screen and sash.

Despite its large size, the fireball vanished with only a snap and did no damage. The other ball lasted longer. Questions exist. In that time, what did it do? Did it explode and cause the temporary blindness? What does it take to cause such temporary blindness? Or was the room dark at night and the witnesses' night vision temporarily impaired?

Dazzling sizzler seeks manhole cover, radio hisses. [218 McNeil] A two-foot crackling and sizzling fireball white as the sun traveled down the middle of concrete drive, covered with two inches of water, then made a right turn and traveled for a hundred feet to a metal manhole cover, where it seemed to go down or disappear. A radio made static.

Red blue-green silver ball floats, radio crackles. [226 Hart] Lightning struck nearby. A multi-colored—red, blue, green, and silver—a fifteen-to-twenty-inch fireball that remained in general contact with surfaces caused the radio to crackle.

Author—How far away was the radio? Although in other cases, a nearby radio emitted static—notably Dmietriev of Russia—there are other equivalent cases were no static sounded.

Fast buzzing ball enters and exits police car. During an intense storm, after lightning hit a power line tower 60 feet from the road, there came a beehive-like angry buzzing and a basketball-sized bluish white fireball rapidly shot down at a parked state police car, hitting the rear. The police radio emitted loud static. The ball temporarily vanished and the buzzing stopped. Suddenly it reappeared, making the angry bees buzzing and then flew 30 feet. With the sound of a high-caliber rifle, the ball burst into glowing pieces that shot for several feet in all directions and vanished.

Four beads on a line make radio static. The car radio began buzzing when four bright white (with a bit of yellow and blue on their edges) fireballs rapidly slid along a power line—not on but enveloping the wire, like beads—with the first being the largest and beach-ball-sized; the smallest, football-sized. After 15 seconds, they slid out of sight. The witness did not say if it was AM or FM or a battery-powered radio. In other cases, radios sometimes hissed or even rumbled, as in the Russian case of Dmitriev.

As noticeable as the damage is—either mechanical as in the prior chapter, or from electrical and thermal effects as in this chapter—there is something that is even more noticeable—a fireball's origination, as we will discover from the many dramatic cases in the chapter that follows.

11

Origination and Departure

Fireballs originate and terminate sometimes in the most unusual locations. They appear on trees, bells, chains, knives, steel I-beams, slate roofs, and inside TVs, earthquakes, tornadoes, fireplaces and chimneys, between the fingers of an open hand, atop cooling towers, on picket and barbed wire fences, from phones and wall outlets and light bulbs, from stereos and electronics, inside vacuums, and transformers—in fact, from almost any object. When they disappear, they do so less creatively and there are fewer choices. Some merge and disappear into conductors, like lines and cables, and become large current transient surges.

Trees Emit Balls

Struck by lightning, trees shoot off glowing objects. If the glowing fragment arcs parabolically, in a curved flight path, then it is under the influence of gravity and is not a fireball—just a flaming chunk of wood or vaporized resin. Your author examined the film from James Tuck taken at Los Alamos, and recognized that this light in its parabolic flight was subject to gravity, and thus probably was not a true fireball. Remember, if it floats free and flies in a directed path, or if it slowly rolls down a tree and then flies and defies gravity, it is a fireball.

Cedar tree spits dissolving red fireballs. [442 Wallace] I saw lightning strike a thirty-inch-diameter cedar tree, quickly releasing many deep red, four-inch fireballs that quickly dissolved in the air or followed conductors.
Big fireball strips bark off tree. [55 Garden] A dazzling bluish-yellow fireball rolled down a large tree (four-foot-diameter) and stripped off its bark, then broke into small multi-colored fireballettes. [See Martin Uman's description of bark stripping (on pages 36-46 in his book all about lightning) of tree barks' torn off.]
Melts transformer, burns off treetop. [446 McCown] Airborne part of the time, an eight-inch fireball melted a transformer and burned the top of a tree.

Hickory tree descender grounds on metal garage. [448 R. Jones] A ten-inch, bright orange-red fireball went down a hickory tree and flew forty yards across a paved street, then grounded itself against a sheet metal garage.

Orange fireball bursts oak's bark, scorches tree and leaves. [183 Wilson] Standing in a clearing, the large oak tree was isolated from surrounding trees. Lightning struck the tree twenty-five feet above the ground and a six-inch orange fireball—not too brilliant and moving slowly—traveled down the trunk and discharged to ground. The bark burst loose at the point of impact and splintered. The fireball singed the lower tree trunk and wet ground leaves, then disappeared after four seconds.

Potpourri of Births

Fireballs can appear in almost any manner and from or near almost any object. From bells and tree chains, knife handles and wet wood, pressure gauges and steel I-beams, showerheads and slate roofs—the possible origins are infinite.

Firebell disengages blue fireball. [108 Leichsenring] A large fire alarm bell mounted on the wall of an upstairs closet in our house connected to our local fire department (my dad was a member). This closet was large enough to contain a window. At the height of a thunderstorm, looking out the window, I saw a lightning flash and heard a single tone from the bell. A five-inch, bluish-white glowing fireball disengaged itself from the bell and floated slowly to the floor. When several feet above the floor, it faded. Cover a sphere with mohair fabric with an extremely long nap, say an inch, then photograph it and project the slide onto a screen using a blue-white projection lamp—that is how it looked.

Tree chain makes fireball. [103Blue] Lightning struck thirty feet away at a large elm tree. To one of its branches was tied a chain that reached within three feet of the ground. The four-inch fireball materialized on the ground below the chain. The event removed a strip of bark from the branch where the chain was attached.

Throws knife into lawn, fireball appears on handle. [411 Meons] During a thunderstorm I threw an old table knife down and the blade stuck into the lawn. At that instant a five-inch fireball appeared on the handle and instantly faded.

Orange-corona fireball slowly follows 3-phase line. [413 Walker] While in a rural section of Jefferson County during the last week of July in 1932, on a typical hot summer day with distant sheet lightning, I saw a six-inch fireball—of translucent orange with an iridescent [incandescent?] glow similar to a corona—travel slowly along or close to a high-voltage three-phase transmission line until it was out of sight.

Wet rotten wood ejects fireball. [143 Snead] When I saw ball lightning, both occasions were in wooded areas of mountains. In both, lightning struck wet and partially rotten wood. A softball-sized fireball with blue edges rose slowly in the air, leaving a light blue trail. After twenty seconds, it slowly decayed.

Fireball atop pressure cooker gauge. [499 Crowe] During a lightning storm, a golf-sized fireball materialized on top of a gauge on a pressure cooker and disappeared after two seconds.

Building's ceiling descender. [ORNL J. Lane] I worked in a process building at the DuPont Chambers Works. Lightning struck the building, burning out all the lines and other things. At the instant of discharge, I looked back into the building and saw a one-foot fireball descend from the ceiling onto the steel structure of the apparatus on which I stood and vanished in one second.

Orange ball visits judge—in the door, out the window. [548 Young] I was in the law office of Attorney Murphy—now a judge—of Florence in Alabama in 1930 during a thunderstorm. After a sudden flash, a 24-inch ball came through the open door, floated through the room between him and myself, and then after four seconds it flew out an open window.

Distant smokestack-lightning induces orange fireball. [89 Ross] My house in northwestern Alabama was within one mile of the large concentration of high-voltage transmission lines, transformers and equipment of the Wilson Dam hydroelectric plant. Severe thunderstorms, sometimes without rain, are common. At a hundred yards directly behind my home was an old dormitory and mess hall left over from construction days. The mess hall heating system had a steel exhaust stack rising up fifty feet.

I sat in my living room one summer afternoon in 1936 with my mother in the kitchen. The back door was open because there was little rain. Lightning struck the exhaust stack. My mother cried out when a brilliant orange, six-inch fireball entered the door and moved in a straight line. I looked up and saw the fireball in the kitchen. It moved through the dining room in a linear movement, close to the wood floor, into the living room. It covered thirty feet in a couple seconds. I could not tell if the ball dissipated at an iron radiator or if it exited out the front door. There was no damage to the carpets or wood floors.

Bolt spits out blue ball. [329 Owings] During summer, with temperatures about 85 degrees, a small dark cloud was accompanied with a few drops of rain. A bolt struck the earth. The blue-and-yellow softball-sized fireball fell from it, and seemed to touch both at the same time, and the ball bounced out of the lightning charge. The ball appeared to bounce about ten times, going "downhill," but followed an erratic path. It scorched a small tree, possibly from the lightning and not the ball. The soil was red clay.

Oven door opens, dazzling spitting ball rolls out. [78 Godfrey] During a storm, something hurled open our stove's oven door and out rolled a temporarily blinding, six-inch brilliant yellow spitting fireball that lasted four seconds. Ozone lingered.

Yellow-green ten-second glider-ball instantly appears and silently vanishes. [Bussell] At four on a cloudy 1976 April afternoon, I walked on Bromfield Street in Quincy, Massachusetts. I saw a six-inch yellow-green fireball, as bright as a 100-watt light bulb, instantly materialize 20 feet to my right. It moved like a man running, gliding across my path within ten feet from me. After a ten second flight, it disappeared silently to my left.

Bolt spits out blue ball. During summer, with temperatures about 85 degrees, a small dark cloud was accompanied with a few drops of rain. A bolt struck the earth. The blue-and-yellow softball-sized fireball fell from it, and seemed to touch both at the same time, and the ball bounced out of the lightning charge. The ball appeared to bounce about ten times, going "downhill," but followed an erratic path. It scorched a small tree, possibly from the lightning and not the ball. The soil was red clay.

TV emits orange ball. The event occurred in a rural area near a great lake. Born in an explosion of debris and smoke, a large orange fireball emitted from the front of a stereo then entered the front of the television and then out back directly through the wall. During its flight, it opened a cupboard door and ripped it from off its hinges, and broke jars.

Furnace emits expanding fireball that melts through glass. A bright white one-inch fireball emerged from inside a furnace flame and flew as fast as a man walking into the room and quickly turned to orange and grew to five inches. It pressed against a glass window, and with a hissing it passed through the pane and flew away, leaving a 0.75-inch oval hole with glass edges not melted, but sharp.

Slate roof emits two fiery bubbles. A roofer who was applying second slate layer was removing debris from a poplar tree's fluff and buds off the old slate. At a yard away, two one-foot fireballs came from under the roofing. The tangles inside the balls looked like poplar tree fluff interconnected. The roofer felt no heat. Crackling, the balls glided down 30 feet like soap bubbles and hit some trees and sprinkled out weak white-rose sparks and after 20 seconds, they extinguished with the report like fireworks. While flying, the balls' gel-like surfaces vibrated.

Four Roman Candle fireballs. Low storm clouds scuttled by as rain pelted. At 2,000 feet away, a large fireball appeared on the grass and shot up a hundred feet and exploded. After five seconds, another identical fireball shot up from another location. This event repeated two more times. The witness, a doctor of physics, reported that the origination seemed to move with the cloud base.

Showerhead sprouts fireball. When lightning hit a nearby tree, a fist-sized blue-green fireball appeared on the showerhead and the bather (aware of optical illusions) looked down and to the side and back—the fireball was still attached to the showerhead, blinking. He turned off the water, exited the shower, and left the room.

Tornados Spawn Fireballs

Tornadoes make fireballs. Tornado lightning is brighter, bluer and far more frequent and intense than ordinary thunderstorm lightning. In a typical tornado, twenty lightning flashes flicker each second, and a single tornado generates over 20 billion watts, and several dozen could replace America's generators.

Before a tornado is seen, lightning interlaces the dark clouds, and fifteen minutes before the funnel appears, lightning becomes continuous. The funnel spins around with gathering speed as it descends from the dark clouds. Once contact with ground is made, enormous pulses of electricity flow up the funnel, which behave like a giant vacuum tube.

Twisters cause curious electrical effects. Not only does lighting flicker inside the funnel, but also loud hissing sounds come from its bottom edge, and St. Elmo's fire may appear on nearby house s and trees. At times a bright blue light or luminous band appears around the funnel at 900 to 1000 feet. Witnesses smell ozone, such as from sparking electrical equipment, and even vegetation is scorched and set afire by its electricity.

One witness saw "fire near the top of the funnel." The light was so intense that he had to look away. Another saw a funnel lit with an "orange-colored fire in the center" that resembled a giant neon tube with orange fire bushing from the bottom, with a terrific light. Another survivor heard a screaming and hissing noise, then looked up into the tornado mouth poised above him and said that the walls of the opening were rotating clouds brilliantly lighted by constant flashes from the lightning, which zig-zagged from side to side. Another witness described a beautiful electric blue light circling a tornado spitting orange fireballs.

Thunderstorms spawn tornadoes whose rotating speeds vary from 100 to 300 mph, with 250 being common, and the maximum (in Oklahoma) clocked at 318 miles per hour. But some are far greater. And they can strike in regions not used to them. On June 9 of 1953, a monster tornado a mile in diameter struck the Massachusetts city of Worcester, just west of Boston, traveled 46 miles and killed 90 people. It mangled steel towers designed to withstand 375-mph winds.

A tornado behaves like a giant resonating vacuum tube and emits specific radio frequencies. In the 1960s, Weller, a disabled trucker who became an electrical engineer determined the mechanism and first discovered that if you darkened your television, turn to the opposite channel, and the screen goes white, then a tornado is near. Although this Weller Effect saved lives, more recent Doppler radar, severe weather sirens and alarms over television and telephone, replaced it. Doppler can anticipate the likelihood of a tornado before it forms, and its speed and have increased effective warning times to 12 minutes.

Tornadoes are electrical. Some are luminescent, at least on the inside wall. Some witnesses who survived being inside a tornado—there are very few who did—one was a meteorologist—report that the inside walls were a sheet of strange electrical fire. If so, the lights were shielded from the outer wall by opaque clouds and debris.

Sometimes trees and objects near tornadoes glow. This is St. Elmo's fire. But this corona does not set fires. But sometimes tornadoes do scorch and burn leaves and char tree limbs that were within the tornado's circumference. The sides facing some tornadoes were totally scorched and burned, but the opposite side, untouched. Neon-like lights, beautiful electric blue lit, and balls of orange are sometimes seen.

Cases of tornado fireballs are not uncommon. Usually these fireballs do no damage. Some witnesses see several fireballs coming from a tornado.

Tornado's blue-corona sphere drifts three minutes near ground. [205 Harris] On a level prairie in Minnesota in 1936, just before the tornado struck, the air was still with considerable sheet lightning. A spherical corona-like glowing ball of electric blue fire slowly drifted just above the ground for three minutes and left an odor of ozone.

Tornado fireballs drill holes in windows. A large number of fireballs shot from a tornado and entered chimneys. Some did not slip through the glass windows—as often occurs and as described in Smirnov's fractal theory—but instead bore circular holes in the glass of several windows. Sometimes fireballs do melt metal screens and burn holes in glass. This was presented to the French Academy of Sciences in 1890.

Tornado pours out fiery spheres. [Cade, pg 36, case 8; George Raveling] A weather bureau meteorologist witnessed a tornado that hit near Rock Rapids in Iowa on a July evening. From the side of the boiling dust-laden tornado cloud, he saw a fiery stream pour out like water through a sieve, break into descending spheres, although some seemed irregular. No lightning occurred.

Earthquakes Emit Fireballs

Witnesses report strange and intense electrical phenomena before, during and after earthquakes. These include fireballs and clearly defined luminous masses that float, vertical pillars and horizontal beams of fire, scattered sprinkles of light, and scintillating sparkles, emanations and bright flames, glowing fogs, and phosphorescent clouds, and a glowing sky.

Inside the many reports mentioning fireballs and strange lights, many descriptions and phrases reappear, such as in this sampling—"Opening fissures emit bluish flames...Flying luminous objects...Swarms of fireballs flying in different directions...Glowing air...Luminous bodies in the sky several nights before the earthquake, and for 20 days afterwards...Nightly sheet lightning luminosity flickered in the cloudless night skies...Just before one earthquake, a fiery cloud moved to a mountaintop in a whirling motion and vanished, followed by a rumbling and a severe earthquake...The night preceding the quake, the sky luminesced and the ground emitted a glow...Many witnesses report white or pale blue illumination and occasionally red and orange." Of course, such electrical glows should come as no surprise.

Earthquakes cause rock slippage and massive shearing of rock that through friction generates and redistributes charge. Before and during intense earthquakes, there are intense electrical events—lightning, sparks, strange flashings, floating luminous masses, fireballs, columns of fire, luminous funnels, glowing masses and vapor, phosphorescence of sky and clouds.

Mysterious glowing lights and fireballs have long been associated with earthquakes. One of the earliest recorded accounts of lightning and flashings was recorded by historian Tacitus of a massive earthquake that struck the Achaean cities during 373 BC. It was 746 years later in 373 AD that an earthquake hit opposite Delphi [Hale, de Boer, Chanton, Spiller, *Scientific American* August 2003] on the south slopes of Mount Parnassos, and lights appeared as gas belched up from the seafloor. Mysterious lights (fireballs) flew through the skies. After this, the sweet-smelling hallucination-inducing ethylene gas that had emerged from Delphi's temple of Apollo since 1500 B.C. only weakly vented up at the temple—and the Oracle (Pythia) lost her power to prophesize. (Recent recreations duplicated the prophetic hallucinations.)

To this day, remnants of the oracle chamber lie over the intersection in the limestone of the Delphi and Kerna faults, from which flowed a spring and the ethylene vented. The navel of the Greek world, Delphi possessed a vast intelligence-gathering network and influenced the course of history, leading to the defeat of Persia at Salarmis by the Athenian triremes, and later launched Alexander the Great—events that totally altered world history.

When fissures opened up in the great Japanese Kamakura Earthquake of 1257, lightning flashed, corona glowed, and fireballs floated about. In the winter of 1672, an earthquake hit Yedo or Tokyo and a fireball was seen. The 1698 Tosa Earthquake caused numerous fireballs.

The 1730 Tokaido Earthquake produced "luminous bodies" that floated through the clear night sky several days prior to the great quake. Strange lights in the night sky, an intense corona emanating from the ground, and a mysterious glow intense enough to make the landscape glow—all these ominous signs warned the citizens of Kyoto; and, the next day in August of 1830, a giant quake struck—and for twenty nights thereafter, a large luminous light floated in the night sky.

During the Kwanto Earthquake of September 1, 1923, a staff member of the Central Meteorological Observatory reported a stationary fireball in the sky over Tokyo. It was November 25, 1930, when fishermen near Mount Amagi in the late afternoon saw a luminous fire globe appear west of the mount and then rapidly fly to the northwest. Many citizens reported fireballs and glowing clouds, and a funnel of electric fire that appeared. Other strange lights also materialized and then faded.

Near the Hanshin region at 5/46 am (LT) on January 17 of 1995, a magnitude 7.2 earthquake hit the city of Kolbe. Fishermen at Osaka Bay saw a fireball follow what they thought to be a fault line until it seemed to collide with Mount Rokko. And when at 1:47 am on September 21 in 1999, a 7.7-magnitude earthquake hit Chi-Chi, many witnesses saw "fire poles" glowing several yards high. Three appeared near the mountaintop and one on the top.

Electrical friction is odd. For example, static electricity is created on the topside of self-stick floor tiles when the bottom paper is peeled off—even during humid weather. These static discharges are visible in low light conditions.

Crystals such as quartz and tournaline (a ring silicate) exhibit piezoelectric and sometimes pyro effects—that is, by mechanically stressing them with sudden tension, compression or shearing force, they generate a difference of electric potential on opposite sides. They are mounted and hermetically sealed in tiny metal containers. In a crystal clock, the quartz oscillates at a precise mechanical frequency and outputs an accurate electronic period that is used to precisely synchronize computers, wristwatches, utility generators, GPS location modules, communications equipment, and surveying instruments, to name just a few.

Gas stove igniter piezoelectric crystals create potential differences or voltages that drive a spark which ignites the gas. Barbecue handheld igniters do the same. Earthquakes are like giant quartz igniters—mechanical force and torque on rock cause charge displacements on a gargantuan scale.

Some cracking crystals generate electrical currents. Some even glow. This glowing extends beyond rocks. Face a mirror, and extinguish lights. After your eyes become accustomed to the dark, put a dry wintergreen Life Savers in your mouth and bite hard and fast, and your mouth will sparkle—the flashes of eerie light coming from the sudden fracturing of the combined mint flavoring methyl salicylate and crystalline sugar.

Earthquake slipping and shearing rocks create large electric potentials. Today, this is not surprising. But what no scientist can explain or duplicate is not just earthquake fireballs, but some odd electrical phenomena. Nor can anyone explain why such phenomena occur several days earlier and up to 20 days after the earthquake. Mysterious electrical effects occur elsewhere.

Dust Devils and Volcanoes

Glows, strange night-lights and fireballs also appear in certain deserts. Wind shears sand, separating charge. Compressing silica grains creates opposite electrical charges at opposed sides, and this charge separation causes grains to attract each other. In some regions, such as the Kalahari, grains sometimes adhere to form filaments half an inch long, and electroscopes verify that such threads are electrically charged. In over 30 acoustic "booming dunes" in deserts and beaches on all continents, mysterious "booming sands"—at 50 to 300 hertz for up to 15 minutes—generate electrically charged sands.

In the 13th century, Marco Polo, while on his Silk Road route, encountered a giant sand dune in the Chinese desert that made singing and whistling sounds. Today, Chinese historians believe it was the Mingsha Dune, which overlays solid granite. Although odd effects are accompanied by strange electrical effects, these phenomena are not well researched nor understood.

Dust devils produce magnetic fields. Desert dust devils of North Africa are 30-feet-wide and 900-feet-tall and emit magnetic fields that change magnitude 3 to 30 times each second, and are detectable over a thousand feet away. Lifted aloft, grains of sand collide, and negative charges transfer to lighter, smaller (often clay) grains, which are lifted higher than heavier ones. These charged grains move at speeds that vary, swirling in varying circular paths. Since moving charges generate magnetic fields—just as do electrons flowing in a coiled wire—they create varying magnetic fields.

Most sands in the deserts—in fact, almost everywhere in the world—begin as continental rock, mainly as feldspar and quartz, which when mechanically stressed generate a potential difference or voltage between their sides. Sand is between 0.05 and 2 millimeters wide, and larger particles as silt, which is rarely lifted by wind. Typical rivers move sandy deposits a hundred miles downstream, while chemicals polish grains to high gloss.

Unlike water, wind blows particles a foot or more above ground, but does not polish them, so they take on an opaque and even frosted appearance. Some beach sands do not originate as fragments washed and in water flow from mountains, but are from limestone near the sea. In the deserts, wind erosion creates dull opaque surfaces that are roughened more than those by water because of greater abrasion, and wind speed exposes grains to more wear. Desert sands exhibit a wider range of grain sizes, because water sifts sediments more selectively than does wind.

Volcanoes spew fireballs. Volcanic gas is hot and corrosive, and through friction and chemical processes, intense charge separation occurs. Volcanoes spew out clouds of particulate matter. Frictionally generated charge-separation generates large electric fields that exceed 11,000 volts per meter. Generally, the larger particles of volcanic ash obtain positive charges; and the smaller ones, negative charges. Winds that blow acidic dusts produce potential gradients; basic dusts, positive gradients. But different kinds of ash can get the exact opposite signs. When hot magma reaches and rapidly heats water, or water falls onto magma, it causes appreciable charge separation.

Some volcanoes sprout lightning and fireballs. Washington's Mount Saint Helens' dormant volcano exploded in 1980. Meteorologist Peter R. Chaston describes how observers a hundred miles southeast saw a group of large fireballs float above the ground, all moving in the same direction. An observer 29 miles north of the volcano saw profuse lightning 700 feet overhead after the cloud passed overhead. It formed huge luminous fireballs the size of a pickup truck that rolled above the ground.

Fireplaces and Chimneys

Some physicists such as P.A. Silberg and Nobelist Pyotor Kapitza felt fireballs are created, attracted to and guided by electromagnetic "tunnels" or wave-guides out in the open. Some elaborated on Kapitza's theory and speculated that electromagnetic waves from lightning might reflect off chimneys' creosote carbon-coated and conducting inside surfaces—which, they allege, would behave like a wave guide—and cause the waves to interact, both adding or reinforcing and subtracting or canceling each other to create nodes and antinodes that the fireballs "ride" in. This may be occasionally true, but not often, and Kapitza later admitted the theory did not work.

Kapitza became famous when he observed superfluidity in 1938 when he cooled helium 4, its most common isotope, to 4.2 kelvins, at which it becomes a liquid. Cooled lower to 2.7 kelvins, it does not freeze but becomes frictionless and flows without viscosity, and passes through tiny pores; and when rotated, it creates mini-vortices that obey quantum rules.

Golka visited Russia several times and spent time with Kapitza and his son, and obtained previously unpublished details of Kapitza's wave-guide theory. Although disproven, the microwave waveguide theory resurfaces in many "improved" versions—but none work.

Why fireballs are attracted to and descend chimneys is another mystery. Some researchers suggest that chimney smoke attracts ball lightning. Although certain chimneys seem to attract fireballs, this is not because the warm smoke is electrically conductive; otherwise such fireballs would prefer lightning rods and skyscrapers.

Furthermore, statistics disprove this folklore that smoke attracts lightning and fireballs. The Department of Lands in Schleswig-Holstein in 1884 conducted the first study and concluded (as did subsequent studies) that there was no significant statistical correlation between lightning strikes and chimneys, other than elevation—the tallest buildings got hit most, not those venting the most smoke.

The hot campfire gases that rose from the top of the closed-bell tent once used by British troops were believed to be conductive and attracted lightning. During the Boer War, many troops observed that some soldiers in the tents were electrocuted during storms. Many more were shocked. Theorizing that lightning and fireballs might seek the smoke from chimneys was not unreasonable to those scientists of earlier times. As early as 1600, it was discovered that flame gases would discharge an electroscope. Conduction of electrical currents in flames was known in the nineteenth century.

But scientists disagreed. Dr. HA Spencer, who witnessed these Boer War cases, first discovered that smoke was not the culprit. Oddly, these "electrocution tents" rarely caught fire. The wood tent poles, which should have shattered from a lightning stroke, went undamaged. The tents often had a few rips, and only soldiers touching the wooden tent pole or wet tent suffered severe shock or death. Others in the tent might receive minor shocks or nothing. However, if lightning had really hit their tents, Spencer knew that few would survive.

Spencer discovered that charge accumulated on the metal cap and wet tent, and the wood pole acted as an insulator, behaving like a Leyden jar or capacitor—two conductors separated by a gap or insulator—that stores excess charge (electrons) on one conductor. This can occur in the strangest ways. While he stood on the ground, lightning expert, Sir Basil Schonland received a powerful shock in a similar way during a storm by touching the trunk of a small thorn tree that was mounted on insulators.

During a storm, invisible silent discharges bleed current of positive ions upwards into the air from each projecting point—such as thorn tree branch tips, from edges of buildings and tent pole-caps. From such single tips, current or charge flow is small; but it can add up—for Schonland's insulated thorn bush's many branch tips, this collectively added up to a significant current, and briefly was three amps. (A typical light bulb draws about one amp.) The British soldiers in their tents and Schonland near his thorn tree were all positively charged (that is, lost electrons), but the insulated wet tents and insulated thorn tree—because they bled positive ions to the air—became very negative from electrical induction. Losing ions creates a negative charge that induces a positive charge in everything near it. The wet tent and metal cap behaved like a capacitor, or two conductors separated by an insulator, with charge on one conductor.

Fireballs that enter or exit chimneys and pop out of stoves and fireplaces are less common today, perhaps due to the demise of coal and wood stoves, and because of newer chimney designs that thwart fireball entry or formation. And because soot is flammable, and because over-dampened fires of softwoods produce excessive creosote on the flue liner (or brick or stone liner in earlier chimneys), some experts propose a questionable theory that lightning could ignite the creosote and create a rare fireball and not a typical flashover.

Attempts to duplicate fireballs in this manner proved unsuccessful. Rich creosote-laden chimneys do occasionally explode from the heat of the fireplace or stove, but these terrifying events are not fireballs. Even if this doubtful theory is true, then these creosote fireballs are rarities and cannot explain the following cases from nuclear scientists and engineers at Oak Ridge.

Chimney fireball circles andirons, explodes. [566 Keavreu] I sat before my fireplace one Sunday afternoon. The thunderstorm receded, thunder rumbled distantly. I had left the damper open because of a fire in it last night, and the fireplace still held charcoal chunks and cold ashes. Two-inch diametric balls topped the andirons. Which were spaced fourteen inches apart. Suddenly a 1.25-inch fireball floated down the chimney moving in an up-and-down motion as it slowly floated towards the sphere atop one of the andirons. It circled at increasing speed and made three concentric revolutions around the andiron sphere, then spiraled away and flashed to the second andiron. With a loud report, it threw ashes and charcoal all over the floor. A small shiny spot appeared on the sphere where the paint and some metal were burned away by the discharge.

Chimney fireball melts hole in glass pane. [533 "CD"] My grandmother always warned us to never stand near a fireplace during a storm. Lightning once struck her house and a fireball came down the chimney, floated across the room and passed through a window opposite the fireplace. It melted a six-inch diameter hole in a glass pane.

Author—Notice the melting of the large hole in glass. Although the Smirnov fractal theory states that fireballs will slip through glass, in ten to fifteen percent of such cases, it instead blasts holes through glass. In one case, the fireball melted a hole in one glass pane but not the other one. Fireballs behave quixotically and exotically, unpredictably and schizophrenically, with many multiple personalities.

Bounces off iron, strikes tree, enters fireplace. [362 Lemons] One hot summer day in the 1930s, I was ironing at my house in Sweetwater, twenty miles southwest of Oak Ridge. A two-inch fireball entered the window and bounced on the iron, then flew over my head through the house. It struck a tree limb directly across the street, in the line of flight. There was no nearby lightning. Several years later, my mother saw a fireball roll across the living room floor and into the fireplace

Chimney emits kitchen-trasher, rolls, flies out window. [519 Hillard] A very large fireball came in through the chimney, tore all the bricks loose to the stovepipe, and burned a ring in an oilcloth on a table. It knocked a pan of rice, a coffee pot, and a kettle of water off the stove. It then rolled through the house on the floor and flew out a window.

Down chimney, out fireplace. [209C Cottrell] My friend reports that his father was seated near a fireplace when lightning hit the chimney. A fireball rolled down the chimney, across the living room floor, then out through an open door onto the ground.

Swiss wood stove spits multi-colored fireballs. If God chose a residence on earth, he would choose this one. At 14 miles southeast of Interlaken and located in the Bernese Oberland chain, the Swiss resort village of Grindewald ("the glacier village") rests in the "Valley of Glaciers," dotted with chalets and overlooked by a long east-west mountain chain both beautiful and treacherous that lies in the heart of the Jungfru region.

Grindewald lies at the foot of the Wetterhorn (3,701 m) and Eiger North Face (3,970 m), and boasts first-rate winter sport resorts. Amongst the most impressive scenery in the Alps, many ski runs wind through pine forests. There is spectacular skiing for all levels, with winter sports that include night skiing, ice-skating, curling, and night ski jumping. The Masters Cup in cross-country skiing are held here. Non skiers appreciate the lively clubs, cozy cafes, and excursions to Europe's highest mountain railway to the "Talhaus" mountain museum, plus group and individualized excursions.

Nature extends the winter wonderland. But in early 1999, a major avalanche cut off road and rail to isolate Grindewald, and the Masters were cancelled. However, in summer, impressive mountain lightning storms here can be localized but formidable.

If you look towards the north face of the Eiger, with Grindewald below, you can look down from the "First" station (2,200 m), where the cable car up from Grindewald ends, and begin our walk up to the Faulhorn, after which the long-forgotten "Faulhorn Hotel" took its name.

Back on a July 1921 afternoon at about five the hotel manager Suzanne Iacci sat in the rustic Faulhorn Hotel near Grindewald. In the dining room, she was talking with her sister Margaret and six guests, some who came to enjoy the July mountain air and hike the alpine trails. Thunder boomed faintly. Dark clouds came in the east.

Suddenly, the small wood burning stove in a corner spit out many very bright multi-colored fireballs—the largest, basketball-sized. (There was no lightning nearby yet.) A blast shook the building and all balls vanished. Ozone and a gray smoke filled the room. Guests nearest the stove were electrically shocked, none seriously. The fire still burned undisturbed and there was no damage.

Seven years earlier, two women had a similar experience during a hailstorm (the fire was unlit) with red grapefruit-sized fireballs. The old hotel, which your author has not been able to locate, they said stood somewhere near the 1.7-mile scenic peak and had numerous lightning rods. The stove, measuring 5 feet, 4 inches tall and 2 feet, 2 inches in diameter, stood away from the wall on a wood platform. The rhomboid-shaped inlet was 0.7-by-1.7-inch. Gases fed into a 4.3-inch diameter iron flue atop the stove.

To extract maximum heat from the exhaust gases, the pipe passed horizontally through several bedrooms above the dining room, then exited up 20 feet above the stove. The stove was not electrically grounded until later, after several people complained of receiving shocks during storms, none serious. [*Nature*, pg. 413, March 15, 1930. William Colebrook Reynolds regarding conversation with Iacci on August 2, 1921]

Red ball enters smoking chimney. At 4:15 on January 11, the first of three orange-red softball-sized fireballs exploded against the corner of a metal-bound chest just inside the open door of a large shed in a walled-in yard attached to the house. A second exploded a minute later against a ladder leaning against the shed. A third came down the chimney (antenna attached but undamaged) against smoke, due to a fire, and entered a first-floor room but exploded in the fireplace. Two men outdoors and two women indoors saw the events. [Joseph M. Wreath of Ballymoney, Co. Antrim, in Ireland; January 11, 1934, reported in *Nature*, Feb. 3, 1934; Pg. 179, by Marshall Holmes of Thirlmore, Innisfayle Road, Belfast Jan. 20]

Chimney fireball sparkles and vanishes. A lightning storm descended. Larger than a basketball and 28 feet away, 15 feet above the earth, and 15 feet from his neighbor's chimney, a silent sun-like sparkling fireball floated motionless for 20 seconds and silently vanished.

Atop Cooling Towers

Because fireballs form on or near many buildings and structures, it is not an anomaly that they also appear atop or close to water towers. There is an account of one nuclear plant-cooling tower atop of which regularly appears a fireball during electrical storms.

Fireball appears atop nuclear cooling tower. [27 Muir] In the company of three other nuclear technicians in 1954, I stood at the street near building [deleted] of the [deleted: exact location inside nuclear bomb-making] facility. I faced plant [deleted] while they faced north. A storm loomed. A bluish-white glowing ten-inch fireball suddenly appeared atop or above the cooling towers between [deleted]. There was an arcing to transformer and power lines across the street with a loud crackling and loud boom. Ozone lingered. Our hair literally stood on end from the electricity. In the past, other Oak Ridge nuclear physicists have noted similar phenomena.

Fireball atop wood water tower rolls down. [419 Sherrod 1940] Intense inter-cloud and to-ground lightning flashed. Clouds hovered close to ground. Intense rain let up. With iron bands around it, a wooden water tower stood on old Maryville Pike in Knoxville. A pale blue two-foot fireball formed atop the tower, rolled down one of the supports to the ground and out of sight beyond a railroad fill. Lightning previously damaged this tower so that lightning protection was added.

As we just saw, the origination and departure is as varied and unusual as is our world. And even then, there are always unusual cases with new surprises. The most unusual, as we just saw, were those that came from trees, were spawned by tornadoes, emitted by earthquakes, emerged from (or entered) fireplaces and chimneys, and—most scary—even formed atop cooling towers of nuclear power plants. But when they form, not surprisingly, many fireballs are associated with electrical effects.

12

Electrical Effects

Electrical effects vary greatly. Although some wall outlets and sometimes light bulbs emit fireballs, occasionally a fireball seeks out a wall outlet or light bulb to enter. Telephones are well authenticated as sources of fireball emission—particularly the older wall telephones. Occasionally, a fireball enters a phone. Fireballs leave line pole transformers, but occasionally one does the reverse. Electronics emit fireballs, but some attract fireballs. Sometimes the mere presence of a fireball inexplicably burns out electronics, sometimes even when it is inactive.

Sometimes, a fireball or possibly some underlying yet-unsuspected phenomenon (a "Hidden Progenitor"?) will cause electrical effects. In one case, a fireball turned on a wheelchair. It was during a lightning storm, when a fireball flew through open patio doors. The charger for an electric wheelchair turned on, a battery-powered game started, and then with sparks flashing inside it, a toy bouncing ball glowed. A child was moderately shocked, which left a red mark on her head, perhaps caused by one of two metal hair clips.

Bulbs Burp Balls

Both light bulbs and wall outlets burp fireballs. Although it is not uncommon for electrical outlets and light bulbs to emit fireballs, it is less common for fireballs to seek out and enter these outlets and bulbs or sockets. Although they usually grow out of outlets and sockets, slowly like glowing bubbles, sometimes they escape abruptly.

Wall outlet burps fireball. [331 Huffstetler] Lightning struck an outdoor transformer and a wall outlet burped a five-inch, blue-green fireball that bounced five feet into the middle of the floor and vanished, lasting two seconds, leaving an ozone odor.

Light bulb spouts blue fireball. [6 Bradshaw] During a storm, from an old-fashioned ceiling drop-socket light in the center of the room, a two-inch blue-white fireball jumped from the bulb and floated through the air for several inches. Within a second, it—ball or upward discharge connecting to descending leader?—loudly popped, followed shortly by a loud crash of thunder outside.

Light bulb emits floating roller. [369 Parkham] Lightning struck near my father's house. Lights went dark. Immediately, a light bulb (suspended on a drop cord from the ceiling) emitted a yellow fireball that floated to the floor and for several seconds slowly floated across the floor, then disappeared. The outside telephone line to the house was cut in two at several feet from where it entered the house. There was no other sound.

Bulb emits shrinking drifters. [98 Osborne] I am in the Oak Ridge Solid State Division of Thermonuclear Experimental Division, but before entering thermonuclear semiconductor research, I lived in an old family home in the mountains of western North Carolina, a region of many thunderstorms. On five occasions, two-inch fireballs appeared at a light bulb and slowly drifted randomly away. A loud pop and nearby lightning flash accompanied each appearance. When a ball drifted away from the bulb, it gradually decreased from the original three-inch diameter, and then vanished after three seconds. They caused no damage, nor did anyone feel shocks, despite being within four feet.

Wall outlet emits red-blue floater. [174 WC Brock] Lightning struck a telephone line and the wall outlet emitted a three-inch red-to-blue fireball. It remained airborne for four seconds and slowly disappeared.

Enters electric wall outlet. On a cloudy day a faintly yellow fireball rolled in through a window—it was open 10 inches—and flew to a ceramic wall outlet. Most of the ball entered one of the holes until only the tail stuck out. The tail then bent over and entered the other hole. At this the receptacle exploded, fire shot out of the wall, and soot came out. Broken into pieces, the outlet was blackened.

Outlet sprouts ball that enters another outlet. From the outlet that the television was plugged into, out came a softball-sized light-blue fireball that emitted small white sparks, like a birthday sparkler. In an erratic path, it slowly crossed over the living room rug into another room, and after its 20-second flight, entered a plug outlet.

Through glass, into switch. A glowing ball that came in through a closed glass window, floated across the room and dissolved into a light switch on the wall.

Socket spits sparking ball that burns bed. When thunder roared, a spark burst out of a socket above a bed, followed by walnut-sized fireball that fell onto the bed and rolled across the mattress for one yard, stopped, made a rustling noise, and many eight-inch sparks spurted from it, resembling an arc welder. It suddenly disappeared, leaving only a small hole in the mattress and sheet.

Phones Spit Balls

Phones that burp fireballs during storms were well known in stories of fireball lore. Although today's tiny and portable cell phones do not burp or spit fireballs, and today's wired phones are less likely to emit fireballs, the hardwired phones still do it, but not as commonly as in earlier days.

Since the earliest days of telephony, mouthpieces sprouted balls of fire that floated or darted. In the mid-1970s, a quaint Midwestern town with a very small, local phone firm served only four-dozen subscribers, mostly small rural old-time farmers. The lines ran across semi-arid land over a twenty-mile area. Oddly, many of the phones were legacy phones from a bygone era, still functioning. Fireballs frequently floated from the mouthpieces during storms, but the old farmers were used to this, and told the investigator that when they used their phones during a storm, they would first point the mouthpiece to the ceiling. If no fireball came out, it was safe.

Five-minute fireball sucked into phone's mouthpiece. [33] It was a hot, humid Baltimore afternoon in early August of 1938, so dark that I turned on the indoor lights. Lightning struck nearby. A car headlight shone (I thought) into the screened window, which opened onto a roofed porch. The light grew in size and dimmed in intensity. Basketball-sized, the fireball floated slowly toward the window and struck the copper screen.

There was a ripping sound and ozone smell. A blue-green edging appeared on the fireball's circumference. It continued through the screen into the room and swung in a slow arc to the hallway. It speeded up as it moved toward an old-fashioned stand-up telephone. As it touched the mouthpiece, there was a bluish flash, which lit up the room brightly. The ball zipped into the mouthpiece as if sucked into it by a vacuum cleaner. The bell on the box under the phone stand jingled.

The screen had a basketball-sized hole with smooth melted edges. The [undecipherable handwriting] were scratched across the folds, and the wallpaper above the phone was scorched. The mouthpiece had a whitened [undecipherable] ring on the top edge, and on the mahogany table was a mat covered by the lace doily, which was scorched. The doily itself was undamaged and the phone was operable.

The lightning hit a tree a hundred yards away, across a street divided by a hedge lined streetcar track system. Blown apart by the storm, unblackened tree pieces were found on the porch—the largest was baseball-sized. The fireball swirled internally much the way cream does when poured into hot coffee. This resembled Brownian motion of a colloidal suspension. The fireball lasted five minutes.

Cedar tree spits dissolving red fireballs. [442 Wallace] I saw lightning strike a thirty-inch-diameter cedar tree, quickly releasing many deep red, four-inch fireballs that quickly dissolved in the air or followed conductors.

Big fireball strips bark off tree. [55 Garden] A dazzling bluish-yellow fireball rolled down a large tree (four-foot-diameter) and stripped off its bark, then broke into small multi-colored fireballettes.

Melts transformer, burns off treetop. [446 McCown] Airborne part of the time, an eight-inch fireball melted a transformer and burned the top of a tree.

Hickory tree descender grounds on metal garage. [448 R. Jones] A ten-inch, bright orange-red fireball went down a hickory tree and forty yards across a paved street, then grounded itself against a sheet metal garage.

Orange fireball bursts oak's bark, scorches tree and leaves. [183 Wilson] Standing in a clearing, the large oak tree was isolated from surrounding trees. Lightning struck the tree twenty-five feet above the ground and a six-inch orange fireball—not too brilliant and moving slowly—traveled down the trunk and discharged to ground. The bark burst loose at the point of impact and splintered. The fireball singed the lower tree trunk and wet ground leaves, then disappeared after four seconds.

Phone fireball streaker leaves trail. [539A Metcalf] A fireball came from the telephone during a storm. With a loud crackling, the baseball-sized fireball streaked across the room, leaving a trail of light and then instantly disappeared, leaving no sign of a burn, no odor, no change of color. Even the phone was unharmed.

Phone emits five-inch fireball, floats by screen. [3 Johnson] Lightning struck a telephone line and the phone discharged a five-inch fireball that floated by a metal screen door and after several seconds, vanished.

Revolving, shrinking pinwheel disintegrates and crank-phone jingles. [487 Hoglund] Like a pinwheel, the six-inch yellow-gold-white fireball tumbled across the floor, leaving no marks. Revolving, it threw sparks and got smaller as it moved. I was ten feet away and noticed no heat. It disintegrated, seeming to run out of fire. As a child in Kansas, before moving to Oak Ridge, I saw a fireball come in an open doorway, float toward an old crank-type telephone and disappear. The phone jingled.

Crank phone shoots balls at ceiling. [46 Foltz] In Iowa and later in western Nebraska, after a lightning flash, the mouthpiece of an old crank-style country telephone shot a red tennis-ball-sized fireball diagonally at the ceiling like a Roman Candle discharge, but in half a second. My dad taught my seven brothers and me to always point the mouthpiece at the ceiling during a storm to minimize being hit by a fireball.

Crank phone spits red-roller. [4] During a storm, an old crank-type telephone emitted a five-inch red fireball that rolled on the floor, leaving behind an odor of ozone.

Crank phone's white dazzler rolls. [22 Hemphill] In north Texas, during a short spring 1936 thunderstorm, lightning struck a telephone line. A fireball followed the line to the receiver. The fireball entered the phone by the entrance cable and departed from the mouthpiece, and rolled out into the room and disappeared, leaving no trace other than the faint ozone. The telephone was a wall-mounted hand-crank with a long mouthpiece.

Telephone ball exits window. [206 Wheatley] After a lightning flash, the telephone emitted a six-inch fireball that flew fifteen feet to the window. It moved slowly, taking a couple seconds.

Hallway fireball discharges to phone. [83 Picklesimer] On both occasions during a heavy thunderstorm, lightning struck close by. In my first encounter, two three-inch fireballs floated through a hallway in the house to a telephone coming two feet from the phone. I was ten feet away. The phone rang and ozone lingered. In my second case, after lightning struck a tower, two twelve-inch fireballs rolled along a high-voltage transmission line. The duo slowly rolled 200 feet and simply disappeared.

Seeks ancient phone and explodes. During a dark thunderstorm in early August of 1944, at 2:30 in the afternoon, hail and rain pelted a Victorian house on a Vermont hill. When lightning struck nearby, a physicist who was in the attic saw a faint sizzling blue-green grapefruit-sized fireball appear three feet above the landing and 12 feet from the inside of the window glass. The ball looked like a colored globe of the old oil lamps, now seen in museums, with a yellow-orange glowing inside the ball.

Spinning about its vertical axis, the ball slowly floated sideways across the landing to the center, paused there, and then slowly floated down the stairway, gradually increasing its speed. It suddenly flew sideways four feet, attracted into the wood cover of an old wooden box mounted on the wall. It contained an ancient telephone and its electrics, which were bulky. Inside the box, the ball exploded and blew the door open. An odor of burnt wood and varnish lingered.

Transformers Emit Balls

There are some reports of fireballs spontaneously coming from transformers during sunny weather. But when lightning strikes transformers, sometimes fireballs float out and fly about.

Transformer's fireball falls and rolls on concrete. [168 Witt] I sat 35 feet from a transformer pole set in the sidewalk and next door to a gas station with a concrete apron extending down to the street. The pole set between the two properties. The ground was dry—rain had not fallen yet. Lightning struck at or near the transformer and an eight-inch white fireball appeared at the top of the pole. It floated down. At the base of the pole, it moved southwest above the concrete for thirty feet and after five seconds, it faded out.

Transformer spouts blue roller. [99 Crawford] I parked with my family outside the Snow White Drive-In restaurant during a storm in 1951. Lightning struck a power pole bearing the main transformer for the restaurant. Immediately, a large bluish-white basketball-sized fireball drifted slowly—ignoring the wind—down the power line strung from the pole toward the building. The restaurant lights stayed on until the fireball reached it. (We were 20 feet from the spot.) There was another crash and the lights went off. It blew out their fuses.

Transformer repeats swooshing slow risers. [360 Blevins] During a storm I was parked near an electrical substation. Immediately after each flash, a swoosh sound and an eight-inch, yellow-to-orange-to-smoky-orange fireball slowly (three seconds) rose vertically from the substation transformer and far circuit breakers. Each event emitted a strong ozone odor.

Transformer's mother fireball spawns ballettes, avoids dynamite. [354 Bennett] On July 20, 1948, on Highway 58 South, three miles from Decatur in Tennessee (I work at Union Carbide Nuclear at Oak Ridge), I drove a jeep for Volunteer Cooperative Power Company. Torrents poured, gale winds blew, hail pelted. Water flooded the ignition, my jeep quit. I parked 75 yards from a power pole. A fireball traveled along the electric lines and hit the pole's transformer. Several baseball-sized sunshine-bright fireballs bounced off the transformer and hit the road before me. These diminished into several smaller ballettes that vanished before they reached my jeep. Was I scared? My jeep was loaded with four cases—two of caps, two of dynamite.

Transformer's ten-foot fireball hangs, then falls. [384 Lane] During a storm, lightning struck a transformer (elevation 2,000 feet) and a ten-foot bright fireball hung suspended at 25 feet for five seconds and then lost form, fell to the ground in a shower of glowing fragments.

Transformer fireball tastes of brass. [557] A small lightning flash in 1926 struck a pole transformer and a 16-inch yellow fireball appeared. After several seconds it slowly decayed from 18-to-1 inch, then vanished after twelve seconds. I tasted a strong metallic taste of brass in my mouth and throat, and the air was filled with a metallic ozone odor. My body felt electrified, but that was my only other sensation.

Transformer spouts two fireballs, float two seconds. [311 Hatfield] In the dark of dusk an electrical storm lashed Harriman, Tennessee. While at my sister's home on Margrave Street, I stood in the front door, facing east, when lightning hit a transformer across the street. There was a loud crackling and three two-inch flame-colored fireballs came from the transformer into the air. They floated two feet in two seconds and vanished.

Became Current Surges

Fireballs that contact a conductor, such as a cable or equipment, and disappear will usually send a large charge into it and create a large transient current surge. In one case, the fireball floated into one of a high-voltage, three-phase power cable and disappeared. Twenty miles away, the surge of current damaged equipment and was detected. No one has reliably calculated the charges and currents of such fireballs, but it is obviously enormous—far exceeding all estimates.

Powerline fireball rolls into fusebox, destroys house appliances. [484 Kenerly] In 1958, a volleyball-sized fireball ran down a 220-volt entrance cable to a four-room house but loudly disappeared when it reached the fuse box mounted to the side of the house. It did extensive damage to an electric stove ten feet from the fuse box and burned out its heating elements. (Two people stood within eight feet.) Next to the stove stood a tabletop water heater with a half-inch hole burned in the front right corner of its outer shell.

In the living room fifteen feet from the fuse box, a table TV badly smoked on its back and sides. It was set in a corner by a picture window, which disintegrated. It completely burned the 300-ohm antenna cable. The black cabinet was not noticeably harmed. The only damage to the TV electronics was two elevator or antenna coil primaries burned open and two AC line-input molded capacitors blown to bits. (I understand electronics and electromagnetics—I am with the Instrument Department at Oak Ridge.) Even if the fireball became a large transient current surge, how could that do all this unusual damage? Circuit-wise, the damage made no electronic sense.

Zapped MOV spawns ball. Testing a superconducting storage magnet storing 1 MJ with a circulating current of 1 kA, a senior engineer unintentionally discharged it through a Harris metal oxide varistor or MOV that was nominally rated 10kJ to create a large flash and bang. To his amazement, from this appeared a glowing one-foot bluish white fireball. After several seconds it rose deliberately, not floating, and vanished. Because MOVs contain silicon dioxide, the engineer speculated that filamentary chains of vaporized silicon oxides might be involved.

TWT power supply produces red golf ball. To supply a traveling wave tube amplifier or TWT, a technician designing a high-voltage high-frequency power supply checked its voltage-regulation control loop. He set up a dummy load resistor bank of vertical power resistors on a Plexiglas base. To monitor one current, he placed the clips in series with two sections, but one clip lead disconnected and curled back to touch one hot 375-watt load resistor, which melted the insulation.

There was a loud clap and hissing from a one-inch fireball that formed and floated near the bottom of the dummy resistor load stand. Reddish orange, the ball rotated about its axis. Oddly, the top rotated in one direction but the bottom, in the opposite direction. The ball continued to spin for three minutes, until the technician turned off the supply, at which the ball vanished.

Electronics Births Balls

That electronic devices, and anything electrical, are associated with producing fireballs is certainly nothing novel and has been reported for over a century. But how does this compromise semiconductor memory, microprocessor chips, and panel display devices? Particularly inside the electronics which control nuclear bombs and airliners. As a weapon, the principal behind fireballs that destroy semiconductor devices is distinctly unsettling—especially in the future hands of global terrorist nations.

Strobe tube ejects fireball, hits man in face. When a technician stacked eight high-value military capacitors for a strobe circuit and connected the supply to the 240-volt line, nothing happened, and he disconnected it and rechecked all components and conductors. Still nothing worked. Puzzled, he manually triggered the SCR and at first still nothing worked. So he tried again. Then a slow glow came from the strobe tube—which was very wrong. The strange glow grew until it became the size of a baseball. It was easy to look at, not harsh. The tube ejected the fireball at the tech—standing two feet away—and the ball hit him in his face. Outside of warmth on his face, there was no injury or damage.

Ungrounded dipole antenna emits ball. When lightning hit a 40-meter horizontal dipole antenna wire, with the grounding knife switch ungrounded, a crackling and sizzling fireball came from it and floated across the room and hit a radiator and exploded after its ten-second flight.

TV screen emits fireball into couch. In the early 1960s, nearby lightning struck and a fireball shot out of the round screen of a small television—which never worked again—floated rapidly into a couch between a couple, seated on it watching the set, and vanished.

Rockets trail wires, create fireballs. Walter T. Sullivan on October 29, 1976 in the *New York Times,* where he was science editor, reported that French researchers financed by the French Atomic Energy Commission and Electricite de France, the French power-generating monopoly, fired test rockets with spools that trailed long wires into stormy skies. On 20 of the 30 launchings, lightning hit the rockets, and on several occasions one-second luminous spheres rolled down the wires.

As we just saw, electrical effects are sundry and ubiquitous, including light bulbs that excrete fireballs and phones that burp balls, not to mention transformers that emit balls which vary in both size and intensity—fireballs that shoot out or float slowly. Nor uncommonly, some then enter a cable or wire and disappear, becoming great flowing charges of enormous electrical currents that sometimes do serious damage. Conversely, some electronics such as television sets occasionally burp fireballs. These are not uncommon.

But those in the next chapter are rare cases—ones that are unusual and dramatic, so much so that they demand attention, and deserve their own following special chapter. When an Air Force electrical engineer repeatedly generated fireballs at Wright Patterson Air Force Base, he got more than he expected.

13

Bell Burps Purple Balls

The following, previously unknown spectacular cases of Harold Brown, Lawrence Leach, and the Oak Ridge Hooker Cells are unique because their equipment was small. They were told to me and I taped their interviews.

Wright Patterson AFB equipment sneezes fireballs. [Brown] At Needs Laboratory at Wright Patterson Air Force Base, in 1954, I generated fireballs regularly, and what they did varied. Once I operated a sputtering regulator (jig) to electrode solar batteries with a 1,000-volt second-anode discharge at 40 mA. Suddenly the equipment sneezed, and from the 3/4-inch pump exhaust came a five-inch purple-pink fireball with sharply-defined edges that pulsated, changing shape from oblate to round to ellipsoidal, but always changing back like a rubber balloon full of slow-motion water. Standing two feet from it, I got a good look. It was as bright as a 20-watt bulb but never changed size or color. Airborne, it glided out the door into the hallway, turned left and floated down the hall, westward. Finally it floated through a glass windowpane like it was not even there.

But we never saw where they went after that, perhaps into some trees. There was always a smell of ozone, but no smoke. This occurred often, and whenever we heard the equipment start sneezing, we knew a fireball was coming out, and everybody ducked for cover. We even nicknamed it "The Sneezer." After five times, the fireballs came more frequently. Finally one fireball did not go out the door. Instead it struck a large instrument panel and exploded, burning out the panel. Sparks jumped out and smoke poured out. After that we were ordered not to use the equipment, which sat idle in a corner.

Author—I talked to Brown about this and other cases. This electrical engineer also invented the white ceramic dual-in-line integrated circuit package or Cer-DIP—familiar to all electrical engineers and semiconductor physicists—but he never achieved the appropriate recognition, and later worked at Hendrix located in Londonderry of New Hampshire when I interviewed him.

Leach and Whitmore

Gas Stove's Metal Box Generates Fireball [Leach]. It was not cold and there was no thunderstorm. The sun shone, the sky was clear, there were very few clouds, and there was no storm that day. I am Lawrence W. Leach and I was at work at the *Falmouth Enterprise* on Cape Cod in Massachusetts, when my wife phoned. She was in the living room when she heard a loud crash, followed by a sizzle—a loud crackling noise. A bright brilliant white-yellow fireball floated past the doorway, about one foot above the floor, gliding from the living room to the kitchen.

Later we determined that the fireball went down through the toilet bowl and down into the washing machine—a Whirlpool—and then through its timer, and burned it out. The wires were frayed. It damaged the gas stove. It fused the socket to the lamp, which was melted to the base. The wire that came out of the stove had a quarter-inch long section chopped out—this was behind the stove. It damaged the television antenna rotor, which would not work.

A technician later said that the transistorized circuit was burned out, and he repaired it. There was no damage to the bathroom. It did not touch the drier. Golka also examined it and determined that the ball formed inside a metal rectangular box, possibly functioning he said like a cavity resonator. Where it came from and were it went to I do not know. It went by slowly enough—like a man walking. It went past a fifty-inch-wide open doorway. It was almost tennis-ball sized and left a sparking trail, which lasted for short time and left an ozone smell. My wife knows what ozone smells like because I used to spark generators in the cellar, where I also run my Lintotype machine and do typesetting for several newspapers.

My wife was 12 to 14 feet from the fireball. It tripped the circuit breaker for the gas stove. The electric clock of Mrs. Myet, the first house down on Fox Lane, stopped at 2:47 P.M.—the same time as our event, but all the other houses were unaffected. Where it entered the clock, her wire cord was completely burned off. Our washing machine was not connected to anything above it and its chassis was grounded through the third ground wire. It has a rubber hose.

The trail of gray smoke did not linger long. It was as if someone burned something in an oven and the door opened and the smoke could quickly dissipate. I later got credit from the washing machine company—even though the machine was in the house when we bought it—and the unit was taken to the dump and replaced. An old friend from Mansfield who was into antique wireless radios—he owned a Spartan radio—also investigated. And Golka came down three times to interview us and examine the site.

Author—This was the end of my interview with Mr. Leach. To summarize the above interview, Mrs. Leach was in her living room on a sunny day when there was an explosion and a fireball floated past the French door. Frightened at this apparition, she phoned her husband, who examined the stove and house. On a clear and sunny day, a fireball came from this long steel box, mounted on a gas stove, probably through or from a tiny hole blasted in it. Your author talked to Mr. Leach, taped the conversation, and examined the hole and crater under a microscope, and in 1974 wrote conclusions in the following detailed report.

"Painted black, the rectangular box-like enclosure was bent from 0.0245-inch thick sheet metal. It is 18.398-inches long, 2.559-inches wide, and 2.125-inches deep. The bottom and one end are open and there are four holes in it, probably for attachment with fasteners and for wires. Under this protective cover ran a 14-gauge wire. Where the wire ran by the tiny hole (and possibility touched the metal), it was chopped (melted) in two with a half-inch section missing.

The hole is located 9.705 inches from one end and 0.490 from one side. The hole is tiny—about 1/13 inch in diameter. Although appearing round to the naked eye or magnifying glass, under my microscope it is oval, with a 0.013-inch major axis and 0.094-inch minor axis. The crater around the hole measures 0.311 inch long and 0.171 inch wide. Its slope is shallow and smooth. Several round beads of copper are stuck to the surface of the crater, and though invisible to the naked eye, they were easy to detect under the microscope. The major axis of the elliptical hole does not align with that of the pear-shaped crater but is rotated 40 degrees counter-clockwise.

Radiating outward from the hole and its surrounding crater is a black soot pattern. On top of the box, however, all that shows is the round hole with the paint neatly chipped off near the hole.

The crater was melted first, and the heating was intense and sudden. The center of the crater melted through first, and the blast blew vaporized metal outward through the tiny hole. Though a crater surrounded the hole on the bottom (the inside portion of the box), the hole on the top from where the gases shot out was clean cut and surrounded only by a little chipped paint—but no crater or evidence of melting. The edge is razor sharp, and the hole was clear-cut, indicating that the heat was tremendous, sudden, and localized to this tiny hole. Your author's attempts under similar conditions to simulate such sudden vaporization and clean holes failed and only led to melted holes with lumps and beads.

Owing to the oval shape of the crater, and a small amount of copper from the wire melted into the crater, it is impossible to determine the precise amount of metal vaporized. But measurements that I took under the microscope come close, and calculations indicate that close to 0.594 milli-cubic inch was vaporized, with a tolerance of five percent.

Although a latent heat of vaporization of 1110 calories per gram of metal was assumed, what cannot be determined was how long it took to vaporize the metal. Surely, it was far less than one second; and, so several assumed times were used to calculate the kilowatts needed. If time for vaporization was 0.1 second, then it took 3.35 kilowatts; if 0.01 second, 33.5 KW; and if 0.001 second, 335 KW. This indicates that tremendous energy was briefly concentrated in the tiny hole."

I photographed the metal box and took close-up microphotographs through a microscope of the crater and hole from both sides. I measured the exact dimensions of the box, drew and dimensioned mechanical diagrams, and used the data in an uncompleted attempt through simulations and mathematical derivations (from microwave theory) to determine if the box and wire acted as a cavity resonator. In 1976, I gave the box to a colleague, another electrical engineer and a journal editor at *EDN*, who gave it to a Northeastern University professor for metallurgical analysis who became gravely ill. Although I possess microphotographs, the destiny of this box remains undetermined.

Golka interviewed Leach and his wife in Falmouth on Cape Cod, Massachusetts—examined the house and found no burn spots—and obtained their account and the box, which I analyzed. I later interviewed Mr. Leach.

The Leach case possesses interesting angles that might shed light on unexplained aircraft crashes, perhaps including mysterious explosions of the TWA Flight 800 in 1996 near Long Island. Unbroken wires generate fireballs that exhibit intense heat and melt through wires, screens, and occasionally—as in the Leach case—melt through metal plates. And the proposed solution of inert gas in fuel tanks is unlikely to suppress fireballs, which have formed inside vacuum tubes and burned their way out.

Out-of-phase 5-KW rotary-converter spits fireball. [544B Kouvenhoven] I am a physician and Lecturer in Surgery and affiliated with Johns Hopkins Hospital in Baltimore, Maryland, and submit this report to Oak Ridge National Laboratory at the request of Ruark. One of my classmates in medical school, who worked as a substation operator in the Interborough Transit System in New York, once saw a fireball when he threw a 5,000-kilowatt rotary converter into the line out of phase. A fireball rolled from the commutator onto the substation floor, out the open door and exploded when it reached the trolley track in the street.

Whitmore's Hooker Cell burps fireball. [51 Whitmire] I was terrified by an accident at a nuclear-bomb facility that occurred at Oak Ridge while we were producing chlorine in a Hooker Type S cell. It consumed a direct current of 10,000 amps and 450 volts, under full power. A 12-inch ball of fire floated out.

Author—The Hooker Cell is an efficient method for generating chlorine. Hydrogen is liberated from the cathode, and chlorine is set free from the anode. The two gases are kept separate from each other and from the solutions by asbestos diaphragms. The Hooker Cell for electrolysis of sodium chloride solution to make chlorine gas was manufactured by the Hooker Electrochemical Company and created 95 percent of the chlorine commercially produced at this time.

Such units had a brine feed line and can be a cubical box with a concrete bottom and a concrete dome-like top, at which chlorine gas exits upward from a gas outlet pipe. A brine inlet pipe(s) spray an aqueous sodium chloride electrolyte brine solution—which conducts electricity—onto and over vertical asbestos-covered cathode plates. Below them are anode copper conductors on concrete and graphite anodes. A positive dc voltage is applied to the cathode and enormous currents flow through the cell. Because chlorine is strongly electronegative, it is never free in matter and is removed.

Aerosol in dc short-circuit floats red fireball. Norwegian engineer Nielsen performed famous short-circuit tests by using a large 10-kilowatt, 12 kilovolt DC generator near a waterfall. A fine mist of water droplets filled the sparking electrodes. Several times a great luminous cloud materialized. From the lower part of this cloud came a red fireball that floated upward for several seconds. He published photographs in his report.

Waveguide initiates and guides farmhouse fireballs. [Dr. Paul A. Silverberg, formerly of Raytheon in Bedford, MA] During lightning storms, fireballs haunted an old Massachusetts farmhouse. It always took the same path. Once inside the house, it would appear near the door, move along a telephone line inside the hallway, and then vanish. The investigating researcher, Dr. Paul A. Silverberg, took measurements and concluded that two trees in the yard with other objects created a wave-guide or electromagnetic tunnel for high frequency lightning waves.

Although he did not know how the balls formed, he was sure they followed this wave-guide during storms—an opportunity to study fireballs. He told the homeowners the cause and said he would return during the next storm to study it. When they invited him later for dinner, he was surprised to see the trees cut down. Ball lightning never appeared again.

Author—Fireballs that re-occur are not rare. In the Hill Air Force Base Liniac linear electron accelerator (discussed earlier), the fireballs also appeared—usually on sunny days—and followed the same path repeatedly for years until the Varian tore down and rebuilt the Liniac, after which the dangerous fireballs never reappeared.

Multicolored fireballs blow hole in kitchen stove. [434 Light] These rolling multi-colored fireballs varied in size from one inch to less, in bright colors of red, yellow, orange, green and so on. They blew a hole in the kitchen stove and poured out for ten minutes.

Electric Machinery and Motors

There are many cases where fireballs are associated with, or formed near or inside, electrical equipment and fly or float out. For legal and financial reasons potentially arising from negligent and strict tort liability, the following cases are ignored and denied. They also create adverse publicity. The only way to learn about them is informally from workers, who usually prefer anonymity—for good reason.

Rotary converter creates fireball. [D. Ernesto Caballero] A Spanish physics professor who later became department head at a major university, was an engineer at the Interurban Company in 1914. He was in the main powerhouse watching the main switchboard because many circuit breakers were opening. The firm had recently installed a large 600-volt DC rotary (synchronous) converter or generator to supply power to the company lines and to the city. A steam engineer spun the rotor.

Apparently, a current was applied to the spinning rotor winding, creating a magnetic field that cut through the stationary stator windings and induced three-phase voltage in them. A fifteen-inch fireball popped of the converter commutator and quickly floated down the switchboard. It passed three feet above the engineer's head and a hundred feet from the converter, then hit an office room ceiling and splashed in every direction. Ozone odor lingered.

Told to your author by a Dupont technician, who saw a similar fireball emerge from an electric motor inside one Delaware-based Dupont building, the glowing fireball did no damage. Dupont is safety conscious. Long ago, Dupont experienced a horrific explosion of their explosives, and they arranged their individual buildings so as not to trigger other buildings with cross-explosions, and enforced such excellent safety practices that the Dupont family lived on site.

The nature of electrical machinery and distribution systems generate many such cases. In theory, our ac and dc power systems use neat sine waves. The reality is different. Transients and nonlinearites abound, and computer switching power supplies are worsening the situation.

Author—Transients and nonlinear waveforms may create fireballs in polyphase systems. Invented earlier by Nikola Tesla, the three-wire polyphase system long-ago replaced Edison's anemic dc systems. Polyphase displaces voltages 120 degrees. Contacting the insulated metal slip rings that completely encircled the shaft were brushes. Sparking can occur at these brush tips and power loss occurs. Since frequent brush-ring maintenance is needed, it was common to operate several machines in parallel so they could share the load. During low power demand, such as in early morning, rather than shut down a generator, it was allowed to float or spin on no-load. Synchronous machines are large and carry huge currents and produce heat.

Today, however, coolants are used and piped through hollow "damper bar" conducting pipes. But in 1914, these machines were not self-starting and could not start turning without a damper bar on the rotor. When it starts up, a current surges and transients flow until it spins at constant (synchronous) speed. Other crazy things like short circuits occur to cause transient currents. These current surges or transients can also cause undesired mechanical oscillations and speed variations.

Today, maintenance is easier. We avoid converter commutator rings. Our large machines use brushless exciters or small self-excited generator field wires mounted on the stator and armature on the rotor shaft. The small DC field current to the excitor generator is adjusted without slip rings and brushes. This is a famous-but-old case and there is no information on how many synchronous converters were paralleled or their specifications.

Chain on power line spits out fireball. After a carelessly thrown metal dog leash accidentally wrapped around power lines, an explosion created a basketball-sized bright white fireball that gently floated to the street. After several bounces, it fizzled out, resembling water splashed onto a hot frying pan.

Locomotive emits fireballs. Alternately driving two locomotives provided by the Boston and Maine Railroad, Golka drove them at 20 miles per hour. With a customized control, he opened and closed a submarine circuit breaker between the megawatt 1,600 horsepower diesel generator and 2,000 horsepower motor trucks beneath the locomotive. With a camera, he repeatedly captured photographs of floating fireballs. Earlier, he and Harry (a radio historian) visited the battleship Massachusetts and diesel submarine Lionfish in Fall River to consider if they could try it. Although permission was granted, he did not proceed and instead chose to use a locomotive.

Chemist Analyzes Ball's Bottled Gases

In a well-publicized case, M.T. Dmitriev was a Russian chemist experienced in plasma research. While taking nuclear-bomb radioactive-fallout measurements (according to James Tuck), he was camped on the bank of the Onega River when he saw a bright fireball with a yellow-white two-to-three-inch core surrounded by two shells. Lightning struck the south shore to his left. A fireball floated near the opposite side of the river above a float. Its inner dark-violet shell was over half an inch thick, and its outer bright blue shell was almost an inch thick. This slightly oval fireball resembled a giant football that measured an inch longer than across.

Moving at almost five feet above the river, the ball traveled at over four feet per second and rose an inch or two for each yard, as it flew closer, directly over a series of seven floats tied on a line and reaching two-thirds across the river in a westerly direction. The river current flowed from east to west. Wind blew from west to east, which had no effect on the ball. As the ball approached, the radio's static grew louder into a rumble. Dmitriev stood outside his tent, at the shore where the seventh float was tied to the bank.

The crackling ball flew over his head and to the forest behind him. Overhead, it slowed to one foot per second and began climbing at a steeper pitch, at eighteen inches higher per yard of horizontal travel. It then stopped and hovered noiseless over a knob of land. It left a trail of acrid blue smoke. When the ball reached the edge of the forest and it collided with six or seven trees, it emitted violet sparks with each collision. As it began to zigzag after the collisions, its initial white color turned to bright red. After 60 to 65 seconds, it disappeared. Dmitriev estimated that although he had not seen its formation that the ball probably occurred simultaneously with the lighting, or ten seconds before he saw the ball, so that total lifespan was 80 seconds.

Dmitriev had with him several evacuated gas sample bulbs with which to take air samples. He held them at arm's length overhead. He took air samples at 55, 60, 65 and 70 seconds after he first saw the fireball. These were later analyzed in a laboratory. From his experience with plasmas, Dmitriev compared the brilliant intensity of the core to a plasmatron torch of 13,000 to 16,000 degrees Celsius, or as hot as the sun. (By comparison, the sun's surface is 5,500 degrees Celsius.)

He estimated that to generate the odor of ozone and nitrogen oxide required a radiation of 1,000 to 30,000 roentgens. A meter reading gamma radiation read 1.2 milli-roentgens per hour between the radio frequency radiations could have affected the photo-multipliers to get incorrect readings.

Although the ball passed over six feet from Dmitriev for six seconds, he felt no heat and his film detection showed no radiation. In laboratory experiments, the ratio of ozone to nitrogen dioxide near laboratory discharges decreases with increasing discharge voltage. The ball values were closer to gases near electrical discharges. The highest ratio of 2.5 occurred at 60 seconds and corresponds to an electrical potential of 300 to 400 kV to ground. The hydrogen-to-oxygen ratio was a thousand times too small for any explosion. The measurements in mg per cubic meter of ozone and nitrogen dioxide, and their ratios were both bell-like curves.

Melts Holes In Glass Leyden Jars

Leyden jars are obsolete capacitors. A capacitor—also called a condenser—is two conductors separated by an insulator, such as mica or air. The first laboratory-made fireballs occurred in 1754, after Professor P. van Muschenbroeck invented the first capacitor—the Leyden jar, a wide-mouth glass jar coated inside and out by two separated pieces of tinfoil, with a metal rod touching the inside foil to store electric charge. Bertholon in 1787 described how British physicists Arden and Constable charged a Leyden jar, and created at the point of discharge a one-centimeter red rotating fireball appeared and rapidly moved randomly.

When T. Forster of London in 1823 charged a Leyden jar, a two-centimeter (0.8-inch) red-hot ball suddenly appeared. It pierced the glass wall with a circular hole and disappeared in a bright flash and explosion. Who invented the Leyden jar or capacitor is debatable. Some historians claim that in 1745, Professor P. van Muschenbroeck's pupil Cuneus first discovered the Leyden jar, which Watson improved.

De Tessan in 1859 suggested that ball lightning was a positive ball of charge in equilibrium surrounded by an insulating layer of air, encased in a negative shell, also surrounded by a second insulating outer shell, and stabilized by some little-understood mechanism. Opposite charges leak through the imperfect insulating shell and recombined, thus producing a glow and ozone. Contact with the ground causes the loss of little charge, but piercing the shell with a metal conductor shorts it out, causing it to explode.

The Leyden jar is a phial or glass jar with its lower section coated with a conductor, such as metal foil. A brass rod reaches through a cork or rubber stopper down to touch the glass. To make a good electrical contact, a chain from the bottom of the rod reaches down and rests on the inner foil. The top of the rod reaches up out of the bottle and has a round solid brass ball and binding post that is polished to prevent leakage of electrical charge into the air.

Ordinary wires were not used to suddenly discharge Leyden jars, because it is difficult to draw a spark with pointed wires since they tend to draw out the spark silently. U-shaped metal dischargers with wooden handles were used. With two brass balls on their ends, they created bright blue snapping sparks. Also, it was a good idea to insert a bleeder resistor in parallel with a capacitor to be discharged.

Great charges are powerful. In industry, your author once experimented with sixty-pound 115-microfarad capacitors by discharging them with various gauge wires. Unless he inserted a high-wattage bleeder resistor—even of low resistance—in series with the discharge wires, then these wires exploded in a blinding flash. Some were totally vaporized—definitely dangerous..

Dozens of Leyden jars can be connected together and charged up in parallel, then reconnected in series, thus multiplying their voltages dozens of times. Capacitors are made with water substituted for one foil. Experimenters once used these water capacitors, with several connected in a box of water and poured paraffin on the water to minimize evaporation. The glass jars were thin hard glass, mostly free of bubbles and harmful salts, with thicker tinfoil that did not blister. Cheap lead glass jars leak and crack, and are easier to puncture at lower voltages. Amazingly, historians and hobbyists still use Leyden jars, and firms still manufacture them.

The jars were charged with static electricity from friction by static machines. A hand crank spun an eighteenth-century friction electric machine. The crank spun sulfurous balls—later ones were made of hollow glass cylinders or disks. They could charge a Leyden jar. The Holtz induction machine had one fixed two-foot disk and the other spinning at 450 rpm, to generate 70,000 volts at half a milli-ampere and 34 watts at 27 percent efficiency. The two glass disks of the Winshurst electrical machine, varying from 14.5 inches to seven feet across, reached 180,000 volts and 0.2 mA with a 36-watt 18-inch spark at 19.5 percent efficiency. Current is directly proportional to the rotations speed. The best machines reach 40-to-46 percent efficiencies. By comparison, modern generators exceed 90 percent. Ganged disks of 16 pairs were made. If each of the disks is wired in series, the voltage is 16 times greater, but if connected in parallel, the current increases 16 times.

A capacitor or condenser is made of any pair of conductors separated by a nonconductor, such as an insulator, such as air, vacuum, oil, wax, mica, or any other dielectric. However, a capacitor also can be only one conductor, such as the metal ball atop Tesla's Magnifier, since the other conductor is at infinity. In James Clerk Maxwell's two-volume treatise on electricity and magnetism, he first mentions the single-plate capacitor.

Maxwell's original eight equations were compressed to four with Yale Professor Joshua Willard Gibb's elegant del-sign shorthand nomenclature, a significant time-saver in advanced calculus. But Richard Feynman in his famous three-volume "red book" noted that this shorthand sometimes confused him, and he recommended that we write the eight equations out in Maxwell's lengthier differential equations, frequently so we do not lose touch with intuition. Physicist Murray Gel-Mann in his book *The Quark and the Jaguar* lists the further compression by Einstein in relativity's two-equation shorthand. Although equally accurate and easier to remember, this concise form that Einstein used is absolutely not intuitive.

Coronas Become Fireballs

A corona does not float but is attached to pointed objects subject to high potential differences. Coronas are not fireballs. However, it is not unheard of for fireballs to come from coronas. In several cases, a part of corona on an aircraft window became a fireball and slipped inside the craft. At other times, corona (which is ionization from high electrical potential differences) came from motors and cables.

Trolley car corona becomes fireball, ignites gas tank. [148 Becher] During a thunderstorm downpour, I rode in the front vestibule of an electric surface trolley car, equipped with the usual trolley collector that obtained its power from an overhead electrical feeder. When lightning hit nearby, it tripped the trolley air-circuit breaker near the operator. A bright, 6-to-12-inch fiery ball appeared before the trolley. It appeared first as a heavy corona on the overhead conductor, and then moved toward the trolley car and in a gliding motion descended vertically to the ground. It was not instantaneous but moved in slow motion. The breaker had tripped and I do not know if the ozone came from inside the car of the arc expelled by the air-circuit breaker. It ignited a gas storage vessel and created a fire. I am with the Safety and Health Physics group in the Oak Ridge Gaseous Diffusion Plant and have no scientific explanation.

As we saw in this chapter, electrical equipment accidentally generates fireballs, and no one knows how or why. At Wright-Patterson, Brown did so routinely, Leach saw it blast a hole through steel, Norway's Nielsen used mist, and Dmitriev measured the exponential decay of fireball gases. Leyden jars often formed ballettes, and coronas sometimes birthed fireballs.

Nearby lightning may extinguish a fireball. Although most balls exit by vanishing or exploding, some find strange ways to depart. But, most mysteriously, before they depart, how can certain fireballs magically absorb the entire charge of a lightning stroke?

14

Fireballs Eat Lightning

In "terminating cases," there is no lightning below the ball (in one case a faint return streamer was visible)—and amazingly, the lightning terminated in such fireballs! After traveling two miles, lightning can only be diverted by a conductor (lightning rod, steel building, car frame) but cannot be stopped even by several feet of rubber—yet some fireballs stop it!

So, does all the charge and energy instantly enter the fireball? And where is the energy stored or dissipated? In similar cases, sometimes such terminating fireballs silently disappear. So, to where does all this massive energy disappear? Thus, once understood, could they store and transport enormous charge and energy? And could such man-made terminating fireballs be made to shoot back the massive energy that they absorbed? Your author (personal prejudice) cannot accept this—but cases exist of fireballs that "eat lightning."

Fireball stops lightning. [63 Ellis] An old barn stood 200 yards away. Lightning flashed over the barn but did not strike it. Instead, the discharge stopped and terminated in a four-foot fireball, the color of a spark, which hung in the air above the barn—with no lightning below the fireball! After one second, the fireball exploded in a brilliant splattering. In thirty seconds, it occurred again! I saw no display before, after or between the two strokes. There were no lightning rods on the barn.

Fireball stops lightning bolt 100 feet up. [418 Panter] Like a bright fire, an eight-inch fireball appeared at the end of lightning, which stopped a hundred feet above the ground, terminating in this ball. There was no groundstroke. [Did all the lightning energy go into the fireball?] This occurred near Thelma Road off E. Term Avenue [spelling uncertain] in Oak Ridge. [How did the fireball absorb all that charge?]

Ball breaks away from lightning channel. [180 Turner] It was a very hot summer mid-day when a quick dark cloud came up. A basketball-sized fireball broke off from a weak lightning stroke and fell toward the ground.

Lightning feeds energy into fireball, explodes. [581 Robinson, TSGT, USAF] Before entering the Air Force (I am an officer at McGuire AFB in NJ), I worked in a private home in Springfield, Massachusetts [85 miles west of Boston], caring for an elderly invalid. I sat on a porch, facing west. The August 1936 morning was hot and clear, with a few fleecy clouds. At noon, westerly thunderheads grew, slowly approaching zenith. At three, the storm line came closer; the west sky grew dark.

After the heavy clouds passed, the storm front and heavy rains hit. It became the worst thunderstorm that I ever witnessed. I faced a row of two-family houses across the street with a perfect view of their yards between and in back. In one backyard stood a 30-foot utility pole, used solely to support the outer end of a pulley-clothesline from both houses.

A split second before the flash, I saw a pale, slightly bluish faint streak come up from the pole. I saw the entire flash that came down, except for the lower end of the lightning—it stopped ten feet above the pole and rapidly grew into a 12-inch fireball, and the entire streak ran down into this fireball, which hung in space for three seconds. In a blinding, deafening crash and enormous explosion, the ball vanished. The pole seemed undamaged.

Lightning ends in fireball high in sky. [506 Price] I was driving on a highway (pasture on both sides) during a storm. Lightning hit the ground 450 yards ahead. A ten-inch bright-orange fireball formed and rolled twenty feet. On another occasion, I saw a lightning stroke come down and stop in a large fireball high in the sky. [How did the fireball absorb all that charge?]

Lightning tip stops in dazzler fireball. [461 Egner] Lightning hit nearby with an instant clap of thunder. The lightning bolt had a basketball-sized bright fireball on the end of it. The fireball glowed bright white for several seconds, then exploded. I was half a block from it.

Lightning terminates in brilliant monster fireballs hung in air. [531 Self] I work at Oak Ridge Labs and am providing, at his request, this report to Dr. Bell of Union Carbide Nuclear. While my family and I were tent camping on August 24, 1969, at Myrtle Beach State Park in South Carolina, lightning flashed and rain began to fall. Without any lightning nearby, a white monster fireball materialized at no more than several hundred feet altitude and was at least 25 feet in diameter. It was intensely white and lighted the surrounding area brightly. It hung in the air for several seconds, and then disappeared without discharging to the ground or anywhere, simply going out like a light bulb. After several minutes, the phenomena repeated almost exactly.

Nearby Lighting Makes Fireballs Vanish

As we just saw, when lightning strikes an existing fireball, it may terminate in that ball and descend no lower. Lightning that strikes nearby—that is, within 300 yards—also can instantly cause a fireball to immediately and silently "wink out" or vanish. But is lightning the cause? Perhaps not. More accurately, the nearby lightning itself may not cause the ball to vanish—but the two events are somehow associated.

What could cause a nearby lightning strike to be associated with the disappearance of a nearby fireball? The dilemma is, as is well known, that lightning's sferic or electromagnetic wave intensity is too weak. Is there some mysterious underlying phenomenon—a Hidden Progenitor—associated with both the lightning and fireball? Perhaps a common parent cause, a deeper hidden cause?

Lightning terminates roller. [516 Maney] A fireball fell through the front screen door, rolled along the floor for two feet, and vanished when lightning hit the house at the end away from the front door. The aerial melted, all wiring in the radio burned out, but this was not near the front door where the fireball entered. The ball did not scorch or damage the floor or metal screen.

Lightning terminates in and feeds fireball. [ORNL Flowers] I was in the northeast corner of the second floor of the [deleted] building of Oak Ridge Laboratories on July 15, 1966 at 2:30 p.m.. An active storm with some hail and high wind large lightning activity was approaching. There were many cloud-to-ground strokes.

Amazingly, one lightning stroke stopped at 25 feet above ground at 300 yards and a fireball formed at the terminal and remained after the stroke. With sound, the yellow-white fireball grew larger up to a three-foot diameter. A power circuit failed and arced over. A glowing sphere trailed a tail the same time but not quite phased with the sphere. The ball did not move. After the flash, the ball's intensity grew and peaked at one second and then decayed way after three seconds. There was no retinal after-image.

Vanishes simultaneously with lightning. At four feet above the meadow, a basketball-sized bright white fireball appeared and moved slowly horizontal. Lightning struck nearby and the ball vanished at that same instant.

Enters Transformers

Fireballs that strike or enter large transformers are not uncommon, but often damage the transformers. In some cases, lightning was not involved, so this was not a complicating factor—it was ball lighting, period. However, for insurance and bureaucratic purposes, it is prudent to never blame ball lightning.

In fact, there is one famous case in New Hampshire half a century ago, where witnesses saw the fireball float onto a large warehouse roof and ignite it. But the fire chief and insurance company wisely refused to attribute it to ball lightning, which officially did not exist.

Line slider, transformer wrecker. [202 Miller] While lowering a window during a storm one night, I saw lightning strike a power line a hundred feet at the rear of my house. An eight-inch blue and yellow fireball traveled 25 feet to a pole transformer. During the several seconds of travel, its circumference gave off sparks. It disappeared into the transformer, damaging it and knocking out the power for several hours.

Exploding line-roller damages transformer. [76 Mann] A yellow-white two-foot fireball bounced slightly and rolled down a line following the line closely to a power transformer, destroying it, and where it disappeared after five seconds with an explosion and some smoke. I work at Oak Ridge National Laboratory now, but I saw this in 1915 while in Gainesville, Georgia.

Hideous hisser rolls into line's transformer. Emitting numerous sparks and hissing hideously, rolling back and forth atop power lines, a large fireball finally rolled to the end of the line where it struck a transformer and exploded in a shower of sparks and cloud of smoke. The lines and transformer were undamaged.

Sundry Exits, Strange Departures

There are an infinite number of ways fireballs depart—some simple, others exotic, even strange. They vanish into cables, knitting needles, doorways, sump pump holes, openings, and even into artillery shells.

H-bomb factory inter-metal jumper explodes. [28 Johnson] A five-inch blue fireball jumped from one metal surface to another—lasting five to ten seconds—in the 3039 stack area of the Thermonuclear Experimental Division on June 24th in 1960 during a lightning storm. Two reports were filed.

Line fireballs enter sumps. [34 Thompson] Ball lightning appeared several times where I worked in a compressor house west of Cheyenne in Wyoming at 7,000 feet altitude. High-voltage lines brought power into the compressor house. Fireballs initiated in the terminal boards rolled twenty feet on the concrete floor to disappear down the water sump, into which discharged the compressor cooling water. Diameter was on the order of several feet.

Enters kitchen door, exits other door. [87 Peele] After a lightning flash fifty yards from the house, a four-inch red fireball formed at the strike point, then floated in a straight-line into an open kitchen door, past me, and out the opposite open door.

Ball breaks into shell-like glowing floating pieces. Lightning hit a telegraph pole and where the wires shorted, yellowish green sparks flashed near an insulator. Inside it grew a six-inch white fireball that rolled along the wire, faster, changing to red. Rotating, it emitted sparks. Reaching the lowest part of the wire, it jumped down to a lower wire and rolled to its lowest point, and then dropped to a large poplar branch, which loudly cracked, broke and bent. At this the ball shrank way. But, within two seconds six more fireballs appeared and ran over the branches and simultaneously vanished. Several seconds after this, a five-inch fireball appeared a yard to the right of the origin of the first ball.

It followed some branches, elastically jumping over all uneven obstacles and shrank to three inches and shot out sparks. It flew down to the road and jumped with three decreasing and apparently damped bounces of 8-, 5-, and 2-inch heights, and then broke into three big, floating, shell-shaped pieces and several smaller glowing pieces. The witness, a chemistry professor, reported that these glowing pieces disappeared, but not simultaneously. After 20 seconds, the last piece vanished.

Knitting needle fireball. Laura Ingalls Wilder, Frontier Girl, described an event in her "Little House On The Prairie" books of an event that occurred on February 7 of 1867. In a farmhouse near Walnut Grove, near the banks of Plum Creek in Minnesota, a fireball "bigger than ma's ball of yarn" came in through a stovepipe and was attracted to her knitting needles and silently vanished.

Shotgun blasted, ball grows. The sky was cloudless. A blue 8-inch fireball, glowing like a 150-watt bulb, floated motionless at two feet above the grass. The student engineer fired a blast of shotgun lead pellets into the fireball, which then increased 50 percent in diameter and then exploded into unequal glowing fragments. Lifetime exceeded one minute. In another case, an Oklahoma farmer shotgun-blasted a fireball, but that one vanished instantly.

Multi-colored ball descends into floor grate. Thunder crashed and an orange and deep-blue fireball with other colors arose from behind a sofa up to four feet above the floor, floated to the floor grate (forced hot air), and the baseball-sized ball went silently down into the grate.

Enters illuminator and explodes like shrapnel. It was half an hour after a thunderstorm on a hot windless sunny day. At 60 feet away and ten feet altitude, a one-foot yellow-blue fireball with a tinge of orange moved up and down. As bright as a 100-watt bulb, it resembled an opaque soap bubble that slowly descended and moved 60 feet in half a minute. It was attracted to and entered a very hot illumination device used in their movie filming. The 220-pound unit exploded and metal shrapnel flew around, and the unit fell to the ground.

Flash, crack, gone. [332-1 Wright] During a 1935 day of thundershowers, I walked to the front door to close it. As I got there, I heard a sharp crack and simultaneously saw a thirty-inch, orange-red fireball fifty feet in front of the house, floating two feet above the ground. It exploded—just a flash, crack, gone. Eight feet from it grew a small apple tree. There was no damage or sign.

Suspended and gone. [365 Cunningham] I visited a cousin during 1960 in Rockwood. Lightning flashed. A basketball-sized fireball appeared, suspended in air and quickly disappeared.

Flashes behind trees. [ORNL Bennett] On April 1966, my mother, father and I were on our way to our grandparents' house in the rural town of West Townsend in Massachusetts, on the Vermont border and forty miles northwest of Boston. We drove on Lunenburg Road. Running 9,000 feet, it runs south-to-north up from Lunenburg and stops at West Townsend Center in West Townsend in Massachusetts. For a birthday party, we were bringing presents. As we passed an open field, lightning flashed. My dad stopped the car. At the end of the field we saw a very large orange fireball—much larger than the moon—over the trees. After two minutes the ball suddenly vanished behind the trees, followed by an explosion and flashes of light.

Pole fireball dissipates like sparkler. [332 Shelley] While I stood near a window during a storm, a basketball-sized fireball flew into a utility pole and slowly exploded like a brilliant sparkler, illuminating the area.

Men chase fireball into well. [578 Scandlyre, Addkisson] Two families lived a quarter-mile from me at Big Emory Community, near Harriman in Tennessee, just fifteen miles west of Oak Ridge. A glowing fireball came down off the ridge in front of their house and went up near and all around one family's house. A neighbor and his son chased after it. It crossed the field to another neighbor's home, where he joined the chase. They chased it for some time and struck at it with their hats. The fireball disappeared into the dug well at one of their homes. I will get you in touch with these families. Wright and son Steve and Hobb Addkisson chased it in 1932.

Hangs in air, drops to table, vanishes. [535] This pinkish yellow ball of fire materialized inside a room during an electrical storm. The one-foot ball hung briefly for three seconds in the air, dropped to a table and disappeared. I had spots before my eyes.

Orange golf ball flies briefly, pops out. [479 Barnard] Immediately after a close bolt, my wife and I both saw a one-inch, pale yellow-orange fireball inside our kitchen during a recent storm. It traveled in a straight line for two feet at eye level. After one second, the ball popped like a cap pistol and instantly disappeared.

My wife faced the ball, which appeared five feet before her and flew toward her. I was to her left near the wall, facing her, and five feet from the ball. The water heater was ten feet to my left and five feet from the ball. Drawing a line between us and the ball and heater would create an imaginary rhomboid shape. The fuse box was three feet to her right and just over four feet from the ball at its disappearance.

Multi-Color Balls Shower Sparks

Fireballs often have smooth surfaces, but many are fuzzy. But why are certain fireballs smooth, but others coated with fuzzy fire, and still others covered with spattering sparks?

Entrance cables bursting sparkler damages cable. [385 Beems] Lightning struck nearby. A basketball-like fireball left a telephone pole and followed the entrance cable of the house. When it reached the house in four seconds, it broke up like a giant sparkler. The lead-in cable was split where the junction was made at the house. It damaged the 220-volt wires.

Affects photographic x-ray film. [348 Wilson] There was no lightning flash. A two-foot red-blue-yellow airborne fireball—whose size rapidly changed—melted objects, exhibited magnetic disturbances and electrostatic effects. It affected photographic x-ray film. There were visual after-effects.

Author—Like Roentgen's accidental discovery of radiation on an undeveloped photographic plate, maybe the case of Wilson at Oak Ridge is also a clue, but maybe not. Maybe a few fireballs emit radioactive processes, but others do not. [For more on radiation burns, read the "Hoaxes?" in the next section.]

Ball over power pole showers sparks. [393 Denton] While traveling in a westerly direction toward Valley Pontiac on the Turnpike in Oak Ridge (near Valley Pontiac County) during an electrical storm, I saw a large basketball-sized fireball appear a thousand feet away above a power pole in the vicinity of this business. After several seconds, the ball turned into a great shower of sparks.

Rolls onto porch, splatters against house, dog jumps. [467 Hamilton] While sitting out a summer rainstorm in our house five miles north of Concord in Tennessee during 1942, we were in the kitchen adjacent to the back porch on which the event transpired. About the size of a basketball, a solid ball of fire like a flamethrower rolled onto the porch from the west side of the house, then splattered against the side of our house where our dog lay sleeping. He jumped straight up. A thunderous noise accompanied the fireball, and we felt that something also was occurring inside the kitchen.

The back porch was open and not screened or boarded in any way and had a wooden floor. It vanished after three seconds. A girl in the house across the road from us simultaneously felt a shock on the screen door that she was leaning against. The second fireball I saw was years later. It looked like the transformer that was hit turned into a molten fireball and rolled down to the ground. It fell slowly. After landing it quickly faded out.

Blue ball hugs trunk, leaves three cigar puffs. [459B Davis] In July during the severe rains and thunderstorms, I drove with my family, coming from Clinton to Oak Ridge. At a point, just past Walker Chevrole, there was a thunderclap and in the front yard of one house, about two feet above the ground, there was a blue fireball on the tree trunk. It quickly vanished, leaving a cloud of blue smoke, like three good puffs from a cigar.

Yellow whip-snapper explodes in blue haze. [495 Shebrooks] I was at my summer home on the Jersey coast in High Point in 1931. The volleyball-sized fireball came the entire length of a metal rail that the man next door had for launching his cabin cruiser, which when not in use was laid upright against his cottage. After the yellow ball traveled the length of the rail, there was a whip-snapping crack. It exploded with a bluish light, leaving a smoking haze and a heavy scent of ozone lingered in the air. I was standing at the window when it occurred and was fifteen feet from the rail.

Slips down into hot air grill. Several seconds after thunder, a multi-colored sparking fireball the size of a baseball and mostly orange and deep blue, floated four feet above the floor, then slipped down into the floor grate to the furnace without any noise.

Descending dazzling ball bursts into six. Following three intense storm cells, the fourth was weak. From an inactive cloud slowly descended a blinding white fireball. Halfway to the ground, it burst into six glowing ballettes that diverged, moving down more slowly. After ten seconds, before reaching the ground, they silently vanished, as nearby lines on a pole hissed and crackled.

Exploder illuminates the dark. During a dark storm, a one-foot fireball floated, traveling 300 feet in half a minute, then violently exploded to create a flash that briefly lit up everything within 50 feet.

Sparks fall from ball. Like a firework during a lightning storm, a fireball floated 25 feet above ground, drifted slightly to the south, sparks falling to the ground, and a small billow of smoke drifted from it. After 45 seconds, it vanished.

Rose-pink pear explodes on English yacht. In a classic case, American paleontologist Professor Oscar C. Marsh was aboard a large yacht anchored in the harbor of Southampton in England—a large city on a ten-mile inlet on the English Channel in southern England and about 70 miles southwest of London. A bright rose-pink four-by-seven-inch pear-shaped fireball materialized at the top of the foremast and descended, and when it hit the deck, exploded in a white glow that knocked down one man. Across the deck zigzagged lightning, which quickly vanished. But below deck, sparks shot down the ventilator into the galley, where it threw about utensils. Crackling sparks leaped from everything.

Hoaxes?

In all honesty, web sites that collect and solicit ball lighting cases cannot investigate donations to their sites. Yes, most entries are legitimate. After all, who in their right mind wants to hoax ball lightning! Still, this encourages a few mavericks—literary minds that are strangers to the truth, and with time on their hands—to practice their deceptive craft on web sites.

Although rare, those hoaxers who mimic legitimate cases really do no harm, but those that embellish with new characteristics can confuse matters. So, here are curious cases (totally rewritten, with identifying information deleted) that have the embellishing aroma of fictioneering. Missing is spontaneous human combustion, which does not exist, and is described elsewhere. Remember, as always, as Carl Sagan wisely warned us, that extraordinary claims require extraordinary proof.

Did falling optical black hole explode tree? A severe storm and tornado alert was in progress. Wind gusts exceeded 50 mph. Appearing to absorb light, a totally black, 18-inch ball that looked more like a black hole with a total absence of any light—and not resembling anything painted black—fell slowly from an overhead vortex futilely attempting to form in the sky overhead. Despite a powerful wind, the black ball floated straight down into a tree branch, instantly exploding. The half-split tree fell across a station wagon in the driveway. Looking directly at the black ball as it fell, the witness saw the entire event. He felt no shock, even though in his wet shoes he stood 20 feet from the tree. A few researchers conjecture that these rarely reported fireballs are optical black holes. Although this strange account is very hard to believe, even if true—how can it be true?

Optical black holes were only explored just before the millennium, but the theory goes back to Einstein. In 1923, physicist Walter Gordon mathematically proved that a moving dielectric medium acts as a gravitational field, and noted the mathematical equivalence between light in a moving fluid and a gravitational field. But to detect this dramatic effect required that the velocity of light be low compared to the medium—something impossible to achieve until recently.

Materials like water and glass slow down light passing through them. Theoretically, an eddy will trap a light if the medium circulates faster than the speed of light in that medium. There was no such medium and Gordon's discovery was forgotten.

Unusual dielectric media now exist that slow light to several meters per second, and finally reduced pulses of light to a standstill inside small super-cooled gas clouds and form a Bose-Einstein condensate, in which all the atoms combine and behave as if they are in a single quantum state.

As a medium, water has a refractive index of 1.333 and travels 75 percent of the speed of light in a vacuum. If moving, the medium drags light. Diamond is 2.4.

Led by Lene Hau, her group produced these effects based on quantum mechanics, as a rod-shaped cloud of sodium atoms cooled to a millionth of one degree above absolute zero, trapped within a magnetic field. But the rod is only 0.2 mm long and 0.05 mm thick, far too tiny to consider it as a candidate for an optical black hole fireball! The event horizon of an optically black hole fireball should be large.

Ultra-slow light promises to open up applications in nonlinear optics, supersensitive optical switches, quantum computers, optical communications, data storage, and as research tools to construct analogs of astronomical objects and phenomenon, such as the unbelievable Hawking radiation, and could be a testable model of quantum gravity, and the like. However, your author doubts that optical black holes explain alleged black fireballs—but who knows?

Lukim, Bajcsy and Zibrov of Harvard University subsequently froze a light beam, without removing all its energy—an exciting discovery that could be used to store and process data and even spur on progress in transferring information from photons to photons in future quantum computers, and create secure quantum cryptography.

By firing a signal pulse through a sealed glass cylinder of hot rubidium gas, then switching off the control beam, they created a holographic imprint of the pulse on these rubidium atoms. Earlier experiments switched on a signal control beam to recreate the signal pulse.

But this time they switched on two control beams, thus creating an interference pattern. The regenerated signal pulse attempts to continue through the glass cylinder, and photons bounce back and forth. However, the signal pulse remains stationary and the beam of trapped photons is frozen for 10 to 20 microseconds.

There is no need to further discuss the topic. If you are interested, there is an excellent article by Lene Hau, "Frozen Light," on pages 66-73 in the July 2001 issue of *Scientific American*.

Based on the Centauro events from the Andean, Bolivian and Tajikistan detectors, some conjecture that micro black holes rain down into our atmosphere and explode. Cosmic rays collide with molecules and ostensibly create unstable ten-microgram black holes that explode in under a nano attosecond into a burst of particles. When the Large Hadron Collider at CERN is able to generate enough energy, it may create black holes on demand, and we will have a solution to the Centauro mystery.

Another odd theory speculates that ball lightning contains liquid light. Is it possible to condense photons into drops of liquid light? When a laser beam travels through a cubic-quintic nonlinear medium, it ostensibly slows down passing light—the stronger the beam, the more slowing—and focuses it into a thin column. A gigawatt beam would condense the photons into 0.002-inch-diameter droplets. Liquid light allegedly possesses properties of liquid water, including surface tension, and forms eddies and whirlpools of light. Physicists such as Humberto Michinel speculate that colliding liquid droplets will be used to form massive arrays of logic gates. If so, and if fireballs contain liquid light, could this explain their quasi-intelligent behavior?

To your author, all this above (the applications, not the experiments) is quite doubtful, and your author remains a skeptic. We are close to the edge where the kooks live. Be very skeptical—not of cases and of valid respectable experiments, like above—but of applications to explain crazy things. The theory and experiments are not crazy—the application is. Other ball lightning researchers may have their own opinions, but I say, "Bah humbug on optical black hole fireballs—in fact, all black fireballs."

Ball allegedly powers lamp. After lightning struck near a house, a blue fuzzy quasi-ball floated in through the bedroom screen and glass window, drifted to the floor, and when it reached the dresser, floated up (staying three inches from it) until it reached the top, and then skimmed across the top of the dresser surface into a lamp. With a pop, it vanished. Turned off, the lamp turned on; and when the witness manipulated the switch and pulled the plug, it remained lit. When the witness left the room and returned with another witness, the bulb was off. Really weird. Hard to believe.

Nuclear radiation skin burns? Recently, while discussing string theory, some physicists mentioned a puzzling 1886 fireball case reported by an American representative to Brazil, published in *Scientific American,* which suggested intense radioactivity and "radiation sickness" syndrome. It describes, how, in terror and believing the world was ending, the occupants fell to their knees, praying, but soon commenced violent vomiting. On their upper bodies, extensive swellings began, especially about their lips and face. The brilliant light did not seem hot, although it left a peculiar odor.

Next morning their swellings subsided, leaving their faces and bodies with large black blotches that did not hurt. But on the ninth day, after their skin peeled off, these large black blotches turned into virulent raw sores and their hair fell out on the sides that faced the light. Trees around the house seemed unaffected until day nine, when they suddenly withered.

There was evidence of indirect radiation. In this 1886 case, it is curious that these radiation aspects occurred, because our present knowledge of neutrons, x-rays, alpha and beta particles, and gamma radiation was then-yet-undiscovered. It is unlikely that mere ionizing radiation could do this damage, and that something generated heat within their bodies—which has little heat sensation—although their loss of hair is difficult to explain with internal burning from dielectric and inductive heating.

Only the side of the witnesses facing the phenomena was exposed and suffered injury. This favors particles and weighs against x-ray flux, gamma rays and neutrons, as this would have caused immediate injury, probably rapid death, and might have cooked them. It is difficult to explain their hair falling out due to internal burns from dielectric and inductive heating. Be skeptical. It is impossible to verify the details—and they are important.

Hundreds of lazy yellow golf balls invade cellar bedroom. An intense Midwestern lightning storm was near. Just before midnight, in their basement bedroom, a couple allegedly saw golf-ball-sized, pale yellow fireballs, more yellow near their edges, lazily and slowly float straight—slight down- or upgrade—slowly towards the north wall, which was also the outside wall. After several seconds they would silently vanish, sometimes with a soft pop. There was between one to three fireballs visible at any one time. Maybe, but sounds fictional.

Twin snowballs brave cold winds. In early January with weather windy and unsettled and foggy, with light snow settling, suddenly twin fireballs floated 35 feet above the ground near a highway bridge. The top ball was white inside, encircled with a lime-green perimeter and was three-to-four feet wide; the bottom ball, two-to-four feet.

Neon blue ball above houseplant buzzes. After a small tree located inside the house began crackling, slowly a tiny neon-blue thimble-sized ball formed several inches above the plant, then crackled free. Buzzing, it floated for several seconds but then rapidly shrunk.

Casts no shadow. A ten-inch bright white fireball came through the front glass door, through the wooden door, then flew up to the ceiling, paused and broke into five equal-sized shapes that danced back and forth against the ceiling. After ten seconds, one rogue piece crept down the wall, going back and forth, looking as if it was searching for something. Oddly, the objects, although bright, emitted no light, and even the rogue ball did not emit light and cast no shadow. One of the witnesses approached this Spider Man's odd "wall crawler," but when he reached out to touch the ball [would you?], it shot up to the ceiling and out the front door, followed by the other four. Maybe this witness got a bad ice cube.

Disintegrates over plant. A bright light yellow fireball the size of a hen's egg burped out of a socket, which had a radio plugged into it, headed toward the man, changed direction and skirted a five-foot-high plant and slowly floated by. After flying nine feet in five seconds, with a sudden crackling, it disintegrated into many pea-sized sparks that enveloped the plant, but left it undamaged.

Fiery whirlwind torches house. In 1869, *Symon's Monthly Meteorological Magazine* (4: 123-124) reported a fiery whirlwind. On a hot Wednesday, on the Cheatham County farm of Edward Sharp, five miles from Ashland in England, an odd laming [sic] whirlwind moving 5 mph came over the adjacent woods and tore off some small branches and burned them. The flaming cylinder grew. When it traveled over some feeding horses, it singed their manes and tails up to their roots. [Unbelievable.]

As the fiery whirlwind moved to the house, it picked up some hay and grew brighter. When it reached the house, it set afire the shingles and in ten minutes the house was engulfed. The tall column moved across a wheat field, setting afire all cradled stacks it touched. After this, the fiery pillar passed through woods, in a straight line to Cumberland, and crisped leaves in a swatch to Cumberland, and crisped leaves in a swath 20 yards wide. Several trees where set afire and continued to burn.

When it reached the river, a column of steam rose up for half a mile. It then died. Over 200 people saw the fiery column and told substantially the same account. The team of horses where permanently spooked. There was a mixing of rotational and vortical phenomena.

Sounds suspicious. Maybe sloppy investigators misinterpreted what irresponsible witnesses misreported, and the rest is hearsay that got past incompetent referees and lazy journal editors.

There are other similar alleged cases, such as that seen in Africa by field astronomers awaiting an eclipse. A truck accompanied them with instruments and recorded an electrical field gradient of 600 volts per meter. G.D. Freier reported this in "The Electrical Field of a Large Dust Devil" in the *Journal of Geophysical Research* [65:3504, 1960]. In some cases lighting that continuously moves up and down in a tornado's funnel and oscillates, reaches several thousand volts per meter.

In the *Journal of Geophysical Research* [65:205-206, 1960], Vonnegut speculated that to fully explain the tornado we may need to first understand ball lightning, and wrote that others made the same connection.

Monster uproots trees. Described in 1873 by T. Beesley in *Symon's Monthly Meteorological Magazine* [8:150, 153], the Newcastle Whirlwind of November 30, 1872. As Beesley was about to leave his house, his gardener asked him to quickly come and see a fireball. But he missed it by half a minute He talked to four witnesses who saw it from different points. They all heard a ball "whizzing roaring around like a passing train." This attracted their attention and they saw a large rotating fireball that flew at six to ten feet above the ground.

The smoke whizzed around and it rose high. A blast of wind accompanied it and carried with it branches. It uprooted trees and broke off others ten feet above the ground. The giant fireball bounced on a rock wall and threw some of the stones ten feet.

Beesley rode the two miles of its path from southwest to northeast, and to the end, where it turned northwest. The debris path was 150 yards side. Witnesses saw the ball suddenly vanish. At the end of the path, the ball cut holes in the ground. Witnesses smelled burning sulfur. One gardener at Newcastle Manor described a noise of a long railroad train crossing a bridge and saw a large "dark ball, as big as a carriage" that emitted red large sparks.

Hard to believe. Be skeptical of this one. Maybe there is some truth here that was disguised by extreme exaggeration, hearsay and sloppy investigation.

Levitating eggs? Some meteorological events hint at odd effects. "Electric Storms and Tornadoes in France" in *Science* [17:304-305, 1891] describes such events on August 18 and 19 of 1890. A Boston tornado reportedly exhibited "antigravity" effects on a small boy, who rose in the air but was yanked down by his mom and friend. Just down the street, a man delivering eggs saw his eggs levitate. Oddly, the man was unaffected, although he was more "aerodynamically dirty." Today, we know that tornadoes exhibit no such ridiculous antigravity effects, but can instantly lift victims and objects, but leave nearby persons unaffected. If true, it seems exaggerated.

It was reported that the first Mormon temple in Nauvoo, Illinois, was destroyed by a tornado that suppressed light. What sort of phenomenon associated with fireballs can suppress emission of radiation?

In the murky field of ball lightning research, where facts are fuzzy and we cannot totally dismiss anything, your author assigns mental vectors (magnitude and direction of an ordered n-tuple) to each allegation. When many vectors point to the same general area, truth may lie therein. Perhaps it is better to mentally assign rough probabilities to uncertain qualities. That way, we are less likely to totally dismiss unlikely facts or events.

In our mind, the aforementioned cases might rate only five percent likelihood. However, some of them could add important facts—but not likely. Very suspicious, the aforementioned cases require extreme caution. Remember to shine the light of rational skepticism on all spectacular claims.

15

Odd Cosmetics

There are unusual shapes. A few fireballs are huge, others are tiny. Some are multipliers, shrinkers and expanders, or sport protrusions and emit horns and rays. Some are fuzzy or are enclosed in concentric shells, or exhibit inner motion. A few imitate glowing soap bubbles. Some resemble necklaces. A few trail tails, and some twins are connected by filaments. Although non-spherical shapes—such as rods, pear-shaped oblate spheroids and sausages (some that even morph or change shape)—are infrequent, they do occur.

Repeaters Reappear at Same Location

Why do some fireballs repeatedly reoccur in a specific location or building, usually during a storm? Some researchers attempted to stake out such sites during storms, but nothing ever came from their attempts. One Raytheon engineer investigated an old farmhouse west of Boston during storms where a fireball appeared repeatedly, following the same flight path into the farmhouse. The grateful owners cut down the trees and the fireball never appeared again, much to the shock of the engineer.

Keeps reappearing near ceiling. [294 Packard] After certain lightning flashes, an eight-inch silent, yellow fireball appears inside my house. It was seen many times by members of my family. It always appears in the same part of the room, two feet from the ceiling, and each time it is airborne and motionless, lasting two seconds.

Frequent fireballs storm by. On the opposite side of Montana's Lewis and Clark National Forest down into the plains starting in early spring, storms with fireballs occur frequently. Hikers at 7,000 feet encounter six- to twelve-inch fireballs that float or bounce along. During storms at night, they float by.

Barntop lightning balls repeat. For as long as anyone can remember, a fireball—not corona—appears regularly after dark during electrical storms atop an old barn, but disappears shortly after it forms.

Repeating stove creates balls for years. Beside an old iron wood-cooking stove, mounted onto a concrete floor near a block wall, when bad storms come, an orange white orange-shaped fireball appears and soon explodes with a loud cracking pop.

Fireballs repeatedly visit telephone company. In March 1971 an electronics instructor visited an AT&T central office telephone switching station—a single-story brick building—to meet with an electrical engineer, who took him on a tour and described a lightning ball that repeatedly appeared each summer inside the building during many lightning storms. Many workers had seen it, and he himself saw it last summer.

This fireball was seen over a dozen times and repeated its flight path on each appearance. Glowing like a 60-watt light bulb, this white three-inch fireball bounced several times across the floor, going from the south-west corner to north-east, bouncing fast on the linoleum floor, paralleling the west side of the building. This side had large windows.

In each repeat visit, this fireball vanished when it touched a flat, metal-grounded bus bar (of the translator frame) that ran horizontally above the linoleum floor. No burn marks ever appeared. The large room was filled with many frames, covered with what looked like a rat's maze of wires. These panels in frames contained thousands of tiny thermal safety fuses.

This building was built on a hill over an underlying rock. Thomas Edison had trouble with the subterranean ledge when he visited here. Fifteen months later he wrote a letter, asking for additional information. No doubt, fearful of adverse publicity and exposure to tort liability, management wrote back that no such thing ever happened. Denial is not unusual, and six years later he received another denial letter from a French linear accelerator particle research facility where a fireball blew a crater in their concrete wall—and was witnessed by another engineer from MIT.

Fear of notoriety and liability, and of proximate causality that establishes torts of negligence and intent, and possibly strict or absolute liability, all encourages selective amnesia that silences witnesses. Many great cases are benignly censored.

Fuzzy Surfaces, Nested Shells

Varied are the surfaces on fireballs. When fireball surfaces are smooth, there is little variation. When not smooth, there are a myriad—perhaps hundreds—of variations, and new versions surface to surprise the investigators. The surfaces can be coated with thousands of electrical sparks only an eighth of an inch high, like an electrical rug. Often the sparks, if that is what they are, can be longer, and lengths up to an inch are not uncommon.

Sometimes the lengths are not uniform—some are short, others long. Some sparks wiggle, others move randomly on the surface. Sometime the fuzzy ball appears to spin. A few fireballs are enclosed within one or two transparent concentric shells, often of a different color. Usually these shells are fuzzy, but some were distinct.

Short projecting points cover orange fireball. [537 Rinkenbach, M.S., FAIC] I am a Consulting Chemist with a Specialty in Explosives. I forward this report to Oak Ridge National Laboratories at the request of Dr. Warren. On June 7 of 1947 in Rockaway, New Jersey, my wife sat on a living room sofa, and I sat at a desk in the corner, facing out a window. Lightning struck nearby and a luminous, eight-inch orange-yellow fireball with short, projecting points appeared in front of a radiator, to my right rear, at three feet above the floor. It moved rapidly toward me, until it was within eighteen inches and in front of a window. My wife screamed and the ball vanished. With my back to it, I never saw it, nor did I feel any heat or electrostatic effects, and smelled no ozone or nitrogen oxide.

Double Shells of bright blue and violet encircle yellow fireball. Three Canadian geological expedition members saw a slightly oval fireball over a river—a glowing bright yellow-white six-to-eight centimeter core surrounded by two outer shells. The inner shell was a dark violet layer one-to-two centimeters thick. Outside this, the outer bright blue shell was two centimeters thick. [T.L. Tanton, *J. Roy. Astron. Soc. Canada* 12, 530 (1918)]

Fuzzy Aussie fireball silent. [Reported to Uman] On his horse, riding alongside the right side of Queanbeyan Road in Canberra, a witness saw young farm workers leading a shorthorn bull on the other side, walking alongside the road. It was a rainless quiet that precedes a downpour in this sub-humid area. He drew alongside the bull.

In an explosion, a golden fireball materialized a foot above the asphalt's center, white traffic line. Its surface was fuzzy like wool threads. It flew 20 feet in three seconds and in an instant silently vanished. Shortly after this an intense rain began. The site was not far from the industrial area of Flyshwick near Murray's Dairy, the capital of New South Wales, Canberra is near the southeastern tip of Australia between Melbourne and Sydney, and is 70 miles inland in the sub humid coastal mountain range.

Tentacled-fireball bounces off rug. Through the window slipped a basketball-sized intense glowing fireball with white electric "tentacles" on it that moved quickly. As it floated from left to right, it slowly descended to the rug, which it hit and rebounded up six inches and disappeared.

Three concentric spark layers enclose fireball. [Oak Ridge National Laboratories. analyst] Although these two types were not touching, sparks of the other color surrounded each other. Sparks were replaced each half second by new ones. Concentric layers of sparks enclosed the fireball. Up close, the analyst saw three inner layers. Although the ball passed half a foot from me, I smelled no odor and felt no electrical effects. It bounced off the stairway landing and silently flew into a wall. The undamaged wall was not hot.

Soap Bubbles

If the "soap bubble" cases are accurate, then these may not be ball lightning, but some other phenomena. The "soap bubble" theory of ball lightning, popular in various forms during the 1960s and 1970s, as envisioned by several leading physicists, including James Tuck of Los Alamos, proposed plasma encircled by a thin plasma membrane of unknown nature. This theory has some cases to back it up.

Sinking fiery "soap bubbles" out-speed draft. [518 Huddleston] I saw fireballs several times over the last forty-five years inside buildings or at least under a roof. I saw one ball come out of the ceiling, bounce three times on a bed or carpet then out the door and disintegrate on contact with the ground. These balls behaved like a giant soap bubble, and looked like a soap bubble—except for their fiery color—and bounced just like a soap bubble on a woolen blanket. They always traveled in an air current or draft—but oddly, faster than the air current itself. They sought a lower level and usually passed through an open window or door to the ground. My observations were in a medium-priced farm home of wooden construction.

Fog "Fireballs"

"Fog fireballs" are not true fireballs. They are cold glowing spheres or mists that maintain their shape and float near the ground. They envelope witnesses, with no harmful effect. Some witnesses report that within the glowing fog, sound was suppressed. The fog fireballs resemble mystery lights, described elsewhere, but are not associated with any specific location, as are those like the Marfa Lights. Unlike mystery lights, fog fireballs may occur during lightning storms.

Moving glowing fogball envelops man. Last night at fifteen minutes to nine, I was ascending one of the sharp hills in my neighborhood [near Boscastle, Cornwall, England] when I was suddenly surrounded by a bright and powerful light, which passed me a little quicker than a man walking, leaving it as dark as before. The light was seen by sailors in the harbor coming in from the sea, as it passed up the valley like a low cloud. [*The Times*, letter from J. Brown, Dec. 1, 1858]

Inner Motion—Tumbling, Boiling, Pulsating

For two centuries, researchers puzzled over fireball contents. Are fireballs solid or fractal gases, perhaps spinning? Or do they internally behave like a viscous fluid or a gas? And is there internal shear? Do they slide on a surface? There should be sliding or rolling friction—but if so, this rarely slows them down. Few seem to behave as if they have translational and rotational kinetic energy. Most bouncers do not lose energy and only a few seem to be deformed with each bounce on the floor. Most rollers and bouncers do not scorch the floor—but a few do.

It is common today to invoke chaos, fractals, cluster aggregates and compare this and ball lightning contents with cream poured into coffee, bees buzzing in a globe, and multiple colors swirling inside the ball.

Inner tumbleweed motion. [8A Starken] I saw fireballs several times, mostly at night in southwestern Minnesota. Twice, just before a local tornado (it had not yet rained), I noticed a tumbleweed movement inside the fireball, moving along a fence beside railroad tracks in mid-afternoon. Both exploded.

Pulsating boiling fireball in rotation. [113 Neely] The weather that spring morning before the storm in the late 1940s was hot and humid. Our home was on a high ridge in West Knoxville. The tallest tree in the yard (70 feet) was a tulip poplar. I sat near a window. A loud crack. Looking out, I saw a five-foot blue-white fireball—luminescent, but not extremely bright—bounce off the poplar from 20 feet above the ground.

Lightning struck the polar and split it down to 20 feet from the ground, and the top fell within minutes. The giant fireball floated 70 yards across a clearing in a straight line, stopped momentarily several feet above the ground, and then loudly exploded. Total lifespan was ten seconds. The fireball was constantly in motion about its center, that is, the surface seemed to boil. The effect was that of rapid expansion and contraction, plus rotation about its center.

The giant fireball spun about an axis while its surface had a boiling appearance. Either it rapidly expanded and contracted, or appeared to do so. Did the five-foot fireball bounce off the trunk of the poplar?

Author—The polar grows to a height between 70 to 90 feet tall with a spread of 35 to 45 feet. At this time in the spring, the poplar has yellow flowers high in the tree, 2.5 inches long with yellow-green petals. Their broken twigs emit a spicy odor and the bark has a light gray-green color with white in grooves.

Through screen, bubble of burning flames bursts. [241 Byrge] Lightning struck a telephone line and ran into the house next door. Mrs. HR was telephoning and got shocked. This damaged the telephone line and lead-in and showered her husband, who was planting shrubbery outside, with sparks of fire. I sat in the living room watching television when heavy thunder clapped, simultaneous with the flash. My son lay on the floor five feet before the TV, which immediately blanked out.

Almost instantaneously, a bright golden fireball came from the dining room, rolling along the floor, and stopped three feet from my son. It was basketball-sized and seemed to burn furiously. The flames seemed confined as if in a glass ball, and remained very round in shape. It lasted for an instant and then disappeared, as it burst, not unlike a bubble, after ten seconds of life. There was no damage, no marks on the floor, nor any residue. My wife was in the kitchen and saw the ball pass through the dining room and could see that the fireball entered the house through the door. It was open, but the screen door was closed. The front door was closed. However, several windows had awnings and were open.

Buzzing fuzzy balls in rotation. [388 Bradford] While lightning flashed there was a buzzing sound prior to the ball, then a sharp rifle crack. Then these balls of fire rolled across the floor in one direction. Some were air-borne part of the time. They traveled in straight lines for three seconds, were over a foot in diameter, and looked like fuzzy balls in rotation. They gradually vanished.

Switch grows bubble with sparks inside. *A* utility power plant engineer in Fairburg described how he and others in the 1950s had repeatedly encountered a fireball inside the basement of their electrical power plant. There were boilers on the north end of the building. When lightning storms hit, a panel downstairs would act funny before a fireball would float from it.

One day, a storm hit. He was in the cellar. A fellow worker yelled "Get up on this wooden table!" so they would not be shocked. The workers knew from experience that if they would stand on the cement floor, sparks would jump from the switches and panel meters, but the fireball would not harm them. This occurred frequently during intense storms. All of the power lines for the town would enter through there.

Then a bubble appeared attached to a switch and slowly grew in size from a golf ball to basketball size. Sparks writhed inside the bubble, which was not typical. It broke free and floated, slowly descending to the floor, where it proceeded to bounce off the floor, losing size each time it bounced until it silently disappeared. This was not the first time such a thing occurred to him. The wooden table was still there when Golka visited, photographed the area, and took the VCR film.

Fairburg is a town of 5,265 that is 60 miles southwest of Lincoln and 80 miles west of Missouri River and in the southeast corner of Nebraska. The state is in the center of the great grassy plains, and slopes upward gently from the Missouri River to the foothills of the Rocky Mountains. Winters are dry and bright with little snowfall.

Spaghetti fireball turns and twists. A blue-white fireball hovered motionless above a stove in a lodge, with individual cabins and gas stoves. The ball's interior was dynamic with beautiful spaghetti that turned and twisted, then silently vanished, leaving no odor.

White lines writhe on surface. During a thunderstorm, a blue-white softball-sized fireball floated in through an open window. A nearby witness saw bright lines "writhing" on the surface of the ball, which floated three feet above the floor, across the room and floated out the other window and popped. Ozone odor lingered.

Fire-sausage with internal motion explodes. During a lightning storm, a woman with her back to the door was talking on the phone. Sensing something, she looked back. An orange, glowing, 18-inch-long sausage entered the doorway. A Jacob's Ladder-type discharge ran along its longitudinal axis, and other discharges glowed inside the fiery sausage. It floated to her bed. She jumped out of the bed, upon which she sat, and it hovered over her bed and then floated back to the door and down the hall, and an explosion shook the windows.

Sparks within pink ball. On a late evening during a Southern shower storm, a couple saw a hot pink fireball that was shades lighter in sections. Brighter sparks ran within it. Estimated to be basketball size, the ball may have been larger and lasted a minute.

Colors churn inside blue-green ball. At one hour past midnight in mid-August, just north of a man's house, a thunderstorm sprouted lightning. When he went outside to look, he saw a blue-white (with purple and gray) basketball-shaped fireball, seeming to contain other colors—suggestive of rainbow colors—churning inside. It silently and slowly floated northerly. After ten seconds it turned violet and silently vanished.

Two violet balls contain chaotic beads. During a thunderstorm, one man pulled a cord from the wall socket and another clicked off the switch. Immediately, from the radio socket and inside the switch came a hissing and two clicks. Then, from the socket and switch, out popped two, two-inch, very-bright violet fireballs. They flew quickly across the kitchen and touched, clicked, and leapt outside by passing through a glass pane. In two more seconds, both vanished on the ground. The glass was unmarked, the socket and switch were charred, but there was no other damage. Oddly, both balls appeared to consist of many tiny one-to-two millimeter beads moving chaotically within the larger balls.

Spinning ball contains red and blue threads. During a thunderstorm with intense wind, an 8-inch fireball with the brightness of a 120-watt light bulb rolled in a spiral trajectory. Within, the ball resembled a tangle of red covered by blue woolen threads. It burned the clearing.

Sausage full of fireballs. Feeling heat on the side of his face, a witness turned to see a dark "cloud" [ovoid?] that emitted a radiating white glow and contain shapes like balls that pushed each other forward and disappeared into it. As it floated along, making a quiet crackling, the changing shape mutated, and the cloud's balls seemed to reflect green and red. The cloud's shape was not a sausage, but was five feet long and three across. It changed from a kite-like shape, with its front edge forward, into a "dustbin" shape full of balloons. Upon reaching tree-top level, it vanished.

Green-blue lines wiggle inside. A bright blue, three-foot fireball with well-defined edges and a glowing center and greenish blue lines that wiggled towards its center came to a window. Although it quietly vanished, two seconds later there was a loud boom. Several seconds later, a transformer outside blew up, shooting out sparks.

Cloud drops balls containing fire-snake. While driving through Illinois in a convertible with its top down, despite the storm clouds and silent sheet lightning, a man saw a basketball-sized fireball slowly fall from the clouds and land 20 feet to the front left side of the road. Bright, the ball contained writhing, lightning-like fire-snakes inside it. It bounced three times and upon the third bounce it exploded violently. Balls fell near the car. There was no damage or effect upon the flowers and gravel.

Fireballs Trail Tails

Not a few fireballs fly past witnesses and leave a brief trail of translucent smoke or sometimes trail a glowing tail. Other cases, such as that of Leach, also did this but are listed elsewhere in this book because of their other properties.

Snake-tailed fireball lasts 15 seconds. [118 Cook] Just after a flash, an eight-inch fireball trailed a 12-foot "tail of smoke" and lasted 15 seconds, airborne part of the time.

Japanese see tailed fireball. [56 Tamura] As a boy in Japan, before I later came to Oak Ridge, I was told of this phenomenon of fireballs in terms of ghosts, as part of a Japanese fairy tale. Years later, I saw a yellow-orange, two-foot silent fireball "fleeting" over the treetop trailing a tail.

Necklaces, Filaments, Twins, and Rays

Although certain qualities are rarely reported, they exist. Few Oak Ridge witnesses described a row of fireballs—the classic necklace fireballs. Few see vertical twin fireballs that travel as a pair, separated with a glowing string or filament between the top and bottom ball. This glowing filament is usually several feet long, and occasionally contains several small glowing ballettes in the filament. Usually, twin fireballs travel at the same speed. Often, the bottom ball drifts downward, stretching the filament thinner until it vanishes. And some fireballs shoot out rays or jets, or exhibit cone protrusions, or sport horns.

Mountain fireballs swing 15 minutes on ridge. At 8,215 feet altitude, rain and snow tormented two Swiss mountain climbers. Continuous lightning flashed. A horizontal row of small yellow globes appeared on a ridge, merged to form a large luminous mass that emitted red and blue fireballs that fell downward and exploded. One fireball materialized and moved sideways along the ridge like a thrown ball, curving downward in a parabolic arc but stopping just before it hit the ridge. After moving back and forth for a while, it disappeared, only to reappear minutes later.

Twin fireball's filament stretches. During a late afternoon storm, at an altitude of 3,000 feet, four German mountain climbers saw twin three-foot glowing fireballs floating 300 feet above ground, five feet apart with a luminous string or filament between them. The silent twins flew (not drifted) toward the northeast. The bottom fireball descended—stretching the luminous filament, which soon disappeared—until it fell into a thicket and disappeared. Forty-five seconds had elapsed, and another minute passed before the upper fireball, now alone, also began to drift downward, but vanished before it reached ground.

Twin rotating red fireballs strike building. Dr. Alexander Russell, an English scientist saw one of two reddish-yellow rotating fireballs strike a building and burst loudly. Residents opened their windows to see what exploded, but saw nothing. The other ball drifted away.

Spinning fireball projects ray, hovers. [110 Keele] Dark April thunderclouds hung low. Looking out the window at an Oak Ridge Lab building, I saw a two-foot fireball projecting a cone or ray. It spun, covering thirty feet as it hovered above the ground.

Hissing haloed fireball emits white ray. [Biji] Irish stronomer Edward Biji opened the door of his observatory. A flash of brilliant white light projected up from [down to?] the ground, with a deep-red fireball hissing atop, surrounded by a moving halo. The fireball emitted a line of white fire one-tenth the thickness of the first ray. This ray shot in and out of the ball, stabbing twice downward before hitting the earth, at which the fireball vanished, lasting five seconds.

Street-sitting blue "devil fireball" spouts two yellow horns. The evening thunderstorm's rain ceased. At 9:50 when the witness stepped outside, he saw a glowing 18-inch blue fireball—with two glowing, yellow horns (or rays) resting in the street, at 30 feet—that soon disappeared in an explosion (there was a thunderclap overhead) that left no trace. Lightning and rain resumed half an hour later, lasting to midnight, but with sheet lightning.

At 12:15 he saw a yellow fireball travel high in the sky, to the north, which was now clear with small clouds. Several others reported fireballs during the storm. [*Nature*; June 12, 1910; "Globular Lightning," pg. 284 by George Gilmore, research student at University College, Dublin, Ireland, Physics Department; submitted and attested to by Professor JA McClelland; sighting on May 14, 1919]

Jet pours from hole in fireball. On a humid hot Parisian day in 1849—one with frequent heat lightning—a large red fireball, resembling a rising red balloon rose 80 feet above the ground and a fire started at the bottom and sparks poured out. As the opening grew to three times the size of a hand, flames poured out. The ball loudly exploded, zig-zaging flashes on all sides. A white light continued to burn for two more seconds.

Sprouts Horns, Rays, Cones, Jets

Ever since fireballs were first reported, witnesses described those that spit out sparks and sprout fiery jets. Although most experts felt these jets were hot, it is not impossible that these jets or glowing rays may be cold plasmas. Because some balls seem to open up a jet on one side, and in some cases the hole (if it is that) seems to grow larger and the jet grows larger until the ball bursts. Usually, witnesses report that the jet hissed, but in some cases no sound was heard.

Although we list only two cases here, there are many listed elsewhere (but for their other attributes) that also sprout horns, ray, cones, jets, and spit sparks.

Spits sparks. During an intense storm, a large fireball floated in and hovered, spitting out electrical flashes, bolts and tongues of fire until it vanished after two minutes.

Collision opens up jet. An orange yellow fireball hit the street, cracking it, bounced up onto the rooftop, punching a large hole in the ball and "something" spewed out of it.

Non-Spheroids—Ovoids, Rods, Sausages

Most puzzling, but well-documented, are the non-spherical shapes. Though not common, they appear in one case in twenty. Ellipsoidal or oval-shaped fireballs are reported more frequently than sausage shapes, which are more common than rods. Oblate spheroids or pear shapes are reported least. Discussed elsewhere, fogs and dispersed forms such as glowing mists and mystery lights are not included herein, simply because they are not ball lightning.

Orange ovoid hovers in oak. [556 Izlar] In 1925, while inside a Wake County (North Carolina) thinly forested area, I saw lightning strike an oak tree 150 feet from me. A one-foot oval or egg-shaped orange fireball appeared. Its size gradually changed as it slowly disappeared after ten seconds on the tree. Scorching occurred.

Tub's fireball becomes sausage-shaped, enters bath drain. [589 Hood] I once lived close to the Ohio border near Cincinnati in Covington, Kentucky. One summer in 1912, I saw a blue-green six-inch fireball. It lasted one minute, came along a telephone line, entered the bathroom of the second-floor of a two-story frame house. It bypassed a lightning arrestor four feet above the tub. It slowly circled the tub with uniform speed—bouncing along the tub's rim (otherwise a steady forward motion)—and after four trips around, silently went down the drain.

It was a perfect sphere, except when it entered the metal drain—it changed to sausage shape! I stood transfixed on the wood floor, four feet from the tub. I felt no electrostatic shock. There was no lightning flash. The telephone line lost its covering. There were light magnetic effects.

Crunching red rod morphs into fireball. [Mack] During an early June thunderstorm in 1959, a rod glowing and narrow—two feet long and one inch across, and emitting sound like crunching glass—floated into a room like a spent arrow. Hovering motionless, it slowly gathered itself into a fireball. One minute later it dissolved, spewing sparks (resembling hot metal) from dust-like stuff falling from it.

Exploding sausage damages chimney. [Hare] During a late British summer afternoon, an intense storm passed and the sky cleared. Measuring six inches across and 15 inches long, an oblong sausage of fire descended slowly from the sky and floated toward a chimney 300 feet away. When 30 feet above the chimney, it shot out a horizontal white flash and orange mist. A blinding flash and explosion damaged the chimney; bricks and debris rained down; ozone lingered.

Keyhole rod grows into giant exploding fireball. [Blumenthal] A loud sizzling sound came from the French door leading to the patio. At the keyhole appeared a bright light. After several seconds, a pencil-sized rod of light came through the keyhole and gathered into a basketball-sized fireball. Sizzling, for several seconds it hovered motionless before the door, then shot for the fireplace, passing over the seated couple and their guests. It hit the fireplace and splattered, leaving a burnt mark.

These skeleton keyhole cases raise questions. Why did the fireball or rod not short out when squeezing through or being formed inside a metal hole? Why no melting? Why did it maintain a narrow shape for so long? A sphere is more stable. And why did it change its shape so slowly?

Shape Shifters Morph

Fireballs that squeeze through a crevice such as a skeleton keyhole change their shape from ball to rod after they emerge from the other side. Sometimes they float like an arrow across the room. However, most regain their ball shape after squeezing through, usually immediately. Historically, ball lightning experts such as James Tuck of Los Alamos felt that the balls possessed surface tension—an analogy being the soap bubble—and attempt to maintain equilibrium and minimize their surface, assuming and keeping a spherical shape.

Ball morphs to pour through a hole. Strong winds kicked up during a hot day and a dark thunderstorm arrived. An 8-inch, white-blue fireball as bright as a 40-watt bulb instantly materialized in the corner two feet above the floor in the room in an old house. The ball rose up 8 feet to the ceiling and then moved to the center of the room to a hanging lamp, leaving a string-like trace on the fabric encasing the cord. It returned to the wall above where it materialized and then went to the window and morphed its shape from a ball into an elongated flat stream that flowed out through a hole in the window. One of the three witnesses said the ball traveled 33 feet.

Morphing fire-cone sprouts twin tentacles. A 20-inch glowing fireball—it looked like the hot orange spiral heating element on an electric stove—with "dotted lines" three millimeters thick, floated in from an open balcony door at two feet above the floor. At first it floated totally stationary, spinning and hissing like a swarm of bees, but then slowly floated towards the wall.

From its spherical shape, it morphed or changed its appearance into a cone-like shape, with its cone's tip slowly aiming at an electrical socket, and it continued to float slowly towards the outlet. When within one yard, it sent out two glowing "threads" that reached out from the cone's tip, aiming at the socket. As soon as the threads touched the socket, there came a loud noise and the entire cone-fireball with its two glowing threads was sucked into the socket. The fireball lasted one minute and there was no damage.

Morphing ball harmlessly passes through pole. Trees swayed in the gale and the sky darkened just before a lightning storm hit. Then the wind stopped. A three-foot tangled glowing mix of brown, yellow and orange fire in a ball shape suddenly appeared 3,000 feet away in a field of grass, with a halo of dusty gray and yellow dust vortices. The ball rolled smoothly through the grass. At times it would pause but then continue rolling—at one time passing directly through an inclined pole supporting a telephone pole, but strangely without any visible effect, and kept rolling. Then it rose up, flying over a house and above several trees, then descended. As it came in through a windowpane—it had already lasted three minutes—it briefly elongated itself into an ellipsoid or football-shape, but quickly returned to its ball shape.

Vibrating wall crawler contortionist. Lightning flashed. An intensely bright six-inch orange and yellow fireball with an undefined vibrating surface "crawled" into a room through the upper opening of a window. Slowly and silently, it rolled across the wall and over a door, to the window on the opposite wall. There it squeezed outside though a two-millimeter crack. The heatless ball—it floated eight inches above the witnesses' heads—left no traces.

Colliding Fireballs

As careful as fireballs navigate and avoid collisions, exceptions occur, and they sometimes collide with objects. Sometimes, after flying normally, they inexplicably fly straight into trees or objects. In one French case, the fireball flew into the brick side of a building.

In another such case near Lake Superior, a fireball slammed into the chest of a young man, who was horribly burned and died. In another case from the mid-1920s in Massachusetts, the fireball chased and flew past the mother and her young daughter and smashed into their farmhouse, exploding in writhing sparks.

Fireballs collide and divide. [440 Cameron] During a storm and flash flood, from our office on the second floor, southeast corner of Building [deleted] of the Oak Ridge complex in the nuclear [deleted], we saw these fireballs—red-orange with gold edges, some orange and green—jumping, dancing and rolling on the power wires just south of the building. They appeared and moved like a string of fire. There were assorted sizes, six to fourteen inches in diameter. They hit each other and divided, the effect like a package of firecrackers. They were thicker in the center. There was a bluish-white flash at the time one of them split.

Colliding with fireball spawns trinklers. When a car drove into a one-foot fireball, it exploded against the rainy windshield, and tiny one-inch ballettes "trinkled" across the hood.

Multipliers Divide and Spawn

Fireballs that break up into smaller balls are not uncommon. Those that break into two or three smaller balls continue to fly, but those that divide into many small "ballettes" are more likely to bounce on the floor or ground. Spawned balls are usually smaller than their parent, but not always, and some that divided were close to the size of their progenitor.

Box car ball begets bouncing ballettes. [ORNL White] I was in Waseca Minnesota, 60 miles south of Minneapolis. A thunderstorm was breaking around me, rain pelted down on me. I dashed for the nearest shelter, which was a boxcar, wide open on both sides—an all-steel boxcar with a heavy steel floor, parked about thirty feet straight across from the depot. The depot agent stood outside the depot under the over-hanging roof. Lightning crashed nearby. Instantly a large two-foot, dark-blue fireball with purple interior struck the steel floor just inside the doorway of the car. It seemed to "breathe" and was blinding. I did not look straight at it, rather sidewise. When it struck the steel floor, it broke into smaller fireballs, with the largest about 14 inches.

They rolled and skipped end to end of the car—some seeming to touch the steel floor, others floating a foot above the floor—and it was hard to keep out of the way of these balls of hell, jumping over some, side-stepping others. After one minute, they were gone. After it was over, the railroad station agent (who saw it) said to me "Man! Oh man! How lucky can you get, I expected to see you burnt to a cinder or electrocuted. You must have rubber soles and heels on your boots." And I did, and all sewed with no nails. I think that was what saved my life.

Broken mother ball multiplies. [195 Weaver] After lighting struck, an eight-inch, orange-white fireball rolled across the ground, and then broke up into several smaller balls that also rolled around on the ground.

Ballettes roll down lawn. [Golka] A woman living south of Boston in Easton, Massachusetts was returning home during the summer of 1991. Lightning struck the grass alongside the suburban road, and eight or so small fireballs appeared and began to rapidly roll down the small incline into the street. The woman stepped on one ballette and it vanished without any other effect.

Rolling big ball divides and recombines. In mid-summer during a thunderstorm, while on the front porch watching a storm, a woman saw a white four-foot fireball roll off the street and divide into three smaller balls, and soon split further into six, soon to recombine into three, and then back to a single four-footer. Throughout its journey, a sharp hissing and a distinct crackling accompanied it. As they rolled along, fingers of lightning leapt out to anything nearby—to parked cars and telephone poles, at which the electric power ceased and stayed off for over two hours.

Spinning white ball turns blue, explodes into green, and then ejects "Phoenix fireballs." Lights went out. Outside, a large loudly-humming, yellow-white fireball—light was spinning inside it—floated over a house and an adjacent 600-foot field, turned blue and exploded. Out of this explosion came a basketball-sized, bright fluorescent green fireball that landed on another driveway near a garage whose switch melted and turned on the lights.

Giants

Big fireballs are three to ten feet in diameter; but monster balls, over ten feet across. These are rare and puzzling. With the exception of "fog balls," few giants over five feet across descend lower than twenty feet above ground level, and most of the monsters occur above a hundred feet. According to CSICOP's expert skeptic Phillip J. Klass, some of these natural objects are mistakenly reported as UFOs by sincere eyewitnesses.

Giant fireball near H-bomb facility. [26 Kuna] While a late June 24, 1960 storm raged, we stood and looked out a west window in room [deleted] of Building [deleted] at the nuclear [deleted], a half mile away. Lightning hit the nearby hill. A ten-to-fifteen-foot fireball hovered and jiggled for a second on the hillside, then vanished.

Gold monster-globe plunges, splits tree, spews fireballs, vomits fire-river. [599 Brooks] During a violent lightning storm in 1937, I saw a ten-foot, light-gold fireball fall and hit a large tree. It split the entire trunk from top to ground. Smaller darker-gold fireballs rolled out the limbs and off to the ground and disappeared into the deluge of rain. A river of fire flowed from the tree down to the ground and down the hill away from the tree. The surrounding ground shook violently. This tree later died and never sent up a single sprout. It still stands dead and withstood weathering exceptionally well.

Huge ball explodes, ground steams. [494 Davis] Returning from a fishing trip, a friend and I walked up a steep hillside from the river, when an automobile-sized fireball appeared before us. The explosion knocked down my friend and left him in a state of shock. It did not injure either of us. The ground around us started steaming and a strong ozone odor lingered

Tree-top green ten-footer moves slowly. [270 Sherrod] During a storm in 1940, lightning illuminated the dark sky. Suddenly lightning lit the south sky and a green fireball appeared above the treetops. It flared up as a ball like the moon, but not as big, and seemed eight to ten feet across, and very hot. A smell of burnt electricity filled the air. It seemed to glow and burn slowly out, as it moved along level above the treetops, and held its size. It lasted ten seconds.

Big yellow-orange fireball at one mile. [43 English] On Tuesday night, June 21, I saw a branched lightning streak from the second floor of a Garden Apartment on Virginia Road, facing north. A large yellow-orange fireball appeared along the ridge of the West Outer Drive at a distance of one mile, so I cannot accurately estimate its size (over five feet). It seemed to remain in contact with the surface and lasted one second.

Giant fireball envelops car, mushrooms up. [239 Tate] I rode in the front seat of a car about six P.M. in the summer, driving on a highway, when a ten-foot fireball enveloped the entire car. Then it mushroomed up and the car seemed to be in the center of it. It quickly changed shape and lasted eight seconds.

Author—Was it a sphere? This is not the first encounter with a car, or of a person standing and being enveloped. Three other Oak Ridge scientists reported encounters with such giant glowing fogs. None of these reports mention lightning. Although mysterious, these "fog balls" seem harmless. But is there underlying physics in common?

Fifty-foot skyball. [257 McBryde] Distant lightning flickered on a July Fourth night in Mississippi. At ten o'clock, a fifty-foot fireball appeared in the sky one thousand feet up and a quarter-mile from us. It moved from north to south at a speed comparable to a prop-type plane viewed from the same distance. It suddenly disappeared silently.

Author—How this witness determined the diameter and altitude was not listed in his report. In similar Oak Ridge reports, the physicists or engineers either used multiple simultaneous sightings from different locations and reference to nearby objects for their frame of reference. Some used their fist over fist, held at arm's length. A typical fist height is nine degrees. Then, by applying trigonometry and basic algebra they would obtain numbers.

Monster fireball rapidly rolls down mountain. [75B Myers] I saw a huge fireball roll down on top of the trees of a mountainside (five miles distant). It covered a considerable distance in two seconds. Elevation varied between 1,200 and 1,500 feet.

Monster fireball changes size. [357 Rinderer] At two miles distance, a giant yellow-orange fireball appeared in the orange-tinted sky over the Cumberland Mountains during the season's last bad storm. It moved slowly for several seconds, changed size, and then silently vanished. There was no lightning flash associated with it.

A 20-foot monster frying fireball roars along cable, splits. [508 Clemmer] Rain sheeted down. Lightning flashed. A sharp crack. It struck one of the TVA power lines. A 20-to-25-foot, red-yellow fireball appeared. The bottom was flat as it followed the cable, sounding like a fast-moving train with a loud frying sound mixed in. It traveled northeast on the cable for 150 feet and hit a cable support and split up in all directions. The roaring stopped. My father and I were 200 yards from the lines. It lasted five to ten seconds.

Orange monster skyball trails white column. [Mr. BE Waye of "Larch Gates and Altwood Bailey of Maidenhead, Berks, UK; *Nature*; June 23, 1945; p. 752] On May 8 at 2:20 a.m. during a thunderstorm, an orange fireball the size of the sun appeared on the southern horizon and moved west, followed by a horizontal column of white vapor. After ten seconds it reached southwest, then vanished in a flash. If there was an explosion, thunder obscured it.

Giant glowing ball. Lightning struck. A giant glowing fireball—it was ten feet across—instantly formed, floated deliberately, and after four seconds, it vanished. It was raining and the witness was 50 feet away. Vegetation was unaffected.

Ten-foot blue ball seen 1,000 feet away. A violent rainstorm blew brush across the road. At a thousand feet ahead of them, the driver and passengers saw a ten-foot ball of bright bluish-green light hung in the air. When they were abreast of the light—it was 30 feet off and now directly across the road—it slowly faded.

Bolides

Bolides are spectacular. Can some larger meteors and bolides be mistaken for ball lightning? Perhaps, but probably few. But when a monster occurs during a lightning storm, this explanation is not very likely.

Mega-monster supersonic fireball or stormy bolide? [582 Guice of Oak Ridge, Tenn. (1962); but exact designation number questionable since Phillips also has 582 (582A and 582B). Did I miscopy? Not important since still can identify each by witness names] During an electrical storm on Sunday night May 27 at 10:33, from my home on Alger Road in Oak Ridge, I saw a comet-like fireball above a hill at 125 degrees [bearing] and 6,300 feet away.

The fireball appeared inside a layer of clouds 400 feet above the hill. The low clouds were over 60 feet apart and 200 feet above me. The ball traveled on a northerly path (191 degree bearing), following the layer of clouds. It came in my direction, but on a path that took it 4,257 feet west of me (at its closest) just before it vanished. It was visible for two seconds and disappeared instantly at 30 degrees elevation. The fireball was 33 feet in diameter with a 130-foot tail.

Author—The witness determined dimensions by applying ratios and trigonometry to measurements he made—inches across at arm's length, and in relation to known trees and objects. He knew his neighborhood and acted quickly. He included a map and diagrams, with extensive measurements in his report. Scientists are unaware of any meteor (certainly not a bolide) that flies horizontal for 4,488 feet—silently without a shock wave at a supersonic 1,530 miles per hour—at this low an altitude during a thunderstorm.

On October 14th of 2001, a brilliant and silvery meteor-fireball flew south to north across the deep blue afternoon skies of southeastern British Columbian and southern Alberta. Starting to glow north of Invermere, B.C. and crossing the Rocky Mountains, the eight-foot 10,000-kg meteor fragmented into three pieces with an explosive burst (exceeding 500 kg of TNT) over the Ram Range above mountains east of Baniff National Park. Park visitors heard a tremendous sonic boom and ominous rumble that triggered pressure sensitive doormats and opened hotel doors. Subaudible infrasonic waves were detected 870 miles away.

Meteoroidal bolides are exceedingly spectacular and so are unusual displays of cometary debris fragmenting in the atmosphere or asteroid fragments. A bolide is a bright meteor that randomly appears—that is, not from a usual meteor shower or comet. Very bright bolides cast shadows over a 1300 km diameter. Meteoroids weighing 100 kg are as bright as a full moon. Unlike shooting stars, bolides can change their vivid colors and break apart into fragments that fly in formation, spiral down or fly in a near-flat trajectory and may flare up at the end of their descent. Bolides are very bright and possess a large ionization envelope, thus are farther away than they seem.

Some witnesses near the impact sites report a brimstone odor. A bolide may leave a "train"—a half-mile-wide horizontal trail—that in daytime resembles a column of dust and at night, a glowing pillar that can last an hour. When bolides are up 60 miles, certain witnesses report hearing noises such as an odd buzzing in their heads. Others report a whirring or crackling, swishing or hissing.

Since sound cannot travel that fast, certainly never through a vacuum, this hissing must be caused by electromagnetic radiation. Low-intensity radar beams can create a similar buzzing, clicking and hissing in observers' heads. Such intense emissions should seriously affect radio reception, but paradoxically do not, and some astronomers suggest that bolides emit "extra low frequencies" (elf). Paradoxically, this also raises more questions than it answers.

And could there be some connection with the rare and mysterious "auroral sounds" heard by some witnesses and ostensibly simultaneously coming from the northern lights or aurora borealis? However, there is no accepted theory that can explain the odd rustling, sizzling, swishing, low hissing and crackling noises heard by certain witnesses when they see the aurora. Others compare it to a small animal scampering across dried leaves. Some witnesses hear a crinkling like a cigarette pack's plastic being crumpled.

The auroral noises are heard for a minute or two, and occur when a powerful display is overhead. Not surprisingly, most reports come from the Arctic Circle. Auroral noises are rare and may be heard only once or twice in a lifetime. There are many cases where several people in a group all heard it.

Experiments [Keay and others] prove that electromagnetic waves—with electric field peak-to-peak variations of 160V/m—with the help of objects near a witness can sometimes transfer a small part of their electromagnetic energy into faint acoustic waves that were heard by some laboratory subjects. Subjects differed in their sensitivity by over a thousand times. Electromagnetic emissions from nuclear explosions have an audio component with similar characteristics. Hissing auroral sounds in the 100 Hz range were recorded with large radio antennas used in ionospheric research in Norway.

Shrinkers and Expanders

Most fireballs do not change size. But of those that do expand or shrink, they can do so slowly or quickly. They may remain steady state and unvarying in size, and then for no known reason, grow or shrink. Some reduce their size until they disappear. A few slowly expand and contract repeatedly, as if "breathing." A few expand substantially.

Blue expanding fireball turns orange. [81 Kennedy] In 1960, I saw lightning strike a transmission line near the [deleted] east parking lot at ORNL in early July. Instantly a green-blue sphere formed and immediately began to expand uniformly. As it expanded toward its maximum size, the edge changed to an orange-yellow (but not outstanding). After expanding to four feet, it slowly disintegrated. After it disappeared, sparks dropped. This was not after-image persistence because lighting struck near the periphery of my vision, was not bright enough, and the sphere lasted for several seconds. The wire smoked but was undamaged.

Color-changing fireball grows. [272 Thomas] During a lightning storm, I saw a one-foot airborne red fireball grow larger to three feet and slowly change to white, and then after thirty seconds it suddenly noisily vanished.

Roller shrinks to zero. [122A Simmons] In Orleans, Indiana, I stood in an open driveway garage door during a storm. Lightning crashed nearby. A bluish fireball came from the base of a utility pole, to the center of the doorway, turned and rolled down the drive to the street. The ball floated near the surface, emitting a low crackling noise. It grew smaller, from a bowling ball to zero (five to ten seconds) as it "rolled" down the driveway to the street, leaving a faint odor of ozone.

Orange center, red-rimmed fireball shrinks away. [201 Scruggs] I walked east on the railroad. It sprinkled lightly. Lightning struck a nearby phone pole and tree. A grapefruit-sized fireball with an orange glow and reddish center appeared and lasted several seconds. It burned my face slightly. I was blinded for several seconds after this.

Like collapsing balloon. [135 Richardson] I saw a yellow-white fireball come from a maple tree—during a storm but not associated with any flash—and through which an overhead electrical and telephone conductors passed. Traveling horizontally along a lead telephone cable for 75 feet and following a lower wire to a pole toward the ground, it disappeared after five seconds as it started down the pole. It diminished from basketball sized to nothing, like letting the air out of a balloon. Color was constant.

Volume rapidly expands 13,800 times. Lightning struck the television antenna atop a house, leaving a blinding golf-ball-sized fireball touching the antenna, and in several seconds grew to double the diameter of a beach ball. The sphere did not move and then vanished. The rapid rate of growth or speed of expansion—that is, the increase in volume—was impressive: the "golf ball" grew larger in five seconds from an inch to over two feet in diameter. The quotient of the cubes is about 14,000. This sudden growth is uncommon but not rare. In another case, after a large explosion followed by a crackling sound, a very small white (yellow center) ball formed near the outside television antenna, but in four seconds it grew to a yard wide, and after another four seconds it slowly flew away from the house.

Expanding "soap bubble" noiselessly glides downward. During a rainy Texas day, while walking across an outdoor second-floor walkway to a college classroom, this student saw a faintly yellow-white six-inch fireball silently float diagonally downwards like a soap bubble, descending from a 30-foot altitude to ten feet above the ground (on eye-level with the student). Suddenly after its ten-second flight, the fireball quickly expanded and grew into a "white-out" or noisy lightning flash and disappeared.

Brightens, expands, explodes. During a lightning storm, a basketball of fire, which gave the air around it a greenish hue, floated 20 feet above the street. Suddenly it brightened and expanded, then after 25 seconds, burst and illuminated the dark post-midnight sky. Two minutes later, the power went out and stayed off for eight hours.

Glowing point grows into fuzzy fireball. Rain ceased, thunder stopped, wind raged. Then the last lightning struck a birch and a small glowing point appeared and for three seconds grew into a 3-inch, bright whitish rose fireball with an indistinct white-blue periphery, shaking and vibrating as it grew. It slowly floated toward a wood stove but made a right-angle turn, and then speeding up, it flew into the entrance hall, and flew against a draft. After an 6-second lifespan, it exploded and damaged the corner of this hallway, but did not char anything. The witness, a doctor of physics, compared its brightness to a 50-watt lamp.

Blue fireball shrinks and melts wire. After a rifle-like shot, the bedroom filled with a "horrible blue white light" and a 14-inch blue-white glowing fireball emerged from a socket and hung there for five seconds, and then began to shrink down to four inches and remained at that size for ten seconds, and then vanished. Part of the socket was broken off and that antenna plug was burned off. A three-inch green spot was on the wall and the socket wire was burned bare in two places. The observer's ears rang for one day.

Grows brighter, then fades out. Lightning struck nearby, but outside. A 10-inch, orange silent fireball appeared in mid-air in the basement at one yard above the concrete floor and the same distance from the rebar-reinforced concrete wall. Remaining stationary, the glowing ball grew brighter until it was as bright as a 20-watt bulb, but then it slowly faded into invisibility after half a minute. Then, at thirty feet away, the gas detector alarm sounded and the plant's heating system shut down. The observer, a physicist, noticed no odor.

Expanding and contracting opaque ball grows translucent. Instantly appearing in thin air 25 feet next to a building, a solid silver three-foot fireball floated 7 feet above the lawn and vibrated briefly before it expanded to five-feet diameter and grew translucent. It then shrunk and re-expanded, repeating this expansion-and-contraction cycle four times, vibrating briefly, each time it expanded again. At the peak of its expansion, the translucency was such that the witness, an information technology technician could see the building's wall through it. At the maximum of its final expansion, the ball loudly exploded. There was no effect on the building or its electrical and information systems.

As we just saw in this chapter, of all the fireball categories with "odd cosmetics," most are unique. As we just saw, they include those balls that reappear in the same location, those with fuzzy surfaces or nested concentric shells, and soap bubbles. And few are so odd as glowing fog fireballs. That includes those with tails or that appear as necklaces and filaments.

There are non-spherical shapes, or which alter their odd shapes. This rich diversity of sundry appearances extends to size—some few are true giants. But locations can be unusual, and some fireballs appear along railroads, on towers and bridges, or even over water. And just like cosmetics that are odd, locations can be unusual

16

Unusual Locations

Fireballs occur in all the expected locations—in homes, in airplanes, outside near trees, and so on. But it is the unusual and unexpected locations that are interesting. When they appear at sea or above a lake and skip along the surface or dive into it, then that is unusual. Because roadways and railroads cross the country, maybe they only get an appropriate proportion of sightings, especially because people are more likely to be on them. Or maybe not, and they do get more than their share.

In either case, some of these fireballs do behave differently. Certainly, towers in the sky with metal framework are more likely to attract the formation of fireballs—such as fire towers and the Eiffel Tower in Paris. And the ultimate unusual location? It is inside a vacuum. For the same reason that fireballs are less commonly seen amongst clouds and atop mountains, those reported at sea are also rare for one reason—there are fewer observers.

Sea

Saint Elmo's Fire aboard ships can morph into a fireball and float free. Some have exploded. Sailors reported glowing objects since the ancient mariners first sailed ships onto the Mediterranean Sea. When sailors saw flaming fire appear and dance and hiss on spars and masts, and indeed, on any pointed object, they christened this strange cold glow as Castor and Pollux—the Heavenly Twins. Christians renamed it St. Helmo's Fire and later St. Elmo's Fire, after Saint Ermo (Erasmus), who was the patron saint of Mediterranean sailors. This cold fire was a benign sign of his guardianship. Others claim it was named after the martyred Bishop Elmo of Gaeta, which is on the western coast of Italy.

However, Julius Caesar made the first written description of this phenomenon when he described how the points on the spears of his fifth legion took fire. Seneca saw it at sea and Pliny saw it on the spears of javelins glowing like stars on a ship, hopping about like birds on everything pointed. Italian sailors also called them the Fires of St. Peter and St. Nicholas, while the Portuguese called them Corpos Santos or Body of the Saint. Visiting sailors from England corrupted this into Cormozant and then Corposant.

Whether or not it brought good luck is still conjecture, but we now know that it is a sign that high differences in electrical potential or voltages exist. In the laboratory, we call it corona. When a fireball occasionally detached from it, and especially if it did damage, sailors took notice and it became part of the lore of the sailors' sea tales.

However, many of the earlier accounts in ships' logs are too vague, and even if correct, we should be skeptical. At other times, fireballs appear without any ionization, and occasionally in fair weather. Since electrical storms occur less frequently over the oceans than over the land, and because more people and potential witnesses live on the land than are at sea, then it is no surprise that fewer sightings come from the sea.

Orange sea ball dances near ship's bow. [244 McNamara] When I was in the Merchant Marine I saw a six-to-eight-inch (basketball-sized) orange fireball during a lightning flash that danced (airborne) around the bow of our ship for five minutes.

Balls roll across ship's mast. [47 Ellis and former deck officer McWhirter] This occurred on board ship during an electrical storm. A light bluish-white, basketball-sized fireball usually appeared on the top horizontal crosspiece and the ship's mast, usually starting at one end and rolling slowly toward the other end. It occurred ten years ago, so I do not remember clearly what happened when it reached the end of the horizontal travel. After four seconds, it suddenly decayed. Others saw it, and I saw it twice. For more witnesses, I suggest you contact "deck" officials on merchant ships.

Stationary mast-tip fireball leaves ozone. [41 Doherty] While sailing through a thunderstorm, I saw a basketball-sized fireball materialize at the tip of the mast, remain stationary and suddenly vanish after three seconds, doing no damage and leaving an odor of ozone.

Lakes

Lakes dot the land. Not surprisingly, some fireballs on land fly over lakes. But when they hit the surface or bounce off the water, this is surprising.

Half-minute fireball skips on lake. [389B Northcult] I saw lightning strike the surface of a lake. The fireball lasted half a minute and skipped around before fading out.

Lightning strikes pond, creates yellow fireball. [64 Miller] While I stood on the shore of a mile-diameter lake, lightning hit its center and a 1.5-foot whitish-yellow fireball flew up to ten to fifteen feet above the surface, traveling 300 feet before it hit the surface, bouncing into the air again and flying another 300 feet, bouncing again, only to disappear in mid-flight after four seconds.

Lightning makes Pick Wick Lake fireball skip. [335B Northcutt] Lightning struck Pick Wick Lake in west Tennessee and I saw an orange fireball with greenish tint bounce and skip around over the lake until it vanished.

Dim fireball slows over lake, alters course, climbs slope. [449 Breeding] While on a field trip to obtain biological specimens for study and dissection in Clairborne County, Tennessee (elevation 1,200 feet) in rolling hill countryside of low hills, woodland and pasture land, I stood on the WNW shore of a circular lake (one-acre surface area). Temperature was 85 degrees with cumulo-nimbus clouds and a bright sun obscured. Light conditions similar to early twilight. Soil in drought condition. No appreciable surface winds. Low humidity. No electrical phenomena visible.

An unchanging golden, basketball-sized fireball (from my viewpoint of 750 feet on the opposite side of the lake) approached from NNE (appearing from my left, flying right) at a distance above ground surface of four to six feet, slowly flying in a straight line following a six-degree down slope. Then flying over the opposite edge of the lake, it slowed somewhat over the water. Next over land, it flew tangentially from its outer periphery to a change in course of 40 degrees, or a WSW heading, then up a 12-degree slope, over a hill and out of sight. It now floated two feet above the surface and followed the earth's irregular contour, but a deviation vertically occurred when it crossed a wire fence.

Motion was easily followed by eye. Speed was not such that the fireball appeared blurred, nor did any point discharges or streamers appear. The entire observation took 12 to 15 seconds. It was not extremely brilliant and approximated the moon on a clear night.

Blue ball plunges into pond, hisses. [403] A powerful electric storm with great wind blew tree limbs across power lines along a highway. A golf-ball-sized pale blue-to-orange fireball flew straight until it hit a pond, hissed loudly like putting a red-hot piece of iron into water, and vanished. An ozone odor lingered for a minute.

Lightning spawns lake ball. [389B KJ Northcutt] I saw lightning strike the lake surface. A green-orange, two-foot fireball skipped around for 20 seconds before fading out.

Dying fireball emits lightning that strikes lake. A bluish-rimmed, white one-foot fireball floated onto a lake, lasting one minute. When it was three feet above the lake, it shot out a bright flash of lightning that struck downward out of the fireball to the water, and it then vanished.

Roadways and Highways

Electrical power lines usually travel along most roads, and fireballs have a predilection for power lines. Because roads and highways criss-cross the continent, it should be no surprise that fireballs are seen along them. People travel on roads, not hiking through swamps and forests, especially during electrical storms, so you can guess that most sightings are on roads and highways, not in swamps.

Lime fire-ring in highway. [194 McGinnis] While driving alone in a car on the highway between Sevierville and Gatlinburg [a 14 mile drive] in Tennessee during a lightning storm, I saw, in front of my car and in the road ahead, a two-foot ring of fire, partly lime-colored, which lasted one second.

Glowing fuzzy fireball crosses highway. [283 Reno] I was driving on the Champan Highway off eight miles from Sevierville in a thunderstorm when I heard a thunderclap and a volleyball-sized, white-orange fireball appeared on the highway. A glow extended two or three inches on its outer surfaces. It traveled across four lanes and vanished.

One-minute road ball misses man. [219 Hickey] While approaching a storm center in South Georgia, after a lightning flash, I saw a basketball-sized orange fireball slowly follow the road's contour—heading straight towards me! But it disappeared just before hitting me. It lasted one minute.

Descends pole, rolls up highway. [192 Thompson] Lightning struck a high line pole and a 24-inch fireball came down the pole onto the highway, then came up the highway about 300 feet and vanished.

Railroads

Do steel tracks or currents flowing in them act like wave-guides for guiding fireballs? Ever since the 1880s the railroads sent signals to their switching stations in the track itself. But it was not until a hundred years later when researchers discovered that power lines near the tracks induce transient currents in the rails and could inadvertently switch some of the then-new switching equipment—changing a red stop light to an all clear signal—with only 100 millivolts. Up to 500 times that amount can be induced in the rails by overhead power lines. The alternating rail currents, in turn, surround the rails with a pulsating magnetic field. However, no one knows why some fireballs follow railroad tracks but do not touch them. Like following cables and lines, fireballs also exhibit a preference for following railroad tracks.

Dazzling ball leaves railroad tracks at signal light. [8B Starken] Lightning flashed during a mid-afternoon drizzle. A flat open country railroad track runs for four miles straight to where I saw the fireball. It looked like an approaching train headlight at night, except it was in mid-afternoon daylight. When I first saw it, it was like an intensely bright spot the size of a baseball. It seemed to expand, grow larger and brighter (not as intense) to basketball size, and then suddenly left the track area at a signal light. It lasted four seconds, long enough to call my dad's attention to it and point it out when in its earlier stages. We agreed it was the height above the tracks as tracks were wide. Our distance from the track was one city block.

Hot dry-day fireballs on railroad tracks. [339 Tripp, Jr.] I saw a softball-sized fireball float for 30 seconds along railroad beds on hot and dry days, but never saw them on cooler days.

Fireballs follow railroad tracks and lines, sets pole on fire. [340 Anderson] At 1:30 while walking along railroad tracks during an afternoon thunderstorm, I saw an orange-yellow, baseball-sized fireball coming down a rail. It was in sight for 150 feet, then "grounded off" at a signal wire. Another time, walking in railroad tracks, I saw a five-inch fireball bright red on telegraph wires jump to a pole, cutting into it at the ground and setting the pole on fire. This one was in sight for 200 feet. All those I observed ran on an object.

Blue-yellow fireball wheels down railroad tracks. [485A Moretz and Taflove of Illinois Institute of Technology, *Science 83*; pp. 8 and 12] The fireball traveled along a short section of railroad track, moved a few yards down one rail in a wheel or ball of blue-yellow fire. It extinguished itself between the rails, leaving a smell of ozone.

Railroad tailed-fireball cracks at junction pole. [565 Cooke 1958] Trailing a tail, an orange one-foot fireball traveled along the wire beside a railroad track during a summer thunderstorm. After two seconds, it disappeared suddenly with a loud cracking noise at a junction pole.

Golden fireball follows railroad track. [563 Heyman 1927] Before moving to Los Alamos, I saw a ten-inch golden fireball move on a railroad track in Maryland but saw no residual effects. Another time I saw an almost-stationary 14-inch ball "frozen" on a rain down pipe. Because of my experiences, I looked up the subject in our university's engineering library—and partly because one professor said ball lightning was an unstable plasma, thus impossible.

Towers—Fire and Eiffel

Fire tower fireballs snap at stools. Manned by observers who kept watch for forest fires, fire towers atop hilltops attracted lightning and fireballs. Some towers provided their wardens with tall-legged stools, refuge from fireballs, which would dart about and snap at the wooden legs of the stools and then escape through a door or window. Watchtowers are a thing of the past and only a few are still in use, manned by nostalgic volunteers. Old towers were made of wood; recent ones, of steel staging.

Pink fireball visits fire tower. [582A Phillips] One August weekend in 1940 while I stood alone on lookout duty in the forest-service fire tower atop Buck Mountain near Mormon Lake in Arizona, a storm and hail hit in full fury. After a nearby flash, I saw a pink crackling fireball (the size of a dinner plate) hovering over the field telephone and smelled metallic burning. I got as far away as I could and sat atop a wooden stool. The fireball moved up and down slightly over the telephone; its color would fade, and then intensify. There was a cat in the lookout tower with me. He made a squawling [sic] noise and I looked at him. Every hair on his body stood on end. And there was a blue halo of fire around the cat for a few seconds. He ran out of the tower and I never saw him again. And when I looked at the telephone again, the fireball was gone, having lasted between two and three minutes.

Fireballs infest Eiffel tower. [Roth] A Boston meteorologist who looked toward the Eiffel Tower from the Rond-Point of the Camps Elyses saw lightning strike the tower. Simultaneously a one-yard fireball moved downward to the second platform, traveling a hundred yards in several seconds and the silently vanish. Next day he talked to a tower official and several workmen, who told him (off the record) that fireballs regularly appeared on the tower.

In Vacuum

If fireballs form and fly in a vacuum, then already-failed theories such as vapor and dustball, and even fractal aggregate theories all come up totally inadequate—they fall short. In a vacuum, all physical objects fall and cannot be sustained by buoyancy in the air, as are balloons and dust particles in the air. If fireballs form or enter a vacuum, then by most existing theories, they should fall.

Fireball flies in vacuum. While repairing a television set, a Russian man turned the set on to search for a defect by replacing tubes. Buzzing came from a rectifier tube in the sweep unit. Inside the vacuum tube bounced a three-millimeter micro-fireball that stopped and then melted a two-mm-diameter hole in the glass envelope and flew out across the room and out the window. The smooth contour of the hole was twisted outward. There are other such reported vacuum tube micro fireballs, briefly listed in various old magazine issues devoted to old amateur radio. If these accounts are accurate, then fireballs can form, fly and navigate in the vacuum of space.

Evaporates bulb filament, crinkles linoleum. After a storm passed, when a mathematician entered a dark room, he felt a strong electric field and the hair on his head stood up. When he switched on the light, a bright white, 10-inch fireball with an 8-inch core appeared in place of the light bulb, detached and floated down to the linoleum and disintegrated into glowing points that crinkled the linoleum. It lasted ten seconds. The tungsten filament had evaporated and condensed inside the glass of the unbroken bulb.

In this past chapter we discovered some unusual locations—above the oceans and lakes, and along railroad tracks and atop towers, where fireballs form. Before we move on to examine popular theories proposed and experiments conducted, we cannot omit speculating on fireball's amazing sensing abilities, examine atmospheric electricity, and discover how St. Elmo's fire sometimes mysteriously births fireballs.

17

Saint Elmo Births Balls

Occasionally, fireballs are somehow—no one knows how—created by man-made corona discharges or by natural atmospheric ionization. These two ionization discharges are identical, but the terminology differs. If created by nature, the electrical corona is correctly called Saint Elmo's fire. If man-made, it is called a corona. Modern usage blurs the distinction and uses the term "corona" in both contexts.

History of Corpus Santos

A natural corona discharge of bluish white glow, this strange cold flame appears on pointed objects, such as the wooden masts of ships. From the early Greek and Roman seafarers, on through the superstitious Middle Ages, on to the restless Portuguese and the intrepid French, Spanish and English mariners, and on down to modern days, sailors all were fascinated and mystified by these hissing and sizzling, strange cold flames. Strange and heatless, dancing so restlessly, blazing so vividly on nights stormy and dark, these mysterious glows were both feared and revered by mariners as omens and portents of futures both good and bad.

Known to the Romans as Castor and Pollox, the Heavenly Twins of Roman mythology, these glowing lights were renamed by the Italians after their martyred Bishop Elmos of Gaeta and Saint Erasmus, the patron saint of Mediterranean sailors. A bold seafaring people pioneering the oceans, the Portuguese referred to it as Corpus Santos or Body of the Saint.

Saint Elmo's fire or corona discharge can appear anywhere that high electrical potentials lurk—in high voltage laboratories, on high tension lines, on trees, on mountain tops. It appears on spear and bayonet tips, on helmets, and in one case in the Boer War, a cavalryman saw it between the ears of the horse he rode and on the tips of all the soldiers' lances. It can glow on the tips of hay. It can occur running across a bicycle's handlebars.

Heavy rain will reduce or eliminate it, although after the rain, it may reappear with diminished brightness. Many sailors believe strong Saint Elmo's fire is a warning, a portent seeming less than threat the coming of a bad storm foretells. Others still believe these cold flames are the souls of dead seamen returning to warn of a coming storm.

Saint Elmo's fire is not ball lightning. It is a luminous mass created by corona discharge from a sharp object that appears near strong atmospheric electrical fields, often when an electrical storm is nearby. Since it cannot float, Saint Elmo's fire requires an electrode, something to attach to. It is commonly seen near tornadoes, but only occasionally near volcanoes and earthquakes

At high altitudes in the mountains, the corona will occur on any protrusion, even rounded ones, and including the hands and heads of people. It can move and run along the masts and spars of ships and even run onto and up a sailor's arm if he tries to grasp the flame. The varying electrical fields, which are in constant motion and invisible, will cause the corona to disappear and reappear in other locations, as if it were an unearthly manifestation of spirits returning from the world of the departed. However, on rare occasions for reasons unknown, Saint Elmo's fire can give birth to a fireball that detaches and floats free, but is no longer Saint Elmo's fire.

High Electrical Potential

To understand fireballs and Saint Elmo's fire, we must know something about ionic currents, which create atmospheric electricity. Ions are molecules or clusters of atoms or groups of molecules that lose electrons and become ions, which are no longer electrically neutral and will now move in an electrical field. Under ordinary conditions, these ionic currents that flow in the atmosphere are carried by the ions.

Radioactive substances and cosmic rays create ions. Every ion travels in the electrical field of the air around it until it combines with something and thus neutralizes, and discontinues being an ion. Under normal weather conditions with weak electrical fields, these ions move slower and do not create any new ions when they collide with other air molecules. They lack the kinetic (moving) energy.

But when the electric field becomes strong, which initially and in most cases will occur near pointed objects (such as wing tips, tree tips, a stake in the ground, spars and mast-tops), then ionization by collision will occur by those existing ions traveling fast enough to energetically hit neutral air molecules so fast that they lose an electron. (If this occurs, however, over a long distance, lightning will strike.)

The electric field is enhanced by the concentration of lines of electrical force ending on a point, at which place a point discharge will occur. If this occurs in nature, it is known as Saint Elmo's fire; when made artificially in the laboratory, as corona discharge. Ions cause atmospheric current and cause Saint Elmo's fire. Ions are molecules that either lose or gain an electron and become charged. There are small, intermediate and large ions. To small ions, which are lighter and more mobile, their charge is important: When they lose it, their molecules no longer congregate and these ions cease to exist and cannot be detected. On the other hand, when a large ion loses its charge, it will continue to exist and will be detect-able—they are called "condensation (or Aitken) nuclei."

An ionizing process, whereby one molecule loses an electron and becomes a positive ion and the electron instantly attaches to a neutral molecule, which becomes a negative ion, creates the conductivity of air, which allows the flow of electrical charge. Charge leaks off insulating objects, through ionic conduction, whereby the object attracts ions of opposite charge and repulses those of like charge. With no wind, ions travel along force lines, the negative ions traveling faster and farther.

A small ion is a single ionized molecule with several other molecules attached to it. It may be the size of a dozen water molecules, which are covalent molecules and can be easily polarized by the separation of charge and have a V-shape, with the base of the V as negative. This means that water molecules will prefer to cluster in greater numbers on positive ions. The positive ions, thus, will move slower in a humid atmosphere. Still, that is about 500 times faster than larger ions.

Large ions move slowly. They are far larger and heavier. In fact, they are as large as the "nuclei" found in the condensation of moisture. (The terms "nuclei" and "nucleus" are tiny particles on which water condenses when the air becomes saturated with moisture, and are not the nucleus of the atom.)

Intermediate ions, comprising 2,000 molecules, exist when the air is drier. But they grow into large ions in humid air. Their mobility increases with temperature and dry air. Atmospheric sulfuric acid vapors, a byproduct of industrial processes, help create intermediate-size ions.

Uncharged air particles exit and are called "condensation nuclei." Contrary to popular belief, condensation does not occur on coarser smoke and dust particles, but only soluble in water. Sometimes before the air becomes saturated and below total relative humidity, condensation may occur on these nuclei.

There are 40 to 1,500 small ions per cubic centimeter; and 200 near the sea, and up to 80,000 previously in large cities, before pollution control. Where there are many large ions, there are fewer small ions. Although there are always more positive small ions than negative ones, they are equal for large ions. On mountaintops, things change: There are fewer large ions and more small ions (up to 2,000) and there are more positive small ions than negative ones. There is an increase in small ions and increased air conductivity.

In fair weather, the earth is negative and there is a positive potential gradient, which varies with time of day and the season. There is a fine-weather current that brings positive charge to the earth. The negative charge on the earth would leak away into the air in ten minutes, but does not because worldwide lightning storms return negative charge to earth to balance the rising fair-weather charges.

The higher air of mountains is more ionized and more conductive. At higher altitudes, so conductive is the rarified air that aircraft in the 1930s could not fly higher because the corona discharge and ionization shorted out their high-voltage electrical systems. Cosmic and solar radiation, and radioactivity on the earth's surface and in its air—these all ionize the atmosphere. As a result, as first described by Nikola Tesla, the higher conductivity of the upper atmosphere—higher because of its greater ionization—makes it and the earth behave like a giant capacitor.

Because the air is thinner on mountains, it ionizes easier and Saint Elmo's fire appears there more readily. This fuzzy luminous glow appears on trees, houses, even people. It can fade in and out, but never float free. Saint Elmo's fire is a sign that high voltages lurk, although the glow itself is harmless.

When a point discharge occurs, it liberates ions that in still air, migrate outward, but in a wind, travel downwind. The motion and movement of ions depends on wind speed. Winds are often gusty, so there are eddies and turbulence, so that there are more variation in space charge and complex, violently varying field patterns. In disturbed weather, the potential gradient is so complex it will give little information about charges in the precipitation or clouds.

Energy needed to ionize an atom comes from being in an electric or magnetic field by electron or positive ion impact cosmic rays, or the absorption of a quantum of solar or electromagnetic radiant energy from photons, or by heating the gas. Electron impact varies. Slow-moving electrons lack energy—they cannot ionize atoms. Kinetic or moving energy increases rapidly as a square of power. Moderate-speed electrons lacking energy ionization potential only excite an atom they hit, so that another slow-moving electron may hit it a second time with enough energy to knock loose an electron.

There is an optimum speed. If the electron is moving too fast it will move right through the atom without removing an electron. Electrons lose energy in collisions that excite or ionize atoms, but little in elastic collisions. The colliding electron that spends energy to ionize an atom into ion pairs may impact some of its remaining energy into the ejected electron or use it to excite or eject a second electron from the atom, makes it double ionized. Electrons often create several ion pairs before they lose energy and attach to a positive ion. Free electrons do not last long.

Positive ions have thousands of times more mass than the light fast electrons. Their mass is not as great an advantage because kinetic energy increases only linearly. And to get them to move faster takes much more energy—which is measured in electron volts—than do electrons. Only when these positive ions move fast, such as speeding alpha particles, can they effectively ionize atoms, in which case they become more effective than electrons.

Photo ionization from sunlight occurs more at higher altitudes. A photon can excite or ionize an atom. Unlike electrons, if a photon's energy is greater than required for ionization, then ionization infrequently occurs; and if it is less than resonance frequency, ionization cannot occur. If the air is excited it will be raised to a higher level in the atom and then drop back, emitting the energy it absorbed as a re-radiated photon. It is possible than inside a fireball that the radiant energy is absorbed and re-radiated millions of times, traveling in a skewed random zigzag path to reach the outer boundary of the fireball.

If true, this "imprisoned radiation" may move in a fractal manner through different gases, give rise to different colors within some of the balls. The number of primary ions produced per second by a beam equals the number of photons lost from the beam. When the wavelength of the beam is very short, as with extreme ultraviolet and x-rays, then the ejected electrons possess very high energies, and thus able to produce more ions by secondary electrons than by the primary radiation.

Thermal ionization occurs in high-temperature gases and flames, as in the high-pressure electric arc. In the flame, the relative velocity of the particles will create ionizing collisions. By definition, thermal ionization includes ionization from molecular and electron collisions and radiation or photons.

The gases and vapors with the lowest ionization potential will be the first and most ionized. Walls, turbulence and other factors will alter thermal equilibrium in the gas, be it a welding torch or fireball. At higher gas temperatures, some of the fast atoms' electrons are excited to various higher energy levels in the orbits of the atoms and there will be multiple ionizations. Some polyatomic (multiple atoms) molecules are disassociated.

These prior ionization processes often occur in steps. An atom excited by one process may be ionized by another, or by several processes. This "cumulative ionization" also includes collisions of the second kind with mixed gases. One excited gas molecule can on contact excite atoms of another gas. Atoms in a metastable state—they cannot drop to a lower level by radiating energy—cause ionization-by-contact. They last much longer (sometimes up to 0.3 second) than to atoms in normal excite sates.

In some discharges, there are as many metastable atoms as there are positive ions. Metastable atoms' action can also equal photoemission. An electron that returns to an ionized atom is more likely to park itself in one of the outer orbits rather than return immediately to the lowest level. Recombined atoms are in high states of excitation and can be easily re-ionized. When the courses of ionization are removed, the ionized gas returns to a neutral state, that is, it de-ionizes. Some ions diffuse; others recombine. This may occur when a fireball fades or suddenly vanishes.

To summarize, when the voltages rise, more electrons escape from atoms. They soon recombine with an ion—another atom that also lost an electron. But as the voltage rises higher, these recombinations occur less. And although these more energetic electrons knock electrons out of neighboring atoms, they travel longer before recombining. It glows on sharp points such as tree tips, towers, ship masts and building that extend from a conductor such as the ground into the high electric fields of storms and cause a breakdown of the air near them by avalanche processes. It is often a 10-to-40 centimeter oval- or ball-shaped blue-white glow.

Although it often sizzles, hisses or buzzes, St. Elmo disappears silently, suddenly or gradually, and may reappear. It remains near a conductor and does not float free. It may move along aircraft surfaces or wires, and may pulsate. It is never dazzling and does not blind pilots but only interferes with radio reception. Occasional fuzzy (2-to-18-inch) "spheres" roll along aircraft and are unaffected by the aircraft movement. Such spheres may last a minute or longer.

St. Elmo's fire plagues Harvard's mountain observatory. During the 1870s and 1890s, Harvard Observatory operated a meteorological station atop Pike's Peak in Colorado. Their diary documents intense electrical phenomena that include hail that electrically crackles, electrified facial whiskers repel and hiss; face, hands and scalp prickle as if pricked by hundreds of hot needles (a cap did not stop it); ground currents constantly passed through the arrester, which crackled incessantly. A cricket-like "singing" (that only occurred when it was very damp) came from three locations on a wire outdoors, but would cease when an observer came near, only to resume when he backed off, finally ceasing when that particular storm passed.

At night the glowing singing light for 200 meters would emit jets of heatless fire. These little bluish, quadrant-shaped jets or rays concentrated on the line surface in a small size, and as they constantly jumped about on the line, the rays pointed in this direction. When touched, the wire gave only a small tingle. St. Elmo's fire coated every metallic surface. Phosphorescent light coated the weather vane anemometer. During an electric storm the telegraph line, wind vane and post (in a deep June snow drift!) hissed. When one observer went outside, his hair stood up and he began tingling—always a sign that high voltages lurk nearby—and he ran back inside.

Clear-weather Saint Elmo's blue-green corona edges wings. [473 Krewson] During the war, I remember a spherical electrical discharge that took place during electrical storms on the wings and control surfaces of the aircraft in which I flew. A corona discharge without the sharp electric blue appearance formed on the leading edge and tips of our airplane wing. It continually varied in depth along the wind, was barely visible and was greenish blue, and lasted for several minutes. We were not flying through an electrical storm but over the ocean near the Hawaiian Islands. There was an atmospheric condition of turbulent winds and cloudy skies, which tend to generate waterspouts.

Quasi Dowsing Fireballs?

For a hundred years, some witnesses described quasi-psychic abilities. Some fireballs reportedly anticipate human reaction and seem to "read" minds, detect and follow buried pipes, sense distant electrical fields, and accomplish strange feats.

Some allegedly sensed the movement of a witness—and preemptively reacted—before the witness moved! Some witness sensed a presence before they saw it. In other cases, fireballs seemed to acquire this "knowledge" of their environment. This puzzles us skeptics. It is easier for us to simply sweep these cases under the rug and dismiss the witnesses.

Flaming ball floats down, bounces on road, rolls to house, makes right angle turn. [543 Eldridge of HK Porter] I work in the nuclear [deleted] at Oak Ridge National Laboratory in building [deleted]. When younger, I lived in the outlying suburbs of Philadelphia. While sitting on the porch with my grandfather one summer afternoon, awaiting my father's return from work, we noticed a storm approaching. The sky darkened. As my father reached the front door, my mother also reached it to greet him.

The rain had not yet hit when we were about to enter the house, when a terrific lightning flash hit close by. Immediately we saw an orange flame ball over a foot across drop from the sky 300 yards up the road. It floated rapidly out of the sky, hit the ground in the road center, bounced a half-dozen times, height diminishing after each bounce, as it bounced down the road. It rolled rapidly and was an orange gaseous sphere.

It followed the middle of the road until it reached our house, and turned sharply to the left in a direction away from us. It continued into the open field across the road. After 15 seconds, it disappeared with a terrific explosion. There was a faint odor. Rain did not come for some time after this. Later we discussed the event and agreed on the details.

The city water main was in the road center, but stopped at the road intersection, which was about 400 feet from our house, where it drops to a lower level and turns 90 degrees into our cellar. The fireball followed this underground pipe.

Author—The aforementioned amazing case of Eldridge is not an anomaly. If the fireball did indeed follow the underground pipe, how did it know it was buried there? And know rapidly enough to make a quick right turn while rolling rapidly? Why did it follow the buried pipe? If fireballs can accurately detect and rapidly follow underground pipes, then there are many public works engineers—and also water dowsers of the American Society of Dowsers in Danville, Vermont—who would envy such "dowsing" fireballs!

Slow ellipsoid "dowser" follows buried line. An ovoid or fat sausage, two feet across and almost four feet long, brightly glowing and resembling a roaring noiseless fire, materialized when lightning struck 225 feet from a garage. Inside, the witness watched the ball float half a foot above the ground, moving towards him. It followed a buried electrical line right up to the patio of the attached house. An electrical shock, not as intense as a cow fence, crawled up his legs.

Author—The question is this: How can some fireballs detect and follow buried cables and pipes, and where the cable or pipe made right-angle turns, so did the fireballs. Excluding dowsing, which has no basis in physics (nor does ESP), how do these fireballs detect buried pipes? There is no explanation in existing quantum mechanics, string theory, or any other legitimate theory in physics—including Penrosian and Bohmian Mechanics.

Blue fireball spouts bright white sparks in radio station. Lightning repeatedly hit an AM radio station-broadcasting tower, and the transmitter cut off or recycled several times, thus saving the equipment. With the eerie sensation someone else was in the room, the disk jockey looked up—a ten-inch blue fireball floated a foot above his head. Bright white sparks shot from it. Silent, except for the sparks crackling, it shrank, and after 90 seconds the sparks ceased. The witness felt that the station behaved like a giant capacitor, with a Celotex dielectric ceiling, concrete floor, a steel roof, and a 150-foot antenna eight feet behind the station, which stood atop a hill, with a copper ground-radial system below and around.

Author—There is an inconsistency in the sensation of being watched by a crackling fireball. This eerie sensation—some describe it as a strange or creepy feeling that they are being watched—is commonly felt by some witnesses just before they see a silent fireball.

But this particular fireball made a crackling noise. This contradiction does not mean he lied, but that an interview would clarify the inconsistency. Perhaps after being detected, the ball crackled and the investigator or witness wrote it up wrong. It happens all the time. Ask any police officer or district attorney.

Tears nail from wall. A blue fireball ricocheted around a kitchen and blasted a galvanized nail out of the wall near a stovepipe. In other cases, fireballs floating past instantly pulled out staples and nails from fence posts, poles and buildings.

Detects nearby electrical field. A yellowish light-blue, tennis-ball with a light bluish white around its rim hovered nearly stationary six feet above floorboards before a television, which was off, and then instantly vanished when the occupant turned on a light switch.

Shape shifter senses and mimics witness. A late March thunderstorm with strong winds snowed but stopped. With a semi-transparent halo, a 5-foot yellowish green fireball, as bright as a low-intensity 40-watt bulb, vibrated and floated five feet above the floor. It was brighter in its center. The witness, with a PhD in chemistry, was curious and tested the fireball. He walked toward the ball, but it moved away exactly the same distance, and when he stopped, the ball also stopped, so he could not get within seven feet. The ball then floated to the door and squeezed through a tiny hole, 3 inches wide, by rapidly collapsing its size to squeeze through, but did not change its color.

This ends our anthology of cases. We now examine characteristics, statistics, questionnaires, competing theories, and a new and innovative theory—the Sagan-Hill Hypothesis. Although probably incomplete, it comes closer to the truth. Theoreticians have come to a dead end, and keep re-treading older theories that failed—and keep failing. Little is new. The Sagan-Hill Hypothesis is the first new major theory in four decades. However, to begin our journey, we must first know something about atmospheric electricity.

PART III
Theories and Experiments

18

Atmospheric Electricity

Ball lightning often occurs near or during lightning storms, and many lightning balls are often somehow related to atmospheric electricity and storms. So, to gain clues into ball lightning, we must examine atmospheric electrical phenomena—especially lightning.

To the ancients, lightning was sparks flying from Thor's mighty hammer and anvil. In truth it is more complicated. Lightning is a sudden movement of charge between two oppositely charged storm areas—cloud-to-cloud or cloud-to-ground. Since air is a good insulator, when lightning breaks through, it does so with enormous power—perhaps 400 million horsepower. After a discharge, the cloud can become charged up again for a second bolt in under 20 seconds. Nearly 46,000 storms occur daily on earth. Each second, the earth is hit 3,000 times. The origin of storms and lighting begins with energy from the sun, our atmosphere, and moisture.

Moist Air Creates Storms

In the morning, the sun unevenly heats land. Some land absorbs heat more quickly. As the summer sun heats this land, it warms the air lying upon it. This heated air expands and occupies more space, so it is lighter. One cubic mile of air at 68 degrees Fahrenheit weighs 5.5 tons, but at 95 degrees, only 5.25 tons.

As this warm moist air rises three to five miles, it cools to the temperature of the surrounding air. At one mile up, the air temperature drops 17 degrees; at two miles, 34 degrees; five miles up, down to between -85 and -100 degrees. This chilly temperature is well below freezing even on the hottest days.

The moist air then condenses into moist droplets that become cumulonimbus clouds—lazy white puffs of cotton slowly drifting across a summer sky. This does not produce a thundercloud. But when the air is moist enough and the updraft moves quicker, these fluffy cumulonimbus clouds rapidly grow, punching upward a huge thunder top, a giant anvil that rises high above the spreading thundercloud. This cloud transforms itself from a peaceful cumulus cloud into an electrical Frankenstein, a giant wind machine tearing apart water droplets in its updraft and charging them.

Although larger drops readily freeze within clouds, the smallest droplets remain liquid, some even when surrounding temperatures fall far below freezing, even down to an amazing -37 degrees Celsius. These super-cooled droplets remain liquid until they are distorted or bump into other droplets, at which they instantly freeze. Time to freeze a given volume of super-cold droplets varies greatly, by a factor of 100,000, an extreme variation that depends on the droplet sizes. Only discovered recently, freezing starts at the surface. The reason is simple. Unlike larger drops, a volume of micro-droplets exposes much greater surface area. The surfaces of the water molecules, upon freezing on their surfaces, release large quantities of latent heat outward, and freezing proceeds inward to the core.

Charged particles clump together in 1,200-foot blobs.. The positive blobs drift upward; the negative ones migrate downward, creating on the cloud bottom a strong charge that repels negative charges on the ground, thus creating a positive electrical shadow that follows on the ground below.

Eventually the cloud's electrical potential becomes so great that the air breaks down, ionizes, and becomes highly conductive. An invisible stepped leader travels downward at 620 to 1,200 miles per hour. It reaches down to 160 feet, pauses again, and then continues zigging and zagging all the way down.

But it never reaches earth. At 350 feet above ground, air molecules ahead of the leader ionize, and streams of positive ions stream upward from high points. Suddenly, it happens. One connects with the stepped leader, there is an audible click, and a giant streamer carrying positive ions races upward. Although this return stroke reaches only three-quarters up to the cloud, currents up to 500,000 amps flow in that instant.

Next, a fistful of dart leaders jump into the main channel, and 3 to 40 successive strokes travel the same path. The strokes last only 2 to 10 microseconds. Elapsed time between successive strokes can be as short as 0.6 milliseconds or as long as half a second.

Tesla wrote that each stroke itself could oscillate, and that if a lightning discharge occurs between two clouds, there can be no oscillation, such as would be expected, considering the capacity of the clouds. But if the lightning discharge strikes the earth, it may oscillate unless the channel, ground and cloud possess considerable resistance. Since the "plate" area becomes smaller with each immediate successive stroke, the natural resonant frequency will increase.

But the classical theory of lightning production is wrong. Atmospheric electric fields cannot grow large enough to trigger lightning. Recently, astrophysicist Joseph Dwyer of the Florida Institute of Technology noticed large gamma ray and X-ray bursts associated with lightning strokes—high-energy radiation seen only in outer space, because lower down the atmosphere slows them down.

After Dwyer factored in production of high-energy radiation into a model—which describes building electric fields of storms—he discovered that release of gamma rays and X-rays diffuse the electric field, which prevents it from growing large enough to spark lightning. Large doses of gamma and X-rays can form in small volumes.

In a storm, water molecules shed electrons that overcome air drag, and some accelerate to near light speed, and knock out more electrons, until a burst of gamma or X-rays releases energy from within the field, thus reducing charge. This sets the upper fundamental limit on voltage within electric fields.

Although speculated by Lord Kelvin that the earth possessed a high negative charge, it was Tesla who first discovered this and that the charge fluctuated throughout the day and seasons. During good weather, a fair-weather current of positive ions flow ground-ward from the skies. This neutralizes the earth's negative charge, and in 30 minutes the earth's negative charge should be gone. But as Tesla discovered, this does not happen, and the earth remains negative. Worldwide lightning continually feeds negative charge (electrons) back to earth. Their negative charge distributes itself over the earth's surface. Until recently, no one was aware of the role of sprites in maintaining the earth's charge.

Since current flows from positive to negative, current flows from earth up to the negative base of the cloud, up out over the cloud tops, and into the earth's electrosphere, finally returning to the earth in sunny regions. At 15 miles up, the earth's electrosphere is so ionized that it is highly conductive and behaves like a giant metal ball enclosing the earth. In fact, even at half that altitude, the air becomes ionized so easily that during the last World War, aircraft without proper electrical system insulation could not fly too high or the ionization would short out their ignitions.

The earth's negative charge varies due to electrical storm activity. When more storms occur—when the sun shines on Africa and Europe or the Americas—the earth becomes more negative.

Inner-cloud lightning occurs when charge inside a cloud that is between the lower negative electrical region and positive top of thunder cloud and its negative bottom becomes too great. Air breaks down and the cloud short-circuits to briefly neutralize a region inside that cloud. The common, but incorrect, term is "cloud-to-cloud lightning." Except at cloud edges, all that appears is diffuse illumination that flares up sometimes in a sheet—and so some call it "sheet lightning." Such inner-cloud flashes are more intense when the cloud base is high because the negative charges in the base find it easier to pass up into the cloud rather than descend to the ground, which is positive.

In the semi-arid American Southwest and in Southern Africa, the storm clouds are higher, so inner-cloud bases are 1,000 yards up, and the inner-cloud and to-earth strikes are equal.

Storm clouds reach potentials of 1 to 100 million volts and earth-surface gradients of 5 to 280 kV/m. Power transmission lines are raised to 15 million volts, which creates problems for the engineers, and whose solutions we will not discuss. A negative streamer from the cloud cathode descends in a pilot or leader stroke at high subsonic to double supersonic speeds or faster. Before the pilot reaches ground, a positive streamer arises from the ground until they meet. In the main stroke, currents exceed 500 kA.

Since a storm's charged cloud base and ground create a capacitor, and if the equivalent circuit was underdamped—that is, its effective series resistance is low—then this series circuit will resonate.

Lightning Resonates

We can calculate the resonant frequency if we consider the lower cloud surface and earth to be a charged circular pair of capacitor plates separated by the air. There are formulas that permit us to calculate the capacitance of the equivalent plates and self-inductance of the lightning stroke of a given diameter. For those readers who are interested, see page 199 of James Cobine's excellent book on *Gaseous Conductors*. Your author confirmed his derivations and calculated that for 100 kV/m that the current is a reasonable 200 kA with a natural resonant frequency of 45,800 Hz. If there is significant equivalent resistance in the cloud, column and ground, the envelope of the waveform will be an exponential rise and decay, without any resonance.

If the initial stroke taps into other charged areas of the cloud, there will be multiple strokes, and there can be up to 40. If we solve the second-order differential equations of a series equivalent circuit using Laplace transforms, we will get an envelope of exponential and sinusoidal waveforms. We will get a response in terms of coefficient conditions in the characteristic equation. There are four possible waveforms—an oscillatory steady-state, an underdamped sinusoid with an exponential decay envelope, a critically damped exponential rise and decay without a sinusoid—that is, no resonance—and a similar-looking wave of an over-damped waveform.

Current flows in only one direction and there is an exponential surge and then a decay. These s-plane (Laplace) responses can be seen on page 147 of M.E. van Valkenburg's *Network Analysis (2nd ed.)*. Unlike early pioneers like Charles Proteus Steinmentz, who used ordinary differential equations in the time domain, today we convert differential equations into the s-domain—where it is easier to solve huge sets of big hairy equations—and then covert them back into the time domain. In practice, someone else writes the software and we use their package, such as *Spice* to model our circuits.

Lightning will fall into one of the three later categories, and a single flash can be several consecutive strokes, but each with a different equivalent differential equation and response—that is, a flash can be a damped sinusoid and the next one, an overdamped non-sinusoid.

In an old New England farmhouse that your author lived in as a child, lightning would strike one of the three tall lightning rods atop the chimneys. A thick conductor at the roof ridges and the rod going to the earth would briefly glow red-hot when lightning hit. Ever since the early Twentieth Century, and perhaps farther back, perhaps back to its construction in 1831, every lightning storm struck the house. In 1947, this ended, and lightning never again struck that farmhouse. Although the old two-story barn behind it was never hit, two prior barns were hit and burned by lightning. In earlier times, if a barn caught fire, it burned. Why lightning never again hit that farmhouse is a mystery.

Conifers are struck less by lightning than broad-leaf trees because they have millions of needles. Pointed objects cannot retain electric charge. An invisible "electric wind" of ions (charged particles) streams from the points. Of the opposite charge of the thundercloud above, these ions stream up to the cloud-base and partially nullify some of the charge there, thus lowering the chance of lightning strikes. This was first observed and tested by Benjamin Franklin in the mid-eighteenth century and is the theory behind lightning conductors. A pointed lightning rod connected to a thick conductor to a good earth also discourages lightning by excreting upward a neutralizing positive electric "wind." When a flash does occur, it will follow the invisible ion stream down to the rod. The conductor then conducts the current safely to ground.

But to gather clues to fireball formation, which is related to atmospheric conditions, we must examine a field often ignored by meteorologists—the electrical conditions and behavior of our earth's atmosphere.

Atmospheric Electricity

The lower conductive atmospheric layer is due to primary cosmic radiation hitting air molecules and creating secondary radiation. Higher up, solar radiation ionizes the thinner air into a conductor.

Because the earth is a conductor, lines of electric force are vertical whenever the earth's surface is horizontal. Equipotential surfaces are horizontal and the potential gradient is vertical. The idea of electric potential, important to atmospheric electricity, was introduced by Lord Kelvin. When the earth's surface is not horizontal, as in the mountains, everything is modified. The potential difference between two points is the work (which is the force times distance) in joules or Newton meters per coulomb or charge needed to move a small charge between those two points. We are only interested in the difference between two points—never any absolute value—so we arbitrarily chose some zero of potential at a convenient reference point or location, then measure the potential at other points by the work needed to move a small charge from that zero potential point. This difference potential between any two points, called voltage.

Where would we select the reference point? It varies. In designing computer digital printed-circuit boards, we select a specific metal trace or conductor. To measure the capacitance of a single metal sphere, Nikola Tesla selected his reference at infinity. In atmospheric electrify and in the study of ball lightning, because we are only interested in the potential between the earth or build or other object and a point in the atmosphere or the fireball, we often select the earth as our zero reference.

There are exceptions, such as a fireball near or inside an airliner, where we select the aircraft's skin as zero reference. If work must be done to force a charged particle from one point to another, then the work is positive. The ratio of the force the field exerts to attract or repulse the particle is the "electric field strength," often just called the "field." Since force is a vector—has magnitude and direction—we pretend it is a tiny invisible arrow. So, we visualize the electrical force attracting a particle as an imaginary arrow pointing in the direction of that force is pushing or pulling the particle. The lengths of the arrow correspond to the strength of that pull or push. The force in metric is in newtons and the charge, in coulombs. The field is represented in newtons per coulomb or volts per meter. Since the electric field has a direction, it is a vector.

This leads us to perform a thought experiment. We examine a one-acre site. By moving a tiny positive charge throughout the site, at one-meter intersections on a grid pattern, and doing so at various heights, we can map the electric field onto a grid, similar to a topographical map. At each meter height, we get another grid map. At each intersection we put the particle and measure which direction the electric force of the field wants to push or pull it, and with what intensity. At each point we will get five measurements—x, y, z, f, and t. These correspond to left-right, front-back, up-down, time, and force of attraction or repulsion.

The potential between two adjacent pints will increase or decrease. The direction of the field is opposite to that of the rate of change or potential, that is, the potential gradient. The field in any direction equals in magnitude, but is opposite in sign, to the right of increase in that direction.

The equipotentials are horizontal, and the fields, vertical. The equipotentials can be visualized as imaginary sheets and the fields as vertical arrows piercing the sheets. The field can bend and twist, and be altered, but at the surface of the conducting earth, it will always be vertical.

Ion Currents Carry Charges

Ions carry charges and produce the atmosphere's conductivities, which are measured in "mhos." Atmospheric currents flow. Ohm's Law is usually obeyed. The arrows of current flows are directly proportional to the potential difference. We rarely use resistance because the reciprocal of resistance is conductivity, which adds when a current of ions flow in parallel or side-by-side. This makes it easier to obtain the specific conductivity by merely adding the individual conductivities. Later, to get the resistance (ohms), we take the reciprocal of the conductance in mhos.

Ions move through the air. They incessantly bump into other particles, as they are attracted to another potential. The longer their mean free path between collisions (MFPBC), the greater their mobility, which would be the case in the thinner air atop mountains. The ions can be imagined as tiny pellets sinking by gravity's force down through a viscous fluid, like oil, where they attain a steady velocity that is determined by Stoke's Law and the principles of viscosity and kinetic theory of gases—which we will not go into here, but which are well known—and can be adapted to us to predict the flow of ions (electric currents) in the atmosphere.

When a sphere passes through an ideal fluid which has no viscosity—that is, it is frictionless—the streamlines are symmetrical about the sphere and they are symmetric for the upstream and downstream. Since there is no viscous drag, the resultant force on the sphere is zero. But if the fluid or gas has viscosity, there will be a viscous drag. A resisting force varies directly with speed. Stoke's Law states that when flying through a viscous gas, such as the air, a fireball should be subjected to a force equal to six times 3.14 times fluid viscosity times its radius times its velocity.

Ions tend to move apart and spread out. This diffusion resembles a gas released into the air, except that the ions tend to move in one direction than in the other. Electrical forces reduce the near free path between collisions for particles carrying charges.

Simple diffusion of ions is important, but "eddy diffusion" by convection or air movement is caused by temperature gradients, and though eddy diffusion is described by formulas similar to those of simple diffusion, it is often an important larger factor in ion movement. Eddy diffusion causes the movement of unbalanced charges through the lower atmosphere. It mixes ions below 8,000 feet, most of the time. This is fortunate, otherwise electric charges would build up in certain local areas, burning plants and constantly shocking unwary people who came near these regions.

Ions move very slowly at about 1.5 centimeter in the fair-weather atmospheric electric fields, which is 100 volts per meter. But the velocity of a small ion due to thermal motion is 27 thousand times faster, or 400 meters per second. Even a light 5 miles per hour wind is 220 centimeters per second, which is 147 times faster. In addition, there will be air eddies and turbulence. All of these forces dwarf that of the electric field.

Although ionic current flow obeys Ohm's Law, at least under normal conditions, there are exceptions. When there are insufficient ions, and voltages increase, so will current—until there are no more ions to carry further current. No matter how much higher the voltage gets, current cannot go beyond this saturation level. In enclosed areas, this saturation current is reached quickly with small potential differences, and this value is far lower than outdoors. After the saturation current is reached, Ohm's Law is invalid. But before Ohm's Law fails, the electrical conductivity is the same both inside buildings and outside. This is so because ions with air infiltrate through cracks.

Ohm's law is not sacred. Even in metallic conductors, mean electron drift velocity is lethargic, and the average magnitude of velocity of the same electrons banging around between atoms is hyperactive; and Ohm's law fails at very high current densities (the conductor explodes) or very low densities, which are overwhelmed by thermal noise. Ohm's law fails not only in atmospheric plasmas, but also inside microscopic semiconductor chip geometries and their conductors.

Any conductor bathed in an electric field has charges induced on it, in addition to any of its own charges. These distribute themselves of the surface of the conductor; and, if the conductor's shape is simple, the calculus is simple; but if complicated, then a computer and finite element analysis software can determine charge distribution and the surrounding fields.

As far back as 1860, Kelvin correctly predicted that the upper atmosphere conducted electricity, and that because of this, that the earth its upper atmosphere with the insulative lower atmosphere created a giant capacitor. Nikola Tesla conducted further experiments on ionized gases and experimentally determined the resonance frequency of the earth and its electrosphere. But it was not until the early Twentieth Century that the reflection of radio waves off these layers confirmed Kelvin's prediction. Since the electrosphere is a good conductor, it acted like a perfect electrostatic shield.

When a portion of a conductor is exposed to the atmosphere, its charge is gradually dissipated by ionic conduction. If Ohm's Law holds, and there is no saturation level, then there is a "response time" or "relaxation times," which is an exponential constant. Near the earth, relaxation times is five to forty minutes, depending on pollution, salty air and so on; but at 18 km altitude, it is much shorter—only four seconds, and higher still in the electrosphere at 70 kilometers, it is ten nanoseconds; and for the earth itself, one microsecond.

The "relaxation time" for this transition depends on the parallel resistance and capacitance or R and C. This will be an exponential rise or decay. How fast or slow depends upon the magnitude of R and C: The larger they are, the slower it will take.

When potential differences vary slowly, Ohm's Law is resistive. But a unit cube of air will behave like a capacitor—two separated metal plates—with fast variations in potential differences. This could occur during a lightning flash, and there will be a transition from the first static state to the second one.

After a lightning flash, the sudden current flow will rearrange the distribution of charges. Some that were on the ground will now become space charges in the atmosphere. On the other hand, if instead of sudden change in the distribution on charges, there is a rapid change in conductivities, the same thing happens, except that there is no immediate electrostatic change, just a sudden change in conduction current.

If there is a discharge, be it lightning or something less, the discharge has a self-inductance, which can be calculated from potential difference and length. [*Gaseous Conductors*, by Cobine; p. 199 and *Network Analysis*, second ed., by Van Valkenburg.] If you know the capacitor plate area and their separation, there are simple network formulas that will determine if the RLC systems will act like an overdamped wave of an exponential decay, or a damped sinusoid—much like a ringing bell fading out.

Using a "lumped parameters" approach, your author recalculated the natural resonant frequency (that Cobine gave) for one lightning stroke at 45.8 kHz from initial assumptions. A lumped parameters approach means that even though we cannot see or define the distributed elements of a system as elements that are simple—like the two metal plates of a capacitor, of a spiral wire coil, and of a carbon resistor all suspended in the air—because in reality they are not there at all—that we can still pretend that the system (the imaginary box of air or the clouds and ground) will behave as if they are made of such metal plates and coils, and then apply appropriate mathematics of network analysis to the simplified circuit. Once solved, these differential equations will then predict how the system will behave. If you are clever and lucky, your assumptions will give you good results.

Rayleigh Jets Electrify Clouds

When a charged droplet evaporates, its radius decreases and the charge is distributed over a smaller spherical surface area, yielding a greater charge density and higher electric field strength. As the droplet evaporates, this field will overcome surface tension and suddenly simultaneously eject one to ten very small and highly charged micro-droplets, leaving the large parent droplet without much charge. Just before the parent drop becomes unstable and emits micro-droplets, the field at its surface exceeds the dielectric (insulating) strength of air of three million volts per meter. This spontaneous decay of parent droplets also occurs in clouds, giving birth to vast swarms of highly charged micro-droplets. In this way, large ions disintegrate into small ions.

Water or drops that splash in an electric field produce ions of both signs, but only those of the sign opposite to the potential gradient are released into the air. If drops in a wind or falling freely do fragment, they break up and the charges separate. Results vary. In an intense field of 30 kV/m, as in a thunderstorm, the separation of charges is 100 times more than in the absence of a field.

The exact mechanism is only recently being discovered. Cloud droplets briefly spout transient microscopic Rayleigh jets of water. When electrostatic forces between surfaces and charges become too great for surface tension, two micro-jets shoot out, carrying off the excess charge, and the parent droplet becomes mechanically and electrically stable again.

Through evaporation, droplets shrink and surface tension and area decreases, and electric forces become dominant. Destabilized, these droplets elongate, and then in about 100 microseconds, they squirt Rayleigh jets of water from opposite sides of these droplets. Then, as these emitted micro-jets disintegrate, they form a hundred tinier daughter micro-droplets that collectively carry away one-third of the parent droplet's charge—but only 0.3 percent of the parent's mass. But there are exceptions: Not all parent droplets are left electrically neutral; and, although those that remain unstable do occur, they are much less unstable.

Charges flee flat areas, try to escape other like charges, and migrate away to sharp points, where under correct conditions they will form corona discharges. Can we measure it? Yes. Charge density on an object varies inversely with its radius of curvature. Although this is only a general observation, we can quantify and measure it. A straight line has no curves. On the other hand, the less straight an object is, the more "curviness" it has. A small circle or curved path has greater curvature than a large one; and, so walking around a small circular path, you will change direction more rapidly.

To quantify this, you will measure the change of direction in degrees or radians and divide this by the short micro-distance you walk in a short time. Then, knowing the absolute value of this derivative at a point, you will know what the curvature at that point is. The reciprocal of this curvature, designated by kappa, is the "radius of curvature." Where the radius of curvature is small, near a sharp point, the surface charge will accumulate, and there the electric field strength will be highest. This is where the "action" occurs.

What occurs when two ions or charged particles of these sizes of the same sign touch? Imagine two tiny like-charged micro-spheres suddenly connected by a thin wire so the charges will redistribute themselves. The largest sphere has more total charge but has the smaller charge density and weaker field. If the potential is high, the ions attracted to it will collide with air molecules and create more ions. Air surrounding sharp conducting points will glow in the dark. Normally non-conducting, the air now becomes a conductor. Now, if the electric field is still greater, the discharge spark from the corona will come suddenly.

Conductivity of the air varies: it is greatest in the morning and in the summer. Fogs are poor conductors. In fogs, conductivity drops because the small ions (which conduct more current than larger ions) attach to fog particles, making them less mobile. This greatly increases the potential gradient and slightly decreases conduction current. Mountain fogs, which are really just the insides of localized rainless clouds, have one-third the conductivity of fair weather. Because the mountain fog is so localized and smaller, current will find a way around these poorly conducting regions. In contrast, current cannot get through widespread fog, giving it a greater potential gradient.

"Point discharges" occur when air ionizes near a point. Field lines concentrate and end on a point, intensifying ions colliding with air molecules and other ions to create positive ions and electrons. Being smaller, the electrons travel farther and faster. If this acts over a short distance, St. Elmo's fire appears; if it acts over a long distance, lightly strikes. Point-discharge current only becomes appreciable at tree top levels for 600-to-1,000 v/m potential gradients. For the trees, it is greater. A point-discharge can occur from a point at 35 m above the ground in fine weather for much less potential gradient: 200 V/m.

It is not uncommon for the point discharge to pulse on and off. If the point produces positive ions or electrons, they move into the point and the opposite ions move away to forms space charge. The point reduces potential gradient on the point, so that ionization by collision reduces or even stops. Soon the space charge ions drift away and the process repeats endlessly. Wind removes ions from the point, and wind changes can alter the charges per pulse.

During storms, appreciable currents slow from even low points. During fine weather, it is insignificant. Point discharges from natural points, such as trees, are important to the electrical conduction of the atmosphere and the transfer of charge between earth and clouds.

Point discharges bring down to earth 20 times as much charge as does lighting during a storm. A typical tree can draw 1.3 amp current. Lightning is important in returning negative charge to the earth, but point discharges are far more important. The earth is losing negative charge all the time. There is some mechanism to return it to the earth. Lightning and point discharges return the charges. Current flows from positive to negative, up from the earth into and through storm clouds, up through them into the electrosphere 40 km or so, then returns to earth in fair weather regions.

As a brief aside, you should know that we use the electron volt or "ev" as a microscopic unit that measures energy, usually of subatomic particles. Instead of measuring in coulombs, which is fine for lightning, it is much too large for the tiny scales physicists often use—much like inappropriately measuring your own height in miles rather than feet—so, we use the electron volt, which is a unit of charge (electron) times volts. When working with atomic problems we use multiples of ev, such as Mev, Bev, and Gev. But for Tesla coils and lightning, we appropriately use the joule, which is 6.25 trillion-trillion times greater than the electron volt.

A lightning discharge supplies a definite current, not a definite voltage, to a wire it strikes. The lightning discharge behaves as if it possesses a surge impedance (resistance plus reactance, a complex number) of 5,000 ohms. Thunderstorms maintain the negative charges on the earth.

By some restorative mechanism—previously unknown, but now recently discovered—the earth remains negative (-3.23 nano-coulombs/square centimeter) in spite of positive electrical charges (+3.23 nano-coulombs/square centimeter) flow from the sky to the earth. If not for this restorative mechanism, thunderstorms' point discharges and lightning, then the conduction current would neutralize all the earth's charge in a short time. If we used land values, it would occur in 48 minutes; ocean values, in six minutes. The actual value is probably closer to the second number.

This restorative mechanism is partly due to the red-flashing sprites that erupt after strong lightning strikes during a storm, several branching jets of current streak upward 55 miles above thunderclouds and radiate low-frequency electromagnetic emissions detectable over 1,000 miles away. Most of these occur over oceanic thunderstorms.

Sprites complete the missing part of the global circuit that sends electric charge from ground to the atmosphere and back. Thunderstorms act like batteries in this global circuit. We now know that lightning transfers charge from ground to cloud. Some charge slowly leaks back to earth during clear weather. Giant jets aided by sprites, blue jets and possibly other phenomena, all provide and alternative path for the ionosphere to lose its charge.

Lightning Induced Effects

Do lightning induced effects create ball lightning? Aircraft flying through charged clouds (even some non-thunderstorm clouds) sometimes trigger lightning, which originates at the craft and extends out in opposite directions. When a craft flies through the stroke, lightning reattaches itself along the fuselage at different points as the plane flies through this "circuit." Because aircraft pass rigorous lighting certification tests to verify their designs and verify that there are no conductive gaps, they are safe, but there may still be brief interference with instruments and temporary flickering of lights. Although aluminum fuselages conduct electricity over the outer skin, composites exhibit high resistivities and do not conduct well, so are embedded with screens or conductive fibers.

Unfortunately, lightning flowing over the surface inductively and capacitively induces transients ("lighting induced effects") on wires beneath the skin, which is attenuated with surge suppressors. (This is partly because, unlike steel, aluminum cannot stop magnetic flux.) Skin around fuel tanks is thick to avoid punctures and is tightly designed. Gas within fuel tanks is inert. All fuel lines, pipes, access doors, vents and fuel filler caps, and engineers are designed to protect against lighting; and new fuels produce less explosive vapors.

Plastic radomes enclose radar and instruments, and lightning diverter conductive strips (bars or buttons) redirect charge flow around the radome. Because small aircraft have aluminum skin, non-computerized engines and flight controls—and also because such craft are smaller targets—they are less affected by lighting, although their navigation lights and propeller tips are susceptible. Because kit-built "experimental" aircraft are of graphite-reinforced composites or fiberglass, they never fly near charged clouds.

Although airliner fireballs have appeared within craft when they were struck by lightning, reporting is unofficial and no records are kept.

If atmospheric effects are troubling, then questions of energy content and density pose greater questions for any proposed theory. And the sudden appearances and silent disappearances tweak the second law of thermodynamics.

19

Thermodynamics

Important to physics, thermodynamics of fireball energy content and entropy should provide clues to unraveling puzzling fireball physics. Paradoxically, instead, fireball behavior asks questions of the controversial second law of thermodynamics. And, why do fireballs appear instantly and suddenly—yet so silently? Where does the energy come from and go to?

Energy Density and Content

Energy content puzzles fireball theoreticians. Because observed characteristics do not fit their models, theoreticians became creative with the ball lightning data and their interpretations. If characteristics do not fit their models, they ignore these qualities. Does ball lightning have low energy density, as some—including advocates of the Smirnov fractal aggregate model—now advocate? Allegedly, the very close proximity of traditional lightning and a ball, they claim, makes it only appear that serious damage was caused by ball lightning—an illusion they claim. And thus, according to such low-energy proponents, this erroneously raises estimates of energy density.

However, true ball lighting data demonstrates a much greater energy that cannot be explained by mere fractal aggregates. It is a mistake to assume that the energy is stored within the ball or receives it externally from conventional means. If you do, you will be forced to deny four facts.

First, in most cases of damage, lightning was not nearby and did not do the damage. Second, so great is damage that significant energy was spent. Third, because of the submerged-in-water cases, the energy source was not external. Sufficient energy cannot be stored inside for a true lighting ball to exhibit the damage reported. Conventional physics cannot explain this. If it could, previous physicists that include two Nobel Prize winners would have done so. Finally, energy of flying is considerable, often flying against winds and sometimes flying at over 300 miles per hour beside aircraft.

Although fireballs rarely damage objects, they do leave physical evidence—scorched wood, melted metal, current surges in wires, and minor explosions. But if we examine those rare fireballs that do serious damage, they provide estimates of energy density.

If we assume the contents of fireballs to be in equilibrium with the outside air pressure (a reasonable assumption), then it is at standard temperature and pressure or STP—barometric pressure at sea level—and also assume each atom inside the fireball is singly ionized (each atom loses an outer shell electron), then the average ten-inch ball should contain 0.858 megajoules of energy. But to explain damage, the fireball must be multiply ionized. But for a ball of air to remain multiply ionized that long is mysterious.

Also, Tesla claimed that his measurements on his Artificial Electric Balls indicated they were at 0.01 atmosphere or lower. Assume he was correct, if not just for his own, but also for true fireballs. If so, then it gets worse, and the values must be multiplied by one hundred—which is even more absurd. It means that atoms must be stripped of more than a few electrons, and remain in this unstable state within the fireballs. Let us look at a few cases.

There exist many cases of damage that permit physicists to make calculations of energy density. Balyberdin, a Russian physicist, calculated in 1965 the energy of a 12-inch fireball that knocked down a shack. The ball came through the roof and ceiling, escaped though a window, and floated away from the building. It touched ground 165 feet away and with a prodigious roar, it blew up. The hut collapsed. The energy density of that ball was seven times that of TNT.

Charles Vernon Boyes, inventor of the high-speed Boyes camera (used to study lightning) and pioneer of surface tension, investigated a red ball of fire "the size of a large orange" that fell into a water barrel at Dorstone, near Ross-on-Wye at Hereford in England on October 3, 1936. The fireball glowed under water and then went out. It boiled off four gallons, and for 20 minutes remained too hot for the human hand.

By measuring the temperature rise in this known volume of water, the energy derived by the ball could be calculated. Heat lost by the ball equals heat gained by the water plus energy needed to vaporized some of it. Since it takes one calorie to raise one gram (cubic centimeter) of water one degree, and it takes 539 calories to vaporize one gram of water, the rest was introductory physics. Boyes calculated the ball's energy at eight million foot-pounds. Since one foot-pound is the energy needed to lift one pound one foot, then eight million foot-pounds can lift one ton 4,000 feet.

In the hypothetical case of a ten-inch ball of singly ionized air, will come up with 0.858—*Scientific American* in 1962 rounded it up to one megajoule—or only enough energy to lift one ton 368.8 feet. In reality, the fireball in Boye's case had ten times more energy and was much smaller. Since the Boye's ball was only three inches across, it had only $1/37^{th}$ the volume of the ten-inch ball. If the Boye's ball had been ten inches wide, it would have 370 times more energy. This assumes energy is stored but does not enter or be generated inside the ball.

C.M. Botley investigated a two-foot lightning ball that dug a five-foot trench 328 feet long in soft soil near a stream. If true, then the calculated energy density exceeded one billion joules per cubic meter.

Arthur E. Covington of the National Research Council of Canada reported in the April 18, 1970 issue of *Nature* that, ten miles from Nelson in British Columbia, he saw a lighting ball shatter a pile into splinters. Covington wrote that at "some distance from the shore a fireball fell from the clouds towards his side of the lake. It did not strike the mountainside but came to a stop, floating above the forest. The main road and electrical transmission line along the lakeside were parallel to but away from the shore, and led down into a smaller open valley stretching from the lakeside into the mountainside. The ball continued downwards, drifting above the road and the transmission line until it reached the bottom of the valley. Here it drifted to the lakeshore and struck one woodpile at a small wharf jutting into the lake. It shattered into long splinters and produced a large detonation."

P.D. Zimmerman of Hamburg in Germany calculated the energy required to vaporize a sufficient amount of sap or water inside the wood to shatter the 20-inch diameter oak log. The Covington ball possessed at least a half million joules of energy, and probably much more, since Zimmerman did not include energy dissipated in the lake. There exist more cases where calculations point to great energy densities.

Fireballs exhibit a wide range of thermal effects—something either too anomalously high or cold. In the case of a 10-cm fireball that submerged in a 10-liter barrel of water that boiled, the calculated energy output was 800 calories per cc. James Barry in *Ball Lightning and Bead Lightning* and J. Roth have calculated from measured cases that the energy density will be up to 100,000 Joules per cubic centimeter. TNT manifests energy at "only" 2,000 Joules/cc. People who touch fireballs note one of three effects—heat, no heat, or cold. Oddly, a few observers who contact fireballs report them to be cool or cold. If true, this suggests that some fireballs might be endothermic. Such endothermic processes or reactions, unlike exothermic reactions, absorb heat.

Since not even if we assume total ionization of air—and assume it is stable and lasts awhile—can any fireball store the enormous energies needed to do what the fireballs do. Something else is involved.

Do Fireballs Obey Entropy?

Entropy measures disorder. The universe evolves to heat death. The Second Law of Thermodynamics states that an isolated closed system's entropy must increase or be constant with time. Although entropy is simple to understand, its mathematics is more complicated than those for energy, and the subject generates heated controversy.

Mathematics treats gas statistically, as statistical ensembles of particle motions, not individual particles—because we are ignorant of the individual particles (and no computer can ever deterministically model them and their turbulent behavior)—and then treat these ensemble's evolutionary properties statistically, obeying the applicable dynamical equations of Newton, Hamilton, Maxwell, Einstein, Dirac and Shrodinger.

Although these equations remain unaltered if we reverse the direction or "arrow of time"—that is, these equations are time-symmetric and work if time runs backwards—in our real world the second law forbids initial states where the particles' vector momentums (masses times velocities) and positions lead to dynamically possible but infinitesimally unlikely future evolutions. Just as broken eggs do not re-assemble and leap back into their cartons, there is no perpetual motion machine, and dead people do not come to life and jump out of their graves. In life all your money will not one more moment buy.

Stephen Wolfram's ankos (described elsewhere) says much about the absolute impossibility and irreversibility of time, as does the late Nobelist Ilya Prigogine. Worse, the reversal of time, like time travel, would create catastrophic contradictions in the deepest logic that underlies our mathematics and physics. It also destroys philosophy and is fatal to religious concepts, including predetermination, free will, heaven and hell.

It is the second law that rules out such improbable and atypical evolutions from the irreversible systems that we model and allows us to treat particles statistically and not individually. In the real world, thermodynamics' second law forbids the reverse causal arrow of time because of the special conditions that once existed in the Big Bang that controlled the origin of time.

Before the first picosecond after creation, the sub-microscopic speck that would become our universe was in thermal equilibrium—a steady-state condition that a system will not deviate much from; and if disturbed will quickly return to thermal equilibrium. Statistical mechanics examines fluctuations deviating from thermal equilibrium; and classical theory agrees with observation—except when quantum mechanical effects become important, as with black body (perfect heat) radiators.

Entropy measures ignorance or lack of information and flows to the higher probabilities inherent in disorder because there are more ways for things to become disordered—a cosmic Murphy's Law. [For different insights into thermodynamics enigmatic second law, see Brian Greene's *The Elegant Universe*.]

Thermodynamics Second Law

The Second Law of Thermodynamics is not strictly a true law, but is based on the arrow of time—observations of events that flow through time from past to present. The laws of physics do not prefer either direction in time. The Second Law, which is time-symmetric—entropy increases only in the forward-in-time direction—is not a true law of physics in the inviolable sense, but is a statistical law that is totally time-symmetric.

In cosmology, cosmic entropy decreases backwards, to the instant after the Big Bang, when the universe exhibited low entropy, which grew with time. History answers the question of the arrow of time. Total disorder ruled the very early universe, and when the inflation field was, in scientific notation, only 1×10^{-27} cm across. Where does inflation come from?

Seemingly simple, the Second Law of Thermodynamics is notorious and generates heated controversy among physicists. Fireball behavior adds more controversy. When fireballs suddenly materialize and disappear, where does the energy come from and go to?

Fireball size-changers are somewhat rare. Although most fireballs do not change size, a few expand slowly; others, rapidly. Of these "expanders," a few slowly become brighter; others do so suddenly; and still others expand but do not get brighter. In some cases, the fireballs rapidly grow larger and brighter. In Oak Ridge case 8B of Starken, the bright baseball-sized white fireball rapidly expanded to basketball-size—and grew even larger.

And there are other cases like Starken. But are they illusions? No. Although in some cases, we might wonder if the ball approached and thus only appeared to expand—and either remained unchanged or grew in brightness—there are many undeniable cases confirming rapid expansion. In the Starken case, the ball's volume rapidly grew 250 times. Others have grown over a thousand-fold in three seconds.

Such rapidly expanding fireballs—whatever their sources of energy, of confinement, and of propulsion—and if they obey the physics we know—must obey the laws of thermodynamics and entropy. The first law states that there is a cosmic accountant but no cosmic Houdini—when energy changes form and heat is converted to or from other forms of energy, the energy is neither created nor lost, and can neither appear or disappear. Like a chameleon, when energy changes forms—chemical, light, heat, electromagnetic, kinetic or moving, potential, nuclear energy, and so on—the total energy is conserved and remains unchanged.

When a fireball expands—be it a gas or Smirnov fractal cluster—as in the Starken case, and if the ideal gas laws apply (they should, but they may not), the expansion should occur because inner pressure grew; and, to maintain equilibrium and to balance the atmospheric pressure and thus keep its own inner pressure constant, the sphere will expand. This means that the temperature in the Starken case had to rise 250 times. When heat is supplied into the fireball, held by some unknown confinement mechanism, the size and volume did not change, then the heat would be converted into internal energy in the form of molecular kinetic energy, thus raising temperature and pressure.

On the other hand, expanding gas performs work. If the heated gas expands, the fireball would grow in size, and the heat supplied would change the energy of the gas and perform work in expanding against atmospheric pressure. In surface-tension fireball models (such as the Tesla-Golka-Bass model, described elsewhere), and which I disbelieve, work is also expended to stretch the ball's surface membrane, just like blowing up a balloon or stretching a spring, where the thermal kinetic or moving energy of the molecules become stored as potential (stationary) mechanical energy in the membrane, just like a stretched spring stores potential energy and can later be converted to kinetic energy and do work.

The Second Law of thermodynamics states that an engine—a fireball or almost anything to a physicist is an "engine"—can not transfer heat to another location up to a higher temperature unless that engine draws in external energy. For example, air conditioners and refrigerators must draw in electrical energy. This law saws there is no free lunch and no perpetual motion, and everything eventually seeks lower energy levels, to slide down thermal hills just like skiers descending a Vermont mountain. You cannot push things up thermal hills, or up ski slopes for that matter, without taking in energy. The best reversible heat engine has the maximum theoretical efficiency that any heat engine can ever possess is called the Carnot Engine. In reality, all real-world engines operate at lower efficiencies.

The ideal reversible heat engine of Carnot goes through the Carnot cycle—a sequence of four isothermal and adiabatic steps of expansion and compression that can be plotted on a Carnot pressure-volume diagram, which resembles a rhomboid or squashed rectangle. Although the engine is often envisioned as a piston in a cylinder, it could be anything, such as a fireball, including those that expand or contract (or one that keeps expanding and contracting, as a few balls reportedly do).

According to most theories, fireballs contain atmospheric gas or fractal clusters, which expand with rising temperature and decreasing pressure, and are considered to be an ideal gas—a good approximation that holds true except at very high pressures or extremely low temperatures. The results are independent of the nature of the working substance, and the Carnot cycle applies widely. For fractal cluster fireballs—if that type does exist—the ideal gas law might need modification, but should apply.

The Carnot heat engine can transfer heat to a lower temperature and perform work (steam or gasoline engine) or work in reverse and take energy fed to it from the exterior to move heat from a higher temperature to a lower one (as in an air conditioner).

Treating a fireball as a gas is a simplification, but is probably basically correct. Although we could consider charged particles, such as ions and electrons of Smirnov's fractals, this is covered mathematically in detail by other authors, including Cobine and Chalmers. The rapid collapse or expansion of a fireball would be approximately adiabatic if no external energy rapidly enters or leaves or is generated or destroyed somehow within the ball. Of course, when a hot bright fireball instantly vanishes rather than violently explodes, this leaves unanswered a puzzling question: where did all the energy instantly go?

If work (force times distance) is performed on the fireball, its internal energy increases, but if done by the fireball, its energy decreases.

Most ball lightning theories assume that standard thermodynamic gas and plasma laws apply—although this is questionable because none of these theories has ever explained fireballs. But let us assume that standard physics does apply. Expanding air cools. When a fireball grows in size, especially if its expansion is rapid, the hot air inside it will expand adiabatically—that is, without heat entering or leaving this closed system. Because of isolation (insulation) the relatively rapid expansion in size, the reason is simple.

When gas molecules or ions collide with other approaching particles, there is a brief elastic collision (they do not lose any energy) and there rebound velocities increase, but if one particle is receding, the rebound velocity increases, but if one particle is receding, the rebound velocity is less. In a region of gas that is expanding, more particles will collide with receding ones—and there will occur an overall slowdown of speeds with less kinetic energy, which is the definition of lower temperature. (In particle physics, interactions without energy input are termed adiabatic.)

The rate of flow is often somewhat slow. A fireball's expansion would be approximately adiabatic and its temperature should drop, some deionization should occur, and its color should change from white to red or orange. But witnesses rarely report this color change.

For a fireball at the instant it vanishes or explodes isocochorically, the volume at that instant does not change and remains constant, while large quantities of energy are removed or added to the fireball. There should be a very sudden rise in temperature and pressure, and the process may be treated mathematically as if the energy source or sink isochorically added or withdrew (depending on where it vanished or exploded). This is similar to the sudden explosion of gasoline vapor in a cylinder, which has inertia (just as does the exploding fireball) and cannot move instantaneously.

Isothermal processes would not occur because of the short lifespan of fireballs; for the temperature to remain constant, the other two variables must change slowly with heat transfer—for internal energy depends on the temperature, as with ideal gases. Ideal gases undergo an isothermal process. Their internal energy will not change.

If energy and heat enter a fireball, the hot gases or plasma should try to expand. This would maintain an outward pressure to balance the inward atmospheric pressure and keep a constant pressure and adjust to maintain equilibrium. Thus the fireball should grow larger. Yet, despite all the enormous energy that fireballs can exhibit, it is odd that they can keep their size. If you are rusty on thermodynamics, restudy it, and you will be puzzled by fireball strange implications for thermodynamics.

Sudden Appearances Silent

There are many cases of silent, sudden and instantaneous, appearances and disappearances. In some cases, a ball that silently appears may later explode and even do damage. It is odd that there is no sound in these silent formation cases. By comparison, when air is heated by lightning from room temperature to 50,000 degrees Fahrenheit, it acquires tremendous kinetic energy. Because it has inertia because of its mass, the air has no time to expand and the gas pressure "instantly" exceeds 100 atmospheres, or 1,470 pounds per square inch.

This high-pressure channel explosively expands into the surrounding air, which is at only 14.7 psi, and compresses it. This creates a supersonic shock wave for 30 feet. The sound pulse or blast lasts under a tenth of a second. If you are within 200 feet of a lightning stroke point, you will hear a click followed instantly by a terrifying crash.

The invisible downward-descending stepped-leader from a cloud-initiated discharge seeks the easiest path. At 200 feet above its target, it becomes "aware" of the target, which sends up an invisible leader. When they connect, a brief snap or click occurs and several current surges flow.

But when fireballs form, there should be a sound. But none is heard even with witnesses very close to the lightning ball formation. The ball instantly appears full sized and does not usually grow or shrink. Although it may not seem to be hot, it can set wood on fire and hiss when it hits water and vaporize it into steam. At other times it may do nothing to wood or water.

Some theories (Tesla, Corum, Golka and Smirnov) claim that when Tesla's Artificial Electric Balls form, the air explosively expands. Now, this may be true only for those electric balls created by Tesla and Corum, but they are not true fireballs. This is obvious from examining the Oak Ridge and other true fireball cases.

Supersonic Fireballs Fly Silently

Why are there no sonic booms? In the rare alleged cases of supersonic ball lightning, at least those that were reported, there was no sonic boom. There should be. All jets, rockets and bullets compress the air ahead of them. Compressing air produces pressure fluctuation, called sound or compression waves that travel through air like circular waves from a stone tossed into a pond and travel outward. The speed of sound at sea level is 760 miles per hour; at 10,000 feet altitude, 735 mph; and at 20,000 feet, 707 mph. At higher altitudes, sound travels slower, but not by that much.

A fireball traveling at supersonic speeds flies faster than the pressure disturbances it creates, and the ball is ahead of its wave front. The sonic shock wave is a cone shape. Depending on the fireball's altitude and atmospheric conditions, there is a "cutoff Mach number" that is between 1.0 and 1.3. Below it no boom is heard on the ground. If there is significant turbulence, wind, clouds and temperature variations, then the cone shape is altered by the movement, and refraction or bending of the sound rays occurs through the varying air densities. The waves are unable to move forward to "warn" the air ahead, and the fireball arrives without warning to create a shock wave—or it should, but does not do so for ball lightning. The big question is—why not? The answer involves airflow.

Airflow Around Fireballs

Airflow around a slow-moving fireball parts evenly around the ball and converges on the backside. The front and rear streamlines are symmetrical mirror images at slower speeds, but lose symmetry at rising speeds, with the recombination further in their rear, downstream. At increasing speeds, rotating oval vortices appear in ring-like doughnut shape. At still higher speeds, those stationary eddies that were attached to the rear of the ball are torn away and carried downstream, with a regular alternating periodicity of whirlpools with alternating rotations.

With higher speeds, the vortices grow larger and unravel to lose their periodic shapes. Finally, at an even higher speed, suddenly turbulence sets in and the streamlines swirl chaotically with little apparent pattern. After passing through smooth flow, stationary and moving eddies, the pattern suddenly grows turbulent.

The fireball is not streamlined. If it were, it would alter its shape into a teardrop shape so that its new shape would mechanically fill the space taken up by the paired downstream vortices, so that the air flowing past without eddying will produce little drag. Why does the fireball not alter its shape? And why is it not deformed even at high speeds by aerodynamic forces on its surface? A flying fireball should make some air noise but does not.

Jet-Rays Do Not Destabilize Balls

Witnesses occasionally describe an odd type of fireball—one with a glowing exhaust, ray, rod or jet pouring out of the ball. Most have only one unicorn-like jet-like horn, but some are devil-headed with two glowing "horns." The horns can point up or down, and one can be larger than the other. Other witnesses report a "hole" or "opening" from which issued a jet of fire.

In one case an opening in the side of a huge red fireball grew until it was three times the size of a hand. When the ball detonated, zigzag lightning flashed from it in all directions and men below in the street were knocked down. Some witnesses describe a silent fireball that began to hiss or roar like a blowtorch after it sprouted a glowing jet. Sometimes this jet-ray grows longer. Sometimes it stabs outward, shortens, and then thrusts out farther.

In one fireball after a storm, seen by an Irish astronomer leaving an observatory, a white ray shot up from the ground and formed a red fireball encircled in a halo, and the ray disappeared, and the ball then shot out a thinner ray of white downward into the ground, stopped and repeated itself. Usually, such unicorn- and devil-headed fireballs soon thereafter disintegrate noisily, with crackling and crunching.

The color of these jet-rays varies, but is often white. One ray was white in the center and red near the outside. Even dazzling fireballs have sprouted jet-rays.

Now, by Newton's third law of opposite action and reaction, shouldn't a jet exhaust that is pouring from a fireball make that ball unstable in its flight? Why do these balls not tumble wildly and zigzag out of control? Just like blowing up a toy balloon and releasing it? They are unlike an arrow or rocket. Finned rockets travel rectilinearly in a straight line simply because the center of pressure is behind the center of gravity, thus making it fly nose first. That is why rockets and arrows have fins.

Bullets have no fins but are gyroscopically spin-stabilized and spun by spiral twist of the lands and grooves inside the rifle bore. Cannonballs and musket balls spin wildly, but this reduces their range and impairs their accuracy, thus hurting their aerodynamic performance. No engineer would dare put complicated electronics inside artillery cannonballs.

But fireballs have no fins, and their centers of gravity and pressure probably coincide closely, making them candidates for tumbling and dynamic instability. This is bad. So, if fireballs do not stabilize themselves like finned rockets—and since they do not pitch, roll and yaw aimlessly like a brainless cannonball—then how do they stabilize themselves? There is one possible answer. It lies in servo-mechanism theory. The science goes back almost a century.

The father of rocketry is Robert Hutchins Goddard. He lived in Auburn, near Worcester in Massachusetts, just west of Boston. In 1926 he constructed and flew the world's first liquid fuel rocket. But he had problems with stability. How could he keep it from tumbling? For his first rocket, he took the direct solution—he put the rocket motor on top of his rocket! It was an ungainly-looking contraption. Goddard went on to larger rockets, relocated the motor to the bottom, and used gyroscopes for stability and navigation. He filed many patents. The Germans studied them.

In its first lumbering minute of liftoff, to keep it from tumbling, the World War II German V-2 rocket had graphite vanes stuck inside its exhaust. These vanes soon melted off. By then the rocket was moving faster, so its four giant external fins took over in stabilizing it. But this changed because there was a better way. In 1958, the finless Vanguard—looking like a pencil balanced on a plume of fire—and its successors, discarded their fins and used gimbals to turn their motors to balance themselves, much like a juggler balancing a vertical stick on his finger. That is why the rocket exhaust flickers from side to side. The servomechanisms compensate, sensing and sending directions to compensate, thus doing a balancing act.

Anyone who studied linear and nonlinear servomechanisms, also called feedback or control systems, is aware of these difficulties. Today, so refined has modern servomechanism theory become that these systems can stabilize even odd-shaped aircraft like the aerodynamically unstable, ugly "woblin' goblin" known as the F-117 Nighthawk stealth fighter. Without modern servomechanisms, this would be like trying to fly a brick—which is why some engineers call it "the flying brick." If servomechanisms can stabilize such odd shapes, surely they could stabilize a flying ball. So far so good. But from here on it gets weird.

Now, obviously, fireballs are not hardwired with sensors and microprocessor-based electronics. That is a problem. If there are sensing and servomechanism subsystems equivalents—and some fireball behavior indicates such mechanisms do exist—then what are these sensing and servomechanisms made of? Cellular automata? Fractal gases?

There is another problem. For any body such as a fireball to move and overcome air resistance and its own inertia—sometimes flying 400 miles per hour close beside an airplane—and to change direction, to accelerate or decelerate—sometimes instantaneously—requires the application of a force to the fireball. And since fireballs flying at 400 mph do not deform, then that force must be applied to every particle within the ball uniformly and equally. It is easy to suspect that some unknown form of propulsion exists.

Fireballs ignore translational and rotational friction. In relation to the equilibrium of fireballs, can we idealize fireball Newtonian mechanics? We can try. We must idealize and ignore the apparent rotation of the fireballs and the forces that occur from air being dragged around by the spinning ball. If Newtonian mechanics apply, then we have not robbed the problem of its reality, for our assumption should yield results close to actual values at lower velocities. Our approximations will yield good results without resorting to complex mathematics. Or so it should be.

Friction occurs when one body—an object, a gas molecule, or atoms—slides near or past another and each body exerts a frictional force on each other, parallel to their surfaces or direction of movement. Some fireballs change their linear velocity and thus change their instantaneous rectilinear acceleration. Most fireballs do not behave as if they were subject to frictional forces. They do not slow down from air resistance or from mechanical friction. They often keep bouncing at the same intervals. A fireball circling, as in an ascending spiral, has an acceleration that is the velocity squared divided by the radius of its spiral. Collisions in the extreme are completely elastic. Fireball collisions with objects are usually totally elastic.

Do Spinners Possess Rotational Inertia?

Some fireballs exhibit inner motion. This may be chaotic, as in the theory of chaos. But some fireballs spin. Are spinning fireballs gyroscopes? A gyroscope is any body revolving about an axle—real or imaginary—that penetrates its center of gravity. If the body is homogenous, this is uniform, this is its geometric center. For example, a spinning bicycle wheel and child's toy top, or the earth revolving around its axis. If it is a fireball that is revolving about its spin axis and if a force attempts to rotate or displace the quasi-axle of that fireball, then the axle will resist the motion that attempts to impose on the ball and will rotate about the axis that is parallel to the force. This turning is called precession.

If the spinning fireball is free from any force applied perpendicular to its spin axis—called a couple—it will maintain a fixed orientation no matter where it flies. This is similar to a gyroscope mounted in a Cardan's suspension, which is free from any couple or force-pair applied to its spin-axis, provided gimbal lock is avoided. It takes a large couple or force-pair to twist a gyroscope.

So great is the twisting, that large gyroscopes were used as stabilizers inside early battleships or dreadnaughts so they would not roll with the stormy waves and could accurately fire their huge naval canons. Today, some binoculars contain gyroscopic stabilizers to reduce shaking. The gyroscopic spin stabilization of wheels makes it easier for a rider to balance on a moving motorcycle.

Gyroscopic spin imparts to bullets a spin-stabilization so they fly nose first and stay tangential to their flight path. Tumbling bullets do not fly far. Just as bad, a bullet that tumbles will "keyhole" when it hits. So, to impart spin, spiral lands and grooves inside the barrel typically twist one turn in nine inches of barrel length. Gyroscopic precession can be formidable. It will damage the bushings on shafts of ships and aircraft propellers.

If fireball contents are gaseous, the precession may be weak. But it is there and how it would affect a fireball's sensing, propulsion and navigation systems is unknown. A revolving swarm of hornets or swallows in flight may spin about an apparent center or centroid, but there is no physical force or connection between each, so the swarm lacks any moment or gyroscopic action. But a solid, such as a spinning disk or spinning bowling ball, possesses such action. Its billions of molecules are held in place.

Now, in between these two extremes are the galaxies, of which there are five categories. But no matter what category, each galaxy spins about an axis, and its billions of stars and dark matter are attracted by gravity. A raw egg will not spin much—its spin is damped because its contents are viscous and absorb the mechanical energy. But when it is hard boiled, that same egg will spin effortlessly. No one can explain why some fireballs that spin continue to spin without slowing down.

I will not delve too deeply into the physics of angular momentum but will only briefly discuss it. For a deeper treatment, see any introductory physics text. The physics and vector algebra are trivial and well known. A spinning fireball has angular momentum. By calculus we derive the rotational inertia of a spinning solid sphere as $I=(2MR^2)/5$; for a very thin spherical shell as $(2MR^2)/3$; and for a thick shell as one with a denser exterior possesses a divisor between three and five; but on the other hand, with a denser core, it is greater than five.

Bodies absorb great impulse forces and elastically rebound but cannot withstand continual forces. Although a baseball hit with a bat is undamaged, if the ball is slowly squeezed in a vise with that same force, it will be crushed.

Angular momentum is moment of inertia times angular velocity. If there is no net rotational torque on the fireball, that is the sum of twisting forces balance out, then the ball's spin rate will not accelerate or de-accelerate. Conservation of angular momentum provides stability to a gyroscope, to a spinning fireball, spin-stabilized satellite or revolving bicycle wheel. They strongly resist any attempt to reorient them away from their spin axis.

The rotational dynamics of a rotating rigid body, such as a hardboiled egg spun on a tabletop, are convenient to derive. But for a viscous core, as with an uncooked egg, the derivations are difficult and values are better determined by experiment. Angular momentum equals the moment of inertia times angular velocity. If there is no net rotational torque on a ball, that is the sum of twisting forces balance out, then the ball's spin rate cannot accelerate or decelerate.

Conservation of angular momentum applies in classical and modern physics. If no external torque (that is a force that spins a sphere) is applied—or their sums are zero—then the total vector angular momentum does not change and remains constant. An ice skater or ballet dancer can decrease their angular speed or spin rate by extending their arms. The conservation of linear and angular momentum also holds in quantum mechanics. Quantum mechanics—albeit, in quantitized form—is more fundamental than Newtonian mechanics. They also apply to cosmology, such as the collapse of a star or the origin of star systems.

To understand the importance of spinning, we must examine the physics of spinning bodies. Some fireballs seem to spin. If so this raises questions. Do fireballs behave like a solid and thus freely spin like a hard-boiled egg? Or are they viscous, like a raw egg, and thus barely able to spin? Do the spinning fireballs somehow only appear to spin—just some type of optical illusion that they spin? Or does only their exterior spin, but not the interior? This creates shearing forces. Or is there internal chaos and turbulence? Does a fireball's spinning exert aerodynamic forces upon the surface of these spinning balls? If so, the balls must compensate for these frictional and turbulent forces inside them. By what mechanism?

All heavier-than-air flying balls that are made of matter—such as thrown baseballs and fired cannonballs—will curve in a parabolic flight, although slightly modified by spinning and air resistance—as gravity pulls them back to earth. Spinning fireballs should behave the same—unless they are exempt from the laws of physics, as we know such laws.

For simplicity, let us examine baseballs and golf balls in flight. Airflow past a non-spinning ball flows symmetrically around all sides and unless moving slowly creates a symmetric turbulence behind the ball. But when the ball spins, layers of air near the ball are dragged around because of the laminar air flow viscosity and internal friction of a fluid or gas passing near a ball's surface in the direction of spin. The surface facing the oncoming air stream meets it and moves the turbulent flow upward to the top, making it asymmetric, and increasing the pressure on the top.

This spin causes a baseball to curve, and golf balls backspin (imparted from impact with the slanted clubface) creates a lift force keeping the ball flying in the air longer. A golf ball often curves upward on the initial part of its trajectory. Likewise, this should hold true for fireballs. This should be stronger for fireballs with coarser surfaces.

For comparison, dimpled golf balls travel twice as far as smooth balls. This was deeply researched by England's Barnes Wallace, who discovered that only dimpled balls—really bombs weighing 11,280 pounds and spun at 500 RPM—dropped from sixty feet, skipped across water. Smooth ones hit and sank.

Then, on the night of May 16, 1943, 19 British four-engine Liberators launched these spinning skip bombs 470 yards away, which spun after they hit the dam, hugging it, to explode at the correct depth, to destroy four German dams.

The rotational inertia—called the moment of inertia—makes a rotating fireball keep spinning about its axis of rotation. A football spins about its long axis as does a bullet, whose spinning stabilizes it from tumbling. All spinning masses possess rotational inertia.

Rotational inertia depends on the shape and massiveness of the object, so a heavy rolling boulder has more rotational inertia than a croquet ball. The farther the mass is located away from the axis of rotation, the greater the rotational inertia. Industrial flywheels have their steel located at the rim flanges. It is harder to get it up to speed, but once spinning, it keeps spinning. A solid sphere rolls down an incline faster than a hollow one, assuming their masses and diameters are the same. A solid cylinder rolls faster than a ring. The object with greater rotational inertia compared to its weight more greatly resists a change in its motion. Conservation of angular (spinning) momentum says that the total angular momentum of an isolated system (feedback) remains constant.

Angular momentum must be conserved—it does not disappear to and appear from nowhere. This is true on a quantum scale, on a galactic level, and on a planetary scale. For example, if you think that the asteroid that hit Mexico and ended the Mesozoic era and helped terminate all dinosaurs—well, it was a relative midget and was preceded by two monster mega-asteroids that, at 3.5 and 3.7 billion years ago, hit the earth. Molten earth matter spit out. In the lower Pre-Paleozoic oceans, early evolution stalled. Massive tidal waves covered all land, except the tallest peaks, and rearranged coastlines. Your author here speculates that maybe some rock fragments that blasted out carried dormant microbes. Eons later did these rocks rain down into the subsequent primordial pre-Paleozoic oceans of Earth to re-seed them and begin life anew on Earth? And did these Earth rocks also seed the early Martian oceans?

The ejected material coalesced into a molten moon. A larger moon circled faster across the sky. But with passing time, and with viscous losses to friction from its molten core, the earth slowed its rotation. As earth spun slower, the days grew longer, and to maintain the earth-moon total angular momentum—because of conservation of angular momentum—the moon slowly moved from the earth at 354 km per billion years. The point is that this law also applies to everyday-sized spinning objects, including fireballs—if there is no inverse gravity.

Those fireballs that spin possess angular or rotational momentum (that is, spinning inertia), which is linear momentum—mass times velocity vector—at a radius. Examples include a spinning top, dreidel and gyroscope. By the way, Newton's second law states that force is not "ma," as is so often incompletely stated, but is more general and is the derivative of momentum. This law carried over into relativity and added to the stature of Newton.

So, the conservation of linear and angular momentum is more fundamental than Newtonian physics—it holds true in quantum mechanics, and its angular momenta are quantized and only take on discrete values. Angular momentum is a pseudo-vector and is perpendicular to the moment arm and force.

Because of rotational inertia, there should exist kinetic energy in a fireball's rotation. For a rotating rigid fireball, kinetic energy should be the sum of its thermodynamic energy, plus its linear kinetic energy, which is half its mass times velocity squared, plus the same form but with angular velocity in radians per unit time written with omega replacing velocity. Mechanical engineering texts on dynamics list formulas for many different shapes.

All this brings up the question of fireball contents. If fireballs are concentrically homogenous, that is, uniform, then the only explanation for their ability to sense their environment and navigate may come from within the ball or from something outside of the ball. What that "something" is remains an enigma to physics. Some fireballs seem "intelligent" and some "investigate" nearby objects and people. Propulsion, sensing, decision-making "quasi-intelligence" and servo-mechanism behavior, and navigation are several of the many enigmas of fireballs that mystify physicists.

The fireball propulsion system, sensing, decision-making "intelligence," navigation and feedback servo systems puzzle investigators. It is hard for physicists to imagine how a fireball turns to redirect its resultant vector force to navigate. Or is it meaningful to use vectors? It has worked for us for so long, just as has our algebra and calculus that we never question, and maybe our math may shape our physics into domains where it should not.

Do Flying Fireballs "Think"?

Since gases are stupid and behave like Gibb's and Boltzman's statistics say they should, no one ever dares suggest that even though true ball lightning sometimes exhibits curious and intelligent behavior, that it is anything more than a dumb ball of hot gas. At best, some say it merely follows electrostatic and electromagnetic fields like a marble in a pinball machine or maybe perform like some non-linear fractal clusters that are chaotically equally dumb.

But fireballs possess sensory, energy generation, navigation and propulsion systems. Since ball lightning is buoyant, how come some fall rapidly from clouds like falling cannonballs, rebound before hitting the ground, and then float lazily against the wind or rapidly shoot back up to the clouds? How does a ball "know" where an aircraft is, follow it closely and not be affected by the 500 miles-per-hour airflow and turbulence, and not just tumble or be swept away in the wind and turbulence? What possible detection, sensory, energy generation, navigation and propulsion system do fireballs possess? No other natural or man-made control feedback system matches their incredible behavior.

For just one of many amazing examples, in their rising-spiral-around-a-person flight path, fireballs seem to act quasi (as if) "intelligent." Many fireballs display some form of "memory." Your author uses these terms not in a biological sense but in the sense that Stephen Wolfram does and Alan Turing did—that is, of cellular automata, software and servomechanisms.

A microprocessor-based system or fireball with sensors that can materialize and fly, then detect electromagnetic, electrostatic, metal, solid objects, chimneys from a distance, distinguish a standing person from a post, and so on, would be an amazing feat of technology. But, for a supposedly hot ball of gas—it is not made of semiconductors and wires—to do these tasks of sensing, control and navigation is not a question for technology, but a mystery of physics.

These and their other detection mechanisms are remarkable. The mechanisms are most remarkable because many of the fireballs intelligently fly either slowly or rapidly to or from the objects in their environment, and at the same time make complicated decisions and maneuvers relative to these objects. Does coherent behavior, such as the Einstein-Podolsky-Rosen-Bohm Effect at the quantum level, manifest itself at the macro level in ball lightning?

If not a rational skeptic, your author could suspect that fireballs might be a transient life form with some form of "alien" intelligence. Perhaps DNA-based life is not the only form of life on earth. If so, then are fireballs non-carbon-based life forms? Not likely.

String Theory Advances

String theory absorbs quantum theory without alteration, so the mind-boggling weird mysteries are not resolved—especially quantum entanglement. Widely separated particles are correlated, even though they are ostensibly operating independently—a weird idea, even for quantum mechanics. And string theory fails to answer this weirdness.

Loop-quantum gravity has advanced, and its strengths complement string theory's weaknesses. For example, its background dependency, where we must assume an existing spacetime within which strings move. But in loop-quantum gravity, spacetime emerges from its fundamental equations, and so its formulation is background-independent. But they are unable to make the same very direct contact on large scales with General Relativity.

Penrosian twisters advanced loop-quantum gravity to make it viable, together with alternatives. Even with the background-dependence of string theory, physics gets a glimpse of what is going on deep inside, despite the background-dependence formalism of string theory, that developed mirror symmetry, that there can be two spacetimes and one physics, and discovered that space evolves in new ways, and that the microworld may be governed by noncommutative geometry.

On small scales, string theory modifies relativity, which proved that spatial geometry is tightly connected to the physics, so that when the geometry changes, so does the physics also become different. String theory microscopically modifies General Relativity, and says that even though the geometry can vary, the physics strangely remains identical.

Although General Relativity accepts the morphing of space, it cannot tear even microscopically or the equations break down. On the other hand, in string theory small rips are acceptable because Einstein's equations possess extra terms that normally are undetectable.

Earlier, Joel Sherk and John Schwartz discovered that one string vibrational pattern was a massless particle possessing spin 2—very bad because that was not part of the strong interaction—until someone realized that there is such a particle that emerges from attempting to quantitize gravity—the graviton. Someone realized that the telltale vibrational pattern of the strings suggested this was a theory of gravity described by General Relativity.

But Special Relativity modified the equations, so that they are now the lowest-order approximation of this vibrational mode, with additional extra terms that are normally miniscule and drop out (except at the Planck length). In fact, the sun's mass bends starlight, as first observed in 1919 by Eddington. When incorporated inside string theory, it adds an undetectable term—a 1×10^{-90} deviation.

These extra terms physically come from the fact that a string is not a point. The heart of General Relativity is Riemannian geometry, which describes space as a manifold—a geometrical space composed of infinitesimal small points. Because string theory considers the smallest element to be tiny loops, Einstein's equations add a correction. But if we reduce the loop size to zero, these additional terms drop out. Historically, if Einstein had reformulated his equations in terms of noninfinitesimal points, he might have discovered string theory in 1920!

There are other approaches to quantum gravity, including loop quantum gravity, which has made recent progress. Many of the different viable theories, each with compensatory strengths and weaknesses, approach the final theory from different directions.

String theory is background-dependent, that is, it assumes an existing space-time within which move its strings. As we look at smaller scales, the deviations grow. Noncommutative geometry, as developed by mathematician Alain Connes, will be the correct language to describe the very small.

Although we are accustomed to labeling points with matrices, a point is only an approximation, and such a poor one at that for small scales that it becomes meaningless. In string theory, its massless particles substitute for points. Although traditionally physicists visualize geometric points, we really describe something moving through points—its motion, which is described by matrices. On very large scales motions between points is all that counts.

For more on this, please read *The Elegant Universe*. It is not my purpose to further cover string theory—Brian Greene already has done it exceptionally well.

Quest for Quantum Gravity

In an epoch paper in March 1998, Adam Riess described how his team of Mount Stromlo Observatory discovered that the universe's expansion is re-accelerating outward, propelled by some mysterious repulsion force, a negative gravity, that is associated with dark energy. Another team, led by Saul Perlmutter of Lawrence Berkeley National Laboratory, also discovered this acceleration. Skeptics muttered that the data was tainted, until Riess proved that explanations like dust absorption do not explain the unpredicted dimness of distant Type 1a supernovae.

In the distant past, when gravity—a very weak force—dominated the dynamics of the cosmos—and after a brief initial inflation (that is, a cosmic explosion after the Big Bang)—gravity slowed down the expansion. Everyone debated whether the cosmos was negatively or positively curved by gravity, contracting someday or expanding forever. From examining the COBE and WMAP satellite data, as it turns out, astronomers now proved that our cosmos is not curved, but is perfectly flat.

Gravity is "a strange bear" and it is easily the least understood of all the four forces. The attempt to apply the Heisenberg Uncertainty Principle to space-time and gravity and other approaches to quantum gravity fail. It is one of the last frontiers of quantum mechanics and particle physics. So far it is a failure. Physicists do know something about what a quantum gravity theory would be like. It would begin at the very tiny Planck lengths, or less. Einstein's space-time geometry, no longer smooth and predictable, would be strangely turbulent and inexact, making precise measurements impossible. Quantum gravitational space-time geometries would fluctuate between flat, toroidal, saddle and other supermicro shapes.

This turbulent foam of frothing space-time micro-geometries would be riddled with micro-wormholes that multiply-interconnect these turbulent quantum geometries that suffer from the usual inherent quantum-measurement impression of fuzzy statistical wave functions (until a measurement is made). In our macro world and the universe's cosmic scale, things can be measurable and within the restrictions of chaos theory are predictable, and with quantum physics (excluding Bose-Einstein Condensation) operators only indirectly, statistically averaging out.

It is impossible to measure all of a specific particle's properties together at once, such as momentum, position, and Q-spin or quantum spin. Measure one property precisely and the other becomes a fuzzy-and-spread-out probability. All the information we can ever know is in the "wave function." Until you make a measurement, the particle's precise measurements are in a limbo. God in the quantum world plays hide-and-seek with our tools.

By contrast, in Relatively, there is no uncertainty. The metric or set of all shortest interconnecting paths can be determined precisely. But in the theory of quantum gravity, precision is impossible and the metric of the universe would fluctuate between different configurations. Quantum wormhole shortcuts, continually in a state of formation and termination, would create a probabilistic range of all possible paths between two points.

If this exists, then could future technology search, locate and select quantum wormholes between two desired points and enlarge and move them to where we want, thus creating an interstellar gateway, perform non-invasive surgery, even deep-mining the earth's core and of other planets, and maybe do black hole mining? Not very likely. But some luminaries like Kip Thorne theorize that some of this could become reality and that many stellar civilizations may be doing so. And others like Stephen Hawking remain skeptics. Does the ability of ball lightning to "defy" gravity augur a new tool to study gravity?

But before we seek to answer these important questions, we must first examine Artificial Electric Balls, failed theories, and even before that, examine the anatomy of this phenomenon through a powerful tool—statistics. Unfortunately, when misused, statistics do mislead.

20

Statistics

Statistics create mathematical anatomies. Here are fireball characteristics, qualities and properties. They are inexact, but for good reason. Since the mid-1800s, ball lightning researchers—notably Brand, Humphreys, Flamarion, and Dewan—collected cases and calculated statistics. Their results often differed from each other by over 30 percent, and the types of anatomy they construct differ. Recent databases, notably those collected in Russia, also diverge wildly. This disagreement is not surprising. Ball lightning is a fugitive from the law of averages.

Most people are not outdoors looking for fireballs during lightning storms. Thus sampling, as flawed as it is, is even more skewed than previous statisticians and collectors of data ever recognized. The observation and sampling is inherently inaccurate in so many ways that we could never even identify all the sources of error in observation and sampling.

For example, few observers are outside and running about atop mountains during fierce storms, where a disproportionate number of unusual, larger and long-lived fireballs appear, but are rarely reported—for good reason. And for even better reasons, even fewer observers are flying inside dangerous lightning storms looking for fireballs, where many more occur than previously thought. And ball lightning does not knock on the doors of field researchers to answer their questionnaires.

Characteristics

Individual fireballs are unpredictable. Exceptions exist. But as a group, they follow general rules and loosely exhibit certain characteristics. At the beginning of this book, we comprehensively listed attributes. But now, in greater detail, let us examine some qualities that emerge from the various statistical studies. As a composite, there are certain aspects that vary, as listed below. The characteristic that is most readily noticed and described is appearance.

Appearances. Shape, size and color constitute appearance. Most are spherical. A few are oblong or football shaped. Others resemble pear-shaped oblate spheroids. These lop-sided non-balls behave similar to spherical balls. Oblong or sausage-shapes are uncommon, but act differently, often slowly pulling themselves into a spherical shape. Surface tension may pull the sausage into a spherical shape, much like a balloon squeezed out of shape returning to a sphere when released. Torus- or doughnut-shapes might exist, but have not significantly been reported. Shapes include: halo, shell that is clear and distinct or foggy, short fat sausage, long sausage, rod, football, lopsided football, and pear-shaped.

One well-documented shape that is rarely seen is Siamese twin lightning balls, where two fireballs travel in unison, one atop the other, with a space sometimes of several feet between them. A glowing thread or filament can stretch between the balls. Rarely do small beads appear inside the filament, but when they do, there are two or three of them. Both balls float together, and often the bottom one drifts downward until the filament stretches thinner and disappears.

Pearl necklace or bead lightning is usually not considered to be true ball lightning. It is a string of balls that remains for a second or two in the channel of an otherwise normal lightning stroke after the stroke disappears. On occasion, some of these balls float free and drift to earth, but this is rare.

Distinct or fuzzy edges. Some fireballs look fuzzy, with small electrical-like discharges covering their surfaces. The descriptions differ. Some surfaces are coated with short fiery fuzz, others with long strands. But many possess smooth surfaces. One witness, Robert Brown, insisted that the fireballs he and other engineers repeatedly saw at Wright-Patterson Air Force Base had no fuzzy edges and was "as clear as butter." Some suggest these might possess less energy or are a different type, but of this there is no evidence. A single or double shell may concentrically surround some fireballs.

Occasionally, sparks or flames shoot from a ball, apparently from a hole that opens in a ball's shell. As this hole grows, the sparks also grow longer; and finally, after two seconds to half a minute, it explodes, sometimes abruptly and sometimes slowly in a muffled roar. Jagged sparks shoot out and anyone near it feels an electrical shock. Experts such as James Tuck of Los Alamos once suggested fireballs had an envelope with surface tension, but this is now believed not true.

Color varies. Although every color in the spectrum is reported, witnesses commonly report red, orange, yellow and white. Since nitrogen gas makes up four-fifths of earth's atmosphere, and nitrogen's characteristic color is red, perhaps there is some connection. Stationary balls that attach to objects sometimes glow bright and seem hotter. The same is true for balls that roll along objects such as gutters or roofs. Color blindness to some degree could affect the statistics, mostly for males, but no one has studied this.

Heat varies—sometimes cold or hot. While many witnesses report heat from nearby fireballs, others do not. Paradoxically, moments later that same ball that radiated no heat can melt metal, set a house on fire, or boil off a tank of water. After coming into contact with a fireball, some people are burned by heat. Results vary. Some are stung, others burned horribly and electrocuted; and a rare few, killed instantly.

Odor of ozone. Many witnesses compare the scent, when there is one, to nitrous oxide or ozone from a sparking motor or electrical deodorizer. Sometimes, after a fireball's passage, a faint smoke cloud lingers—a mist that may appear brown in transmitted light and white in moist air. Others smell nothing. Determining what gases are inside lightning balls provides clues to what occurs inside. Such measurements were taken in Russia in 1967 by chemist Dmitriev. Experienced in plasma physics, he analyzed his own sighting (described elsewhere) and estimated up to 30,000 roentgens of atomic radiation were needed to generate that much gas and that this fireball's temperature was 13,000 to 16,000 degrees Centigrade—assuming that the ball behaved conventionally.

Noise: Size loosely correlates with loudness. Even up close, they are often silent, but some hiss, sputter or sizzle. A few roar. Some sound like frying bacon. Many fireballs are inaudible, thus giving them an unnerving, eerie, unearthly, floating quality. Or up close, some may sound like a high frequency, and some like a sizzling high-voltage discharge.

Most balls vanish instantly without a sound. They "wink out." A few explode, some more violently than others. Among the loudest, a minority explodes like thunder, even on a sunny day. Smaller ballettes often pop like a cap pistol—but not always, and some midgets are loud and destructive. Larger fireballs that explode can do so with a prodigious roar. Some sizzle. Some that suddenly shoot out fiery rays, streamers or horns are more likely to start hissing or crackling when they do this.

Some fireballs spin, others roll. Balls that attach to rain gutters appear to roll along the surface. A few do not roll but instead slide. Scandinavians call such rolling fireballs *rullende lyne* or rolling lights. But grapefruit-size balls in free flight sometimes seem to spin or have turbulent internal motion. A few contain randomly moving tiny ballettes, beads, or other shapes. These internal lights have varying colors and shapes, and may move randomly, chaotically, or revolve with the larger ball that contains them. In some cases, sparks scintillate within the parent fireball. Some floating fireballs appear to spin.

Some fireballs change color, shape and size. Although they are the minority, it is not uncommon for certain fireballs to change size. Brand reported that those that grew smaller equaled in number those that grew larger. The change in size can be great or small, and its rate or speed of change can be slow or fast. This is puzzling.

If fireballs store energy, then shouldn't gradual dissipation of this energy make them gradually shrink? Do fireballs generate their own energy? Lighting balls that smash into the ground sometimes split into several smaller balls; and, in one case, split into seven smaller balls. In another case, a large fireball exploded, leaving a smaller glowing, luminous object behind.

Shapes sometime change. Most fireballs are spheroids. But other shapes occur. Sausage-shaped fireballs or "fire sausages" often gather themselves into a globe. This can occur after a ball squeezes itself through a keyhole, but some fire sausages occur as standalones. Some are oblate spheroids, others short and fat sausages, but a few are long, thin rods of fire. One ball continually changed back and forth between spherical and ellipsoidal.

Color changes occur. All colors are reported, including gray and the rare death-black, about which I am a skeptic. Blue can be faint or a beautiful sky blue or deep sky blue. In one instance, a fireball turned from red to white before exploding. German investigator William Brand detected no correlation between color change and anything else. Some report that the inner light was as-if filtered by a layer of radial plasma in an outer covering.

TheRussian SKB (Stakhanov-Bychkov-keul) Data Bank on Ball Lightning show that distribution of fireballs by color does not at all depend on location. And their analysis demonstrates no correlation between the color, lifetime and diameter.

Flight paths vary. Several flight patterns occur repeatedly. The fireball may float motionless, remain stationary and attach to an object or roll along that object, bounce along the floor or ground, fly in a straight or curved path, wander about, fall from clouds (often quickly, like a stone) and hit the ground and explode or bounce back to the clouds or level off and fly slowly. Some balls hit the ground and break into several smaller ones. Some rise slowly.

Others float above or occasionally below a line, but rarely through it like a bead on a string. They follow telephone and electrical lines, clotheslines, rain gutters and roofs. A few spiral upward around a person inside a room, then upon reaching head level will dart off out an open window or up a chimney and explode upon reaching outside. One witness touched the ball when it reached his neck level and the fireball exploded, hurling him backwards.

Floating fireballs generally avoid contact with objects, but some are attracted to enclosures, such as homes, which they enter by coming through chimneys, open windows, cracks, and even keyholes. After floating about, they vanish, explode, or leave the building through an opening and explode. Some balls burn holes in roofs. A few float through solid walls or doors.

Witnesses describe fireballs as possessing "intelligence" and moving as though they could "see." It isn't so. Anyone who wants to start a cult based on fireball intelligence cannot be stopped, but don't send me any of your literature or say you read it here.

Lightning balls are immune to strong winds and go where they please. For example, they squeeze through a crack or keyhole, and then move about the room, while carefully avoiding objects. However, at other times, they clumsily bump into objects and explode, or emit sparks with each collision.

Speed varies. Documented speeds vary from stationary to 400 miles per hour, and some may be supersonic. Lighting balls materialize beside airliners. But typical speed is far slower, about that of a man walking briskly. Balls that bounce across floors move faster than those that float, and generally, those with shorter lifespan move faster. Ball lightning that falls can do so quickly like a stone, then glide or float.

Origins and disappearance. Fireballs usually appear suddenly from thin air like magic. They often disappear suddenly. Many witnesses do not see fireballs form because they are looking in another direction. But once it catches their attention, they watch it carefully until its disappearance. Ball lightning can form when lighting hits nearby or within 300 feet. Beads can form in the lightning channel and drift free. It can form with a storm in the distance. Others appear in sunny weather. There is some connection with lightning, but lightning is often not nearby nor is it necessary. Witnesses say the balls form instantly.

The balls often disappear silently but some explode in a violent flash with an ear-splitting explosion—or go mildly like a cap pistol. Most balls explode suddenly, but some disintegrate in flames, shooting out sparks, in a process that can take several seconds or last a minute. At this instant, some nearby witnesses may feel shocks. If lightning strikes nearby or hits the ball, it instantly vanishes, also in a few cases, flaming pieces floated free.

Life spans vary. A few fireballs vanish in under a second. Others last three minutes. In rare cases, balls lasted fifteen minutes. Half of all fireballs last from one to four seconds; one quarter, five to fifteen seconds; and five percent survive beyond 17 seconds. Lightning balls that last longer than two minutes are rare.

Slanted Statistics Inevitable

Because collecting raw data and sampling methodology are unavoidably mediocre, all ball lightning statistics are mediocre. Witnesses fail to reliably sample appearances. For example, few stand outside on mountains during electrical storms and even fewer fly through storms. Some suggest we compensate by weighting the statistics. But no one can reliably predict what people are not seeing!

We are unaware of it and we fail to realize, but we only see a very small portion of our local environment. Think about how little you really see. Are you watching everything near you, behind you, in the other rooms? In your yard at every second? Simultaneously? Your roof? On the street? Well, if you are, you are a super-god. The truth is—you see less than one percent of your local environment.

Usually this makes no difference, because noting important happens that quickly behind our backs. However, many fireballs appear and vanish silently, lasting only a couple seconds, and so most go unnoticed. Even with witnesses around, they may not notice a fireball. For example, just ask Eddy K [listed elsewhere] of Taunton, who saw one inside Truesdale Hospital in Fall River, Massachusetts, and cried out, "Did you see that!" But in that half-second, his nearby friends were looking in other directions.

All statistics are mediocre—even those that were gathered from a small detailed minority of reliable reports. Detailed reports from skilled observers are desirable, but inadequate for a complete statistical study, and it is cumbersome and often impossible to accurately assign compensatory weighting to individual cases. As statisticians lament—"Garbage in, garbage out" applies to all ball lightning statistics.

In practice, mediocre reports tend to skew statistics. Good data submerges in the sea of bad data. Old data? Memories fade. Witnesses vary. It is difficult—even inherently impossible—to separate the missing and intertwined parameters in cases so that data can be accurately tabulated.

And even if we could get only perfect witness reports, our sampling would still be flawed. For example, how many people live atop mountains? Not many. And of those few at the top of mountains, are they more likely to be outdoors during storms? Not likely. So the statistics would under-report ball lighting appearances atop mountains. And how many witnesses are flying though storms clouds? Not many. These biases in sampling will skew all ball lightning data in many ways that it is very difficult to predict. Few observers see flying fireballs that occur at higher altitudes. There are so many other biases that researchers do not even know what to look for that is missing in the cases, let alone how to accurately compensate for it.

Also, even expert witnesses are likely to estimate differently. A trained police office, for example, will be more accurate at observation than a scientist. People who are partially color-blind will not report certain colors correctly. Some are hearing impaired. No questionnaires ask about this. And no one knows how much this skews statistics.

Do fireballs follow below or above aircraft? There is no way to tell because no one can see them close to aircraft. Do fireballs pass through the skin or cracks in aircraft? No one knows. Some slipped in through the side windows, and passengers saw these. Not all flew down the aisle, and a few flew in other directions within aircraft. It is difficult if not always impossible to separate electrical and thermal effects.

Data is only as good as the observer and investigator. And these results will be flawed because observers cannot accurately sample the cases. Thus we lack information and cannot accurately weight the different cases or their internal variables—or, if we could do so accurately at times, we would rarely know how accurate were our weighting errors.

To summarize, it is difficult to accurately categorize most cases, so each is relevant to more than one category or section. Rare are the cases that have all the categories. This leaves big holes in the statistics. The result is this: it obfuscates your statistics. Few adequately mention this dirty secret. Most are strangers to the truth. Second, because certain embarrassing facts do not fit existing physics, they tend to do what they do in the statistics to avoid facing the puzzling facts.

Vision Afflictions Affect Statistics

Color blindness affects some fireball witnesses. Colorblindness is usually a genetic condition that affects males more often, and has to do with color receptors (called cones) in our retina but may be caused in the cone or brain. No one is sure.

Colorblindness is a misnomer. Complete colorblindness is rare—and those victims suffer from poor visual acuity and great sensitivity to light. But most of us who are colorblind possess simple color-vision deficiency and have social troubles, such as mis-matching our clothes. We dare not shop alone for clothes or we risk selecting a purple shirt, rather than a blue one, or wearing green pants instead of brown. Matching ties is socially dangerous. We may select pink instead of white. And, yes, we make poor interior decorators and dangerous fashion designers.

An inability to distinguish between red and green colors is the most common colorblindness. It affects more males. Depending upon severity, estimates vary from 5 to12 percent, but for women only 0.4 to 1.5 percent. Colorblindness is inherited. Most colorblind people do not live in a black, gray and red world because it is a condition of degree, and many see a muted version of red and green.

Although the most frequent colorblindness involves distinguishing between green and red, sometimes it is not so simple. Sometimes such witnesses can tell the difference, but it depends on color intensity, distance to the object, and the brightness of the lighting. Many cannot distinguish a green traffic light from a streetlight, or a yellow signal from a red one. How frequent does this occur? How much does it alter the ball lightning statistics? No one knows. For ball lightning witnesses, only male witnesses (one in ten) might misreport colors.

Certainly, other factors other than colorblindness alter the perception of color. Occasionally, a fireball seen through a cloud in or near a tornado may be white, but appear pale blue through the cloud. Some tornado fireballs blink on and off, or go dark for a while.

Peripheral vision always has low resolution. You can read words within two centimeters from the center of your gaze. We are unaware that only one percent of each image that we see is received and processed at high resolution. We remain oblivious to this feature of our eyes because our eye movements are rapid and enable us to focus on what is of immediate interest. For a glowing fireball, however, peripheral vision is quite good because lights get attention. But glaucoma, which may be in only one eye, might impair detection. Outside of blindness, an ophthalmologist told me that detecting a 20-to150-watt fireball should be fairly easy and no problem. Having conducted tests with various lights and glass-frosted globes at various distances and angles, your author agrees with that ophthalmologist.

Psychology Alters Statistics

Psychology affects witnesses. Individual perceptions, variations in retinal persistence, psyche tache effects, optical illusions, perceptual effects and persistence of memory—these all are important, but difficult to investigate. Law enforcement knows a lot about witnesses' recollections.

Advocates of the Vallee Hypothesis and psychiatrists like John Mack study UFO abductees. Contactees experience an unknown internal phenomena—one that excludes external reality and has nothing to do with extraterrestrial craft. These may be true hallucinations of which medicine knows very little, and it deserves research. It is not our concern in this book.

From personal experience, I doubt if true hallucinations—that is, seeing things that are not there—play much part in ball lightning sightings. Your author suffered at least since age four from moots—annoying floaters or floating specks inside each eye—and since age fifteen from occasional ocular migraines that—although they last only fifteen minutes, occur rarely, and are harmless and painless—and may occur after intense study. They are considered ocular "hallucinations." He would see floating black lights and jagged lightning fixed in place, but that slowly drift and change shape. Each eye sees a different version. Although rare, they are annoying—but he never mistakes them for reality.

Then ten years ago, after a day of intense mountain winter hiking, a curved light began zipping across my upper right eye in semicircular arcs—at first assumed that were gnats (in winter?)—but they continued for six months. An ophthalmologist diagnosed it as detached aqueous vitreous, considered harmless, and after a year it vanished. But never did I "see things," or mistake anything.

I talked to two ophthalmologists on optical illusions and how witnesses see ball lightning. Your author attempted to duplicate various optical illusions, such as looking at light reflections in glass panes, to try and see if witnesses make mistakes. From this informal research, he concluded that few fireball witnesses make mistakes.

Does fraud affect statistics? Very little. Unlike flying saucer sightings, which have a truly sensationalistic aspect, ball lightning is mildly interesting—nothing more—and false reports are rare. For example, CSICOP skeptic investigators Phillip Klass and Martin Gardner justifiably excoriate flying saucer sightings, but never those of ball lightning. There is little thrill from lying about ball lightning sightings. No one cares. And, witnesses are unlikely to be mistaken about ball lightning.

However, this does not mean that ball lighting has totally escaped the attention of inventive minds, especially on the Internet. Today, it is certain that the ease of making uninvestigated reports to web sites attracts some fraud. Be skeptical of any report with anomalous qualities that are too different—such as optical black hole fireballs. Do not totally dismiss anything, but question everything.

The danger is that dishonest and mistaken reports may describe false qualities and contaminate statistics.

Statistics Paint Spongy Portraits

Most ball lightning statistics are spongy. Values vary, sometimes widely. This impreciseness is caused by several reasons, much of it inherent in the way sightings occur. No one questions the integrity of the methodology of gathering and mining data. Observations are imprecise, memories fallible, interviews rare. Most cases are amorphous jello-like blobs, lacking solidity and filled with missing information—that is, full of "data holes." Interpreting data for ball lightning statistics is like making a stew—messy.

There is an unhappy story of a graduate MIT student who attempted to categorize the qualities of ball lightning cases into a computerized database. The idea was good, but the implementation was a disaster. It was like nailing jello to a tree. He quit. I attempted to do so for the McNally study. I quit. Attempting preciseness where there is only imprecise data only gives an illusion of preciseness. Never claim greater preciseness than your data justifies.

Why do detailed fireball databases and statistics provide mixed results? For several reasons. Most eyewitnesses are not expert observers. Few saw the entire event. Most have poor memories. Furthermore, eyewitnesses' ability to estimate accurately varies widely, and there is no way to determine this, let alone compensate or normalize the data. So, if a researcher keeps his categories wide and loose, the results will be more inclusive. But because his categories are so inclusive and broad, they will not provide enough detailed information or resolution. But, on the other hand, if he makes his categories smaller to provide better resolution, he does just the opposite because the uncertainty rises, sometimes dramatically. Worse, the gathering of cases—even at Oak Ridge National Laboratories with nuclear scientists—provides differing results for many reasons.

Sometimes investigators (all of them at times) were insufficiently thorough, and data was missing—often due to no responses to certain questions—and nothing now can be done to remedy this. Second, witnesses often couch their descriptions in qualifying phrases. Finally, unless the witness was an expert estimator, such as a surveyor or police sharpshooter, then he was unlikely to make accurate estimates.

Most statistics loosely agree. Interpolated—a fancy word that here means "eyeballed"—from prior studies, the following rough statistics are typical.

Speeds vary. Usually, fireballs travel at low speeds, of a man walking briskly (2-4 mph). In a few cases, fireballs fell from the clouds faster than a stone. Some travel at 400 mph very close beside aircraft. A few seemed supersonic, albeit silent.

As for **demise**, or how they leave, it varies. Vanished silently, 80%; exploded violently, 15%; exploded softly, 2%; shot out sparks before vanishing, 2%.

Shape: ball-shaped, 90%; donut (torus) shaped, 0%; oval oblong pear, 5%; other or undetermined, 5%.

Color: red orange yellow, 85%; white, 10%; green purple blue, 5%; black, rare.

Size: under 3 inches, 20%; 3-6", 40%; 6-12", 25%; 1-3 feet, 10%; larger than 3 feet, rare.

Lifespan: under 1 second, 5%; 1-2 sec., 30%; 3-4 sec., 30%; 5-8 sec., 10%; 9-16 sec., 10%; 17-120 sec., 5%; longer than 2 minutes, 5%.

Flight path: stationary 10%, horizontal 60%, vertical (up or down) 5%, complex flight, 25%. Did it float? Airborne 90%, attached 5%.

Any successful theory (as repeated earlier) must account for all of many contradictory properties. To partly reword a few properties, they are—(1) **Propulsion**: ability to float against a strong wind or follow closely beside an aircraft in flight and defy gravity in a controlled way. (2) **Longevity**: frequently lasting a minute or more. (3) **Shape**: mostly spherical; some ellipsoidal, rod- or pear-shaped. Edges fuzzy with discharges or clear smooth edges. Height varies from 1/8th inch to one inch. (4) **Changes**: size or color changes are not rare.

(5) **Brightness**: some are dim; others, bright; a few, dazzling so intense that witnesses cannot look at them. (6) **Bouncing**: along floors, street, grass and rarely off walls or trees. Rarely passes directly through walls and doors. Rarely a momentary flat spot or deformation appears on the ball upon each rebound. (7) **Submerged**: Falling into a rain barrel, glowing when submerged and evaporating steam. Rarely do they submerge into a lake or sea. (8) **Energy content**: contains far more energy than singly ionized air, but energy manifestations oddly vary—behaving like a lamb one second and a lion the next. (9) **Openings**: apparent holes open in some balls and sparks or flames shoot out.

(10) **Slipping: through keyholes**, solid walls, and doors, ungrounded metal screens, and glass-sometimes without apparent effect or damage, but at other times puncturing or melting holes in glass and metal. (11) **Spinning**: internal

turning or turbulence of some sort and some contain tiny ballettes that move about chaotically. (12) **Flight path**: seem to control gravity. Exhibits unusual and deliberate movement; rarely a random balloon-type drifting. Flies against the wind and is totally unaffected by strong winds. It may seek enclosures such as entering a chimney or window, roll along lines or gutters, spiral ascent around persons in a room, explode upon leaving an enclosure, and fall from and rebound to the clouds. (13) **Colors**: the entire spectrum—red, orange, yellow and white being most common and black being very rare. (14) Usually **not attracted by lightning rods** or wires, but when it does enter a cable or conductor, it creates a huge current. (15) **Weather**: generally during lightning storms and sometimes with a nearby lightning stroke, but this is not necessary and this association is not as great as some researchers allege. In fact, contrary to belief, most stormy fireballs do not occur near a lightning flash. They also occasionally materialize during sunny or cold weather.

There is a problem. Properties are confirmed by many cases. Any accurate theory must explain every property. All theories fail. And, you will soon discover that exceptions exist. Trying to nail down the elusive qualities of true ball lightning into neat categories without exceptions is more frustrating than trying to nail jello to a tree. No matter what laws you derive from your cases, you will soon come across new cases that contradict your generalizations. It is impossible to make sense, in terms of accepted physics, of these contradictory qualities. Ball lightning researchers who value their reputations ignore these contradictions.

Here is a final warning on gathering your statistics. You can compile and list all the field data that you collect, and take great care to select representative samples. Yet, after all that costly expense in resources and time, you will still come up with a mountain of pyrite—that is, fool's gold—and results not much more accurate than if you had invested only a modest amount of effort in collecting, sampling and interpretation. As we discussed earlier, this is because of "holes" in the data and inaccuracies inherit in observation and collecting data. Actually, your statistics may be grossly inaccurate (skewed) even if you spend a fortune in data collection and interpretation. Remember, witnesses see ball lightning accidentally. Few people go about in the middle of storms looking for fireballs.

What people rarely see, even if it is common, they will rarely report, and we will falsely conclude that it is rare. And this error skews the statistics. For example, few statistics consider mountain and cloud-to-cloud or to-ground or inter-cloud fireballs. The reason is simple—few cases are reported. Fireballs do float between clouds.

Most fireballs are dim—no brighter than a hundred-watt bulb—so they could never be visible from the ground or an airliner. So, those that appear inside the storm clouds could never be seen. Balls falling from clouds generally do not fall, but can plummet like a falling cannonball, so unless someone is looking directly at it, they would never see it. Upon reaching ground level such balls break up or level out into a slow horizontal flight, or occasionally bounce back up into the clouds. Occasionally, a falling fireball will split into two or three small lightning balls that float about.

Customize Questionnaires

In 1975, your author devised a four-page questionnaire. Before doing so, he first collected many possible categories and then selected questions. If you decide to gather cases, first create your questionnaire. Here are some candidate topics for your questionnaires. I list them without any order. Keep in mind, if you select too many questions, then witnesses will refuse to complete your questionnaires, or be confused and make mistakes.

Here are candidates for your own questionnaire. Arrange your selections in any way you like and select formatting and layout to input the data into your data software to make it visually appealing and simple, and minimize your difficulties in later deciphering some mysterious handwriting. Once again, the following are not iconoclastic iron rules, sacredly handed down and chiseled in stone tablets from God, but only suggestions for your perusal, as follows.

Damage. Injury, deaths to people and animals…Fireplace…Skips/bounces…Avoids/seeks contact…Tornados/earthquakes…Origination—light bulb, wall outlet…Transformer spews…Fuse panel…Melts wire, fuse, metal, and cable…Did it penetrate a metal or plastic screen, single- or multi-pane glass (which has central plastic sheet), with or without leaving any damage or evidence, such as a hole—which can be melted edges or clear-cut without melting—in the glass? How big was the hole? Round or oval? Did it crack the glass?…Ball grew larger or smaller…Trailed a tail and length…Did the given quality change once or more…Melted or fused wire or metal…Phone (type and location) emitted or spit out ball…Rolled down/up steps…Floated or drifted up/down/sideways…Glided up or down…In or near or enters or exits clouds…

On or in ships at port or sea, and balls above sea…Multipliers (spawners and dividers)…Line roller/slider…Ground or floor roller…Gutter or roof roller…Fence roller…Hisses or spits sparks…Railroad tracks…Mother ball broke into smaller ballettes...Glowing fog fire...Shape (ball, football, sausage, pear, cube...Tails...Enter/exit...Passed through screen, glass pane (type), wall...Melted something...Shrunk to zero...Location (inside enclosure or building, outside)...

Damage (physical, heat, scorch, melt, move something)...Size (static and unchanging, grows or shrinks at start or end)...Color (mix?)...Intensity/brightness (dazzling, so bright could not look at it...Surface (smooth, fuzzy, small or large fuzz)...Halo with clear or fuzzy edges...Size...Lines unattached, phone, electrical, high-tension, clothesline...Draw diagrams. Measure or estimate dimensions...Ball got larger/smaller...Sounds (silent, hisses softly or loudly, growls)

Explosion (pops, mild poof, mild, loud, very loud)...Odor (ozone, sulfur, odorless)...Captured on film or VCR...How did it move?...Slid or rolled...Rotation/spinning...Internal motion...Falls from cloud...Rolled on street, sloped or flat roof, gutter, floor...Ground, grass or street was wet...Saw through a fuzzy/clear window...Did you see BL? Who else did?...BL qualities/ categories...

Draw and label crude rough drawing sketch pen & ink and map...If you do not have color pens, indicate colors...Motion...Location...Attached...Origin—actual/observed...Demise—actual/observed...Demise: Did it become a current surge, explode mildly or violently, divide, shoot out horns and jets, wink out, leave field of vision, otherwise did not see its demise, faded away...Fell/rose rapidly...Pulled out of a power dive and flew level and slow....Unaffected/ affected by wind...Wind speed

To determine wind speeds, you might insert the following Beaufort Wind Scale into your questionnaires and ask witnesses to select wind speed. The following Beaufort Wind Scale lists—Wind Force categories of 0-12, followed by (speed in mph), and then a description.

WF 0 (under 1 mph) calm/smoke rises vertically

WF 1 (1-3 mph) light air/smoke shows direction of wind, but wind vanes fail to move

WF 2 (4-7 mph) light breeze/leaves rustle, wind vanes move; wind felt on face

WF 3 (8-12 mph) gentle breeze/leaves and twigs move constantly, and light flags move

WF 4 (13-18 mph) moderate breeze/dust and loose paper flow about, and small branches move

WF 5 (19-24 mph) fresh breeze/small trees with leaves sway

WF 6 (25-31 mph) strong breeze/large branches move and difficult to use umbrellas

WF 7 (32-38 mph) moderate gale/whole trees move and pressure when walking into wind

WF 8 (39-46 mph) fresh gale/twigs break from trees and difficult to walk

WF 9 (47-54 mph) strong gale/chimney pots and slates blown from roofs and slight damage to structures

WF 10 (55-63 mph) whole gale/considerable damage to buildings and trees uprooted

WF 11 (64-75 mph) storm/widespread damage

WF 12 (over 75 mph) hurricane/devastation

Remember, ask witnesses the famous 5W and 1H questions—who, what, when, where, why and how—of course, not necessarily in that order. Modify and repeat them in different contexts.

If you make field visits, get estimates and take measurements. What was the wind direction and speed? Practice estimating distances and heights, then measure them, and you will get good with practice.

Most cases lack sufficient details for reasons you cannot avoid. In many cases, time has elapsed and witnesses may apologize for their memory. Some are willing to tell you but not fill out questionnaires. Others may prefer to write detailed accounts. There are so many variations—far more than you imagine when you begin gathering cases—and you will soon discover that no one can convey such complex experiences with the written language. You will face a conundrum. To create greater complexity in your classification systems beyond a certain critical point only creates confusion that discourages witnesses and trips them up. You do not want to encourage errors, especially those that you cannot verify.

Any classification system—no matter how large, complex and well designed or crafted is incomplete. A complete classification system would be so huge and unwieldy as to create confusion and discouragement. And, chances are it would still have holes that need patching.

Avoid confusion. It may be clear to you, but what about the observers? Be careful how you design and word your questions. Do not ask, "Was it a surface, volume, doughnut, or ring type of glow?" Few witnesses know what that means. Their answers will reflect their confusion and you will later be even more confused.

Ask few imprecise or fuzzy questions, such as "Did it last long?" or "Did it disappear quickly or gradually?" or "Was it fast or slow?" One exasperated witness wrote in response: "Two seconds, but if that is fast or slow I will leave up to you." Carry a color wheel or Pantone Letraset Color Paper Picker strips—there are 500 color selections for witnesses to select.

Tape record witnesses. Transcribing tapes is hell. Most journalists make loose written transcription from their tapes. Unfortunately, convenient speech recognition remains cumbersome for personal use and impossible for transcribing field or phone interviews. Which is fine, because the way witnesses speak will not resemble anything you want in print.

Shorthand helps. Try Pitman or Gregg. (Your author studied Gregg.) If you become a fast transcriber, then you may want to try the older versions. Alphabet shorthand systems are easier to learn, but on the other hand, they are slower. They may exist, but your author is unaware of handwriting-recognition "electronic notebooks" or optical character recognition scanners that recognize shorthand and automatically convert your shorthand "outlines" directly into alphanumeric ASCII. This would be easy for electronics to recognize the simple dots, lines, circles and curves of Pitman and Gregg and convert it into ASCII and store it. Or you can read your shorthand into speech recognition software—they are getting better.

To locate ideas for my questionnaire, I located an Air Force questionnaire used for anomalous unidentified aircraft sightings and significantly modified it. I tried to invent a classification system, but gave up. When I crafted my questionnaire, I recognized that no one knew all the characteristics of the phenomena and that my questionnaire would remain incomplete. So, I crafted it to be flexible and partly open-ended to catch these yet-unknown qualities. I anticipated that I could later key the witnesses' responses into a computer data base, but soon recognized this would be difficult to do correctly—and would be too loose—because of the numerous "data holes" or missing data and due to my limited time.

You will discover some witness penmanship is truly mysterious. Get a good pair of magnifying eyeglasses, a handheld magnifier, and a printer's magnifying lens. You will need them. Facing terrible handwriting, you should first read it through once slowly to get the gist, then go back and decipher the cryptic words. Pencil in the words as you decipher them. It just takes time. Sometimes, while deciphering the Oak Ridge questionnaires, I felt like Champollion deciphering Egyptian hieroglyphics. Have a sense of humor. You will need it.

As we just saw, ball lightning statistics are controversial because they are inherently imprecise. Statistics aside, the most controversial scientist in ball lightning is Nikola Tesla, who at Colorado Springs and Shoreham attempted to precisely resonate the earth with Zenneck waves and then also generated Tesla Artificial Electric Balls.

21

Tesla Balls

In early 1972, your author first witnessed a ten-foot, three-coil Magnifier, whipping out continuous, six-million-volt, lightning. In 1972, this Magnifier was one-fifth the size of the giant later operated by Robert Golka in Wendover and Leadville. There was a fourth coil—a single-turn "reservoir coil" atop the ten-foot secondary to suppress corona.

Tesla suppressed electromagnetic emissions, which suffer from the inverse square law of dissipation, and was electrically resonating with Zenneck "surface" waves, at the earth's natural resonant frequency of pulsed 8-hertz. Tesla's Magnifier rectified between 12.5 million volts at 1,200 amps and 20 million volts at 2,400 amps, and low frequencies between 20 kHz and 150 kHz, into 8 hertz and higher multiples of the earth cavity's natural fundamental.

Later in 1985, James Corum speculated that since Tesla was a foremost expert in x-rays, he might have taken an indirect route to charge the metal ball atop the tall vertical mast. At Shoreham in Long Island, he placed a four-foot-diameter opening at the top of the hemisphere. Did a huge, single-electrode x-ray machine of unusual configuration inside generate x-rays that shot upward, perhaps to charge the dome?

Tesla was an x-ray pioneer. Earlier, he x-rayed himself at forty feet! Later, noticing the harmful effects, he discontinued this. Others were not so lucky. And Tesla may have been first to discover x-rays, but—partly because of delays caused by a fire in his Houston Street Lab in New York—he failed to identify x-rays, so lost the Nobel Prize to Roentgen. Tesla was x-raying himself at up to 40 feet, but after suffering injuries, he discontinued. The first x-ray death was Clarence Dally at Edison's laboratory from cancer in 1904.

As described by Lesland Anderson and Corum, Tesla pioneered a large, single-electrode x-ray tube using the Bremmstralung process, using a unique configuration, and patented it. Investigating single terminal incandescent lamps, Tesla investigated x-rays with advanced tubes that lacked large electrodes but created x-rays through vacuum high-field emission or Bremsstralung (slowing down or breaking radiation).

This is the high-energy secondary x-ray emission that is generated when a fast-charged particle (electron) passes through matter, so that energy lost by the decelerating particle emits as electromagnetic radiation, and the primary particle is absorbed. The released energy is highly penetrating. Question—Did Tesla use these x-ray-discharges to trigger ultra-high voltages by charging the capacitor at rf rates, then discharge them at ELF rates?

Unexplored Territory

In 1970, seeking an ideal location for a 51-foot Magnifier tri-coil, Golka suspected that the soil should have low receptivity in ohms per meter, which is the reciprocal of conductance. He examined data and discovered that sea water had the highest and thus best conductivity in mhos per meter, with the highest dielectric constant. Marshland that was forested and flat was almost as good at eight milli-mhos, as was fresh water. Pastoral land of medium hills and forest did well. Industrial cities did poorly. He traveled the country, taking measurements and collected soil samples. At this time, your author studied soil receptivity and the chemical and electrical properties from nuclear bomb test research and conducted electrical tests at various frequency spectrums to develop equivalent circuits to model the soil over those ranges.

In the summer of 1973, a full-scale three-coil-Magnifier was erected on the open Bonneville Salt Flats. The bone-dry summer desert landscape, good ground conductance, and isolation all made it a good choice. And it rarely rained, so the giant Magnifier coils could remain outdoors. This would dramatically reduce costs and dissipate dangerous ozone. Disaster struck. Torrents drenched the desert. An unprecedented deluge left the 51-foot Magnifier immersed in a one-foot-deep, salt lake that corroded metal, ruined tools, degraded insulation, sunk hope, and flared tempers.

When he spoke at Brigham Young University, one attendee was a physics professor—the inventor of the Topolotron, a Tokomak-like fusion reactor—Dr. Robert Bass. He joined Project Tesla and modified Tesla's Artificial Electric Ball model and used this math-heavy paper as his basis for a proposal to construct a prototype fusion reactor that Bass named the *Pyrotron*. But the proposal never received a Federal grant.

In 1974, he selected an unused Air Force hanger at the deactivated Wendover base on the east side of Utah. By coincidence, 30 years earlier, the B-29 Enola

Gay was fitted with the first atomic bomb (it was assembled here in three parts) before beginning its long flight to Hiroshima and into the destiny of history.

This time, hopefully, a new age of low-cost and pollution-free unlimited fusion energy for world peace would be born here, and most swords of war could then be melted into plows for peace. Hopes were high; enthusiasm, even higher. But behind the scenes, hidden and casting an invisible dark portent seeming less than threat, cruel destiny secretly laughed and soon stalled progress.

Ball lightning at that time was believed by leading physicists to be a stable plasma of ionized gas—and just might provide clues to the first fusion reactor—with unlimited near-free fuel of deuterium and tritium, extracted from water. At that time, two gallons of water would provide all electrical needs for Boston for 14 minutes! For 45 cents! Whoever succeeded and patented this fusion reactor could easily become the wealthiest man in history.

In this historic hanger, he succeeded in recreating Tesla's three-coil Magnifier. Although Tesla had trouble suppressing fireballs, without access to all of the secret notes, Golka had trouble creating electric balls. In some film frames, electric balls appeared briefly, but only when ten-feet vertical streaks of blue discharges appeared.

Declared surplus in 1976 and turned over to the town and later made into a municipal airport (Decker Field), the former historic base ceased to exist. Later, he stored his equipment in Butte, Montana, and then had it shipped to an unused industrial mining building in Leadville, Colorado, just 85 miles northwest of Colorado Springs. Suspecting that the lower altitude of Utah was the problem (it wasn't), he moved to the higher altitude of 14,000 feet and its proximity to Tesla's 1899 site.

Several individuals informally joined Project Tesla. Dr. James Corum, who became immersed in Tesla's research, visited the Tesla Museum in Beograd, in Serbia, and constructed his own coils of smaller dimension.

For two summers in Leadville, Golka tried to resonate the earth with Zenneck Waves. He pursued x-ray and rectification aspects in an attempt to convert the 50-kilohertz frequency discharges into 8-hertz extra-low frequencies. He erected a tall outdoor mast.

In late 1989, the carborundum mine was unexpectedly restarted and he had to vacate the large building. He shipped his equipment to a warehouse in Kentucky. Funding was always on a shoestring budget, with grants and funding, personal financing and donated equipment, and volunteers keeping the project going.

Helical Resonator

While flying back home, the Corum's read an article on pages 13-15 and Volume 7, number 3 in the *TCBA News*, and discovered Tesla's secret. They tried much smaller coils to generate Tesla Artificial Electric Balls. The secret? A second or dual resonator in the primary. Then by running one percent of the Colorado Springs power, they created small electric balls of 1.5-inch diameters. By November of 1988, they succeeded.

Your author examined many engineering patents, papers, speeches and lectures, then totally rewrote and re-edited them, and then added his own observations—partly as follows. The Corum's discovered two techniques that generated electric balls. First, a classical Tesla Coil configuration used two resonators. In the second, they discovered the smaller resonator and used the other lower frequency resonator as the secondary of a spark gap triggered by an LC-tuned oscillator.

Classical Tesla coil lumped-circuit analysis yields inaccurate results—bad for small coils, awful for large ones—so that errors grow geometrically with the coil sizes. Golka and the Corums discovered this when they began coil building or "coiling." They discarded earlier analyses for the giant coils. At first, they used crude calculations and trial-and-error that gave them the experience to make rough estimates, to build the equipment, try it, then refine it to interactively converge on the optimum circuit—certainly not an efficient procedure. However tedious, this procedure worked—with it Golka successfully created a ten-foot coil and later duplicated Tesla's Colorado Springs fifty-foot resonator coil at Wendover in the mid-1970s.

By the mid-1980s, the Corums, after scrutinizing Tesla's patents, realized that a distributed circuit analysis was needed and that the Extra Coil behaves like a slow-wave open-helix resonator. Indeed, Tesla patented the quarter-wave resonator in the mid-1890s, so he was quite familiar with this by 1899. By using Master Oscillator, he drove the Extra Coil, an open quarter-wave helical resonator with a high VSWR that reached 12.5 million volts at 1,100 amps, continuous wave discharges, with power delivery rates unequaled until only recently.

In the 1986 International Tesla Symposium at Colorado Springs, the Corums in a historic paper made their revelations. Their solution, once explained, became immediately obvious to everyone with knowledge of electromagnetics and transmission lines.

Most descriptions of the classical Tesla Coil ignore the distributed nature of resonators and erroneously treat them as lumped circuitry, ignoring that the current and voltage distribution of the resonator for a quarter-wave sinusoid with its maximum and minimum at its base, and the reverse at its top. The Corum's used characteristic impedance and slow wave transmission line theory to create their model and applied them with accurate results to Tesla's 1899 results and to those of modern coil builders or "coilers."

Many coilers attempt high voltages to drive devices that generate high-current relativistic electron beams. Using their distributed parameters model, the Corums accurately predicted the waveforms reported and discovered that these modern coilers fail to reach optimum performance. Their spark durations are too short for the coil coupling coefficients. Unlike Marx, Van de Graff and other generators, Tesla's equipment not only delivers high energy, but also allows heavy-duty cycles—high repetitive rates and greater power throughput.

The heart of the Tesla Electric Ball Generator (EBG) was a quarter-wavelength-tall, slow-wave helical resonator coil. Magnetically coupled to a high-peak-power (70kW at 67kHz) spark gap oscillator, the Corum's resonator tightly coupled, as did Tesla. Using only one percent of the power that Tesla's EBG used, Corum's EBG received 3.2kW at its high-voltage electrode to yield 7.5-meter RF discharges. Although Golka's earlier EBG was a hundred times more powerful and a duplicate of Tesla's Colorado Springs coils, it was the Corums' resonators that would first repeatedly create numerous TAEBs because of this operation.

Spark gap frequency was 800 pulses per second or 48 kilopulses per minute; duration, about 100 microseconds. With a 72-mircosecond on the secondary, the induced incoherent polychromatic oscillations require an equal time to rise and set up a standing voltage and maximum voltage atop the resonator.

Artificial Electric Balls

At the top of the resonator, thick insulated copper wire—specifically, the tip of a rubber-coated cable or wire number 10 to enhance pumping the discharge [*CSN* pp. 173-4]—there is mounted a pointed carbon electrode or short wire point atop the high voltage electrode which discharges, suddenly releasing the resonator's energy and delivering the sudden and powerful discharge required to produce lighting balls. Like Forrest and Witten's diffusion-limited aggregation or DLA, this method vaporizes carbon and metal particles from the rubber and wire in burst to form the DLAs. The condensed cupric dioxide vapor forms an aerogel much as does silicon dioxide.

Tesla's primitive rubber insulation of 1899 was even more effective—it contained significant carbon. Attaching a candle atop the high voltage resonator increases the electric balls. Smirnov states that creating a porous fractal bubble is significant to create an electric ball. Soot in candle flames forms fractal structures. The Corum's believe that Tesla's apparatus created massive quantities of ozone and other reactants and gaseous chemicals rapidly absorbed into the charged porous fractal bubbles and the high plasma temperature initiates the multi-step combustion.

Using a dual resonator as Tesla did, they generated swarms of various-sized—millimeters to several centimeters diameter—Tesla balls that lasted from a half to several seconds. They photographed Tesla Artificial Electric Balls on various media. Like Tesla, they had difficulty photographing them, but made video recordings. In some of the films that your author examined, the numerous tiny ballettes look like swarms.

From examining the photographs, your author suspects that Tesla balls are very spherical. This suggests strong surface tension in formation. Limb darkening and near-solid appearance suggest that they are opaque. Other photos show a few sparks striking balls. In one photo, two streamers hit both twin balls.

With the dual resonators, the balls appeared close to the high voltage resonator and shoot out from the discharge streamer, darting down and up. This is not bead lightning, as some critics might allege, because bead lightning forms when the spark or lightning channel breaks up into short beads—or sometimes sausage-like fragments. Occasionally such beads or glowing sausages float free, but this is uncommon.

From examining video recordings, they determined that microballs (6 mm) originate near the electrode and then are struck by streamers, and then glow larger. After being struck by the streamer that quickly disappears, the ball would float there and be hit by the next streamer and glow larger. In one case, a microball was hit six times, and grow in size after each strike. A tiny six-millimeter microball can grew to a five-cm ballette in one second.

Some of these seem to spin, with "dark patches like sunspots." Some look transparent; others change color and burst. A line of balls, sometimes a dozen, may issue from the terminal and then be struck by a discharge that briefly connects the balls. On video frames, "ballettes" materialize between frames, and on successive frames, discharges strike them.

Some photographs show several in the channel. Since the spark gap oscillator operated at 800 breaks per second, many discharges pulsated each second, jumping between and striking the balls in sequence during the shutter's operating time. This optical illusion on film gave the false impression of streamers simultaneously giving birth to electric balls.

To "pump the spark," Tesla used a rubber-covered #10 wire. When this was replaced with a charred wood dowel, and using ASA 400 film with a slow quarter-second shutter speed, the Corums saw a row of a dozen electric balls hit by one streamer.

An amateur astronomer, James Corum compared TAEBs and star formation. Was Corum slyly suggesting investigating a possible dark horse key to fusion? Totally rewritten, and with substantial additions by your author, the analogy is beautiful.

The famous oval, egg-shaped Ring Nebula M57—the showpiece of all planetary nebulae—in the small-but-prominent constellation Lyra, west of Cygnus—forms a thin red outer and inner yellow shell or halo, with a blue center that resembles a blue egg yolk in a yellow fluid with a red shell encircling its parent, a hot sub-luminous dwarf star. This "smoke ring" is actually a shell. It is so thin that several stars behind it shine through it. Intense ultraviolet light from the dwarf stimulates the ejected shell into luminescence. This ultraviolet penetrates the ejected gas; and different frequencies are absorbed, indicating that different regions of the ionized gas are excited to different energies to fluorescence. This causes color stratification.

When a moderate-sized star, up to forty percent larger than our sun, burns up its hydrogen, it greatly swells up into a red giant; and then when the core collapses, it squeezes the center. This generates heat that escapes to the outer layers, igniting a new thermonuclear reaction that blows off its surface into a shell that we see as a ring around the collapsed core of a white dwarf, now shrunk far smaller than the sun. What is left is a planetary nebula.

Through video signal processing and enhancement, they saw what they thought were weak luminosity variations faintly visible across the balls. A glowing shell surrounded some, and James Corum (an amateur astronomer) wrote that one resembled the Ring Nebula (M-57) in the constellation of Lyra. Such luminosity of the envelope otherwise only occurs with the extremely hot O- and B-type stars. He spent much time comparing the formation, life and extinction of Tesla's electric balls to stars in the draft of a paper of his that I read before it was published. With this in mind, we need to examine stars.

Stars differ in brightness and color. Next time you are outside in the countryside, look up. Although most stars seem silverish, look closer and you will see some stars faintly tinged with a bluish, yellowish, reddish, and even greenish color. When the orange Arcturus and blue Vega are in the sky at the same time, you will notice their distinct contrast. Sirius (Alpha Canis Majoris) is white; Rigel

(Beta Orionis), blue; our sun, yellow; Albebaran (Alpha Tauri), organe; and Betelgeuse (Alpha Orionis), red.

Because stars have different surface temperatures, they have different colors—blue is hotter than white, which is hotter than yellow, in turn hotter than orange, and in turn hotter than red. Each color has its own spectrum and is classified into seven spectral classes of OBAFGKM—and sometimes WRN—from blue to red. The sun is a G-2 star.

The Hertzsprung-Russell (H-R) diagram is an x-y graph—with absolute magnitude along the vertical axis and spatial type or surface temperature in Kelvin on the horizontal axis—that plots each star as a point on the H-R graph. These scattergrams always occur when stars are plotted on log-log double-logarithmic graph paper, with their absolute brightness on the vertical axis and their spectral class on the horizontal axis. Like several swarms of tiny gnats, coordinate points are dots that cluster into seven spectral classes.

The super giants cluster along the top; the giants to the upper right, the white dwarfs to the lower left, the red dwarfs to the lower right, and the vast majority of dots from main-sequence stars form the upper left diagonal down to the lower right. The dots on the H-R scattergram cluster into certain regions and not others, and the shape resembles the distinctive shape of a giant propeller. Stars farther away look dim, even if they are very bright. To equalize or normalize them, to make them "equivalent" to make a fair comparison, absolute magnitude compares the brightness of stars as if they were all placed ten parsecs from earth. (Linked to parallax measurements, one parsec is 3.26 light-years.)

Tesla balls begin as red dwarfs, then morph through different colors and sizes, finally to arrive at the blue-white giant stage—then explode like a nova or cool back to the blue-white giant stage. (Such colorful, beautiful analogies of visual imagery!) They compared the Tesla balls to the Great Nebula in the constellation Orion, M42, which hangs in Orion's belt, but is barely visible as a dim patch of pearly light and in a region where stars are being born, about 1,000 light years away. This is strictly not correct, of course, because it would be different if it were only at 10 parsecs from earth, it would occupy a large section of the sky.

I still wonder—was James Corum suggesting that Tesla balls could provide a key to stable plasma confinement?

Secret of Multiple Coils

Tesla also used an attenuative method. He frequently used one primary coil to drive several helical (slow-wave transmission line) resonator coils. In the unpublished Chapter 34 of John J. O'Neill's biography of Tesla, Tesla stated that lightning balls occurred when two frequencies—a strong higher one is imposed on the lower frequency wave of the main circuit. This drives the lower frequency with a longer wavelength to near-instantaneous discharge. This sudden and enormous rate of energy flow cannot remain with their wires and is abruptly discharged with incredible violence to the surrounding space.

Tesla used several synchronous oscillators for a frequency-hopping spread-spectrum communications system, which he patented. The use of such multiple coils is evident in the photographs in Tesla's Colorado Springs Diary laboratory notes. From reading O'Neill's unpublished chapter, it became apparent. By following these instructions, the re-wired two synchronously excited coils (resonators) at 67 kHz and 156 kz, yielding 2.4 MV and 200 kV, respectively. While the low-frequency coil took 72 microseconds to build to its peak, the higher-frequency and smaller coil builds up to VSWR or Voltage Standing Wave Ratio distribution to a lower peak, but does so more quickly.

The smaller coil arcs to the top of the larger coil. The larger coil looks out at its load, and though it otherwise sees high impedance, it now suddenly sees a rapid drop of its load impedance. The larger coil then rapidly dumps its standing wave energy into the reduced load impedance, which is at the arc. The large coil's energy bursts out of the top electrode (or other nearby projections) in a blast of intense current of inconceivable violence.

When high-voltage coil streamers struck glass windowpanes smudged with carbon, balls would form between the electrode and pane, and apparently appear to pass through the pane when viewed by witnesses, but not on video. They seem to come from the opposite side of the glass, slowly float horizontally for about 16 inches or occasionally further, and flare up, explode or burn out.

Here is a possible explanation. Tesla's Artificial Electric Balls seem to slip through glass when electric force lines pass through the pane. These balls' positive ions follow the field but cannot pass through the glass, a dielectric, and thus cluster on this side. Electrons from the air cluster on the opposite side. When the electric ball comes to the pane, it heats the glass, which breaks down and becomes slightly conductive and performs like an electrode. (This may leave a permanent alteration in the glass, so field researchers might detect invisible alterations in undamaged glass.) At this, the electric ball forms on the other side and it floats from the window.

Inside chimneys, there is a carbon-rich environment in the form of charcoal, ashes and carbon-dust films. When struck by lightning—at least according to the Smirnov theory—these chimneys create a carbon aerogel fractal electric ball.

Atop the high-voltage terminal coil, the copper wire and charred wood electrodes emit intense current pulses that create fractal bursts which immediately absorb ozone and nearby chemical reactants. Smirnov believes that charged streamers ignite aerogel-fractal structures.

Recreating Tesla Balls

Tesla Artificial Electric Balls (TAEBs) are not true fireballs—and might provide some clues—but maybe not. In his 1899 *Colorado Springs Diary* (lab notes), Tesla described on pages 368-370 how he created Tesla balls. From the description given by Tesla and Corum, we can describe TAEBs. Although they do not behave like true fireballs, they share similarities. Unlike theoreticians and authors (Stenhoff) who claim energy content is minimal, Tesla and Corum proved otherwise.

Tesla ball size is limited by how quickly energy is delivered. Typically, they start as quarter-inch spheres in streamers near the low-frequency resonator and slide up to a streamer knot where they begin to grow. The streamer fades, and the ball floats there, but will be hit by the next streamer and grow larger. The electric balls can be struck up to six times by subsequent discharges—at 800 breaks per second or 1.25 msec between strokes, although the actual times between being hit usually will be much longer.

When an Artificial Electric Ball is struck repeatedly, it grows in size. One of Corum's fiery red Tesla Electric Balls expanded to four inches in diameter in one second. Some of them spun, but others were transparent—with sparks shooting through them—and some changed color as they evolved.

Putting a wax candle at the low frequency resonator increased fireball size. From these experiments with a small Magnifier, Corum concluded that Golka's Wendover giant Magnifier could have made far larger balls. As it was, Golka's Wendover Extra Coil repeatedly created bead lightning. I possess several film frames showing Tesla Electric Balls briefly appearing in a lightning channel, while several vertical glowing blue sparks (ten feet long) appeared on the Extra Coil—oddly in three, bluish-white, vertically-parallel straight lines.

The Tesla balls were spherical and exhibited surface tension during their evolution. The faint limb darkening and solid appearance indicate that these balls are optically thick.

The streamers not only repeatedly strike each Tesla Electric Ball several times, but jump from one to another. The Corums took infrared photos, which show Tesla Balls much brighter (hotter?) than the streamers. One streamer may strike up to a dozen Tesla Balls, each drawing its energy from the streamers that strike each one.

Tesla said that by interaction of two frequencies, that a stray higher frequency is imposed on the lower frequency oscillations. These discharges expand a portion of air, which cools and condenses, creating a vacuum, and become spherical. Air starts to move in and just as the sphere begins to contract, a second streamer strikes it. Certain powerful streamers occasionally carry several hundred amps in Tesla's Magnifier. When a subsequent discharge strikes through a spot of highly rarefied air trying to contract, it tries to expand, but cannot. The Tesla ball cannot cool by expansion nor give off heat by convection or even radiation.

They discovered that charred wood, the rubber tip of a cable, or carbon deposits facilitate electric ball formation. A discharge may strike material that volatizes or vaporizes with difficulty and subsequently to another material that does so easily, and the ball will form here and float back. These balls move slowly. But if they encounter anything, they explode.

The entire energy inside the Magnifier will not take the normal quarter period to convert from static to kinetic energy, but instead will be converted in a brief transient—at a rate of millions of horsepower. For comparison, consider that Tesla's power input exceeded 15 kW and reached 60 kW for short pulses, and that he claimed instantaneous peaks exceeded 5-million horsepower or 3.7 gigawatts.

The Corums simulated the parasitic oscillations by recreating in miniature. Input power to Tesla's system exceeded 15 KW and for short-pulsed intervals may have exceeded 60 KW. Tesla claimed instantaneous peaks over five million horsepower or 3.7 gigawatts. By contrast, the Corum's input power was only 3.2 KW, but also created two-centimeter (one-inch) balls. In other experiments on Richard Hull's 7-KW-input coil Magnfier, short primary-gap extinction times permitted total energy transfer from the primary and self-resonating in the secondary coil, and yielded a 30-MW instantaneous peak power in the driver tank circuit.

And, this is important. Note how Tesla Artificial Electric Balls differ from the directed, non-drifting behavior of true fireballs, as described by Tesla. In the unpublished chapter 34 of O'Neill's biography, Tesla stated that the electric ball "floats like a bubble, easily carried by air currents, they may last…from a fraction of a second to many seconds…they stay fairly close to the ground…Suddenly for no known reason, the ball explodes doing as much damage as a bomb, if close to structures…It is not pleasant…to have fireballs explode in your vicinity for they will destroy anything they come in contact with."

Tesla Electric Balls float, carried by air currents. But true fireballs do not float. They are rarely carried by air currents. They materialize out of thin air inside closed Faraday Cages. Although important as clues, Tesla Artificial Electric Balls unfortunately fail to match the behavior of true fireballs.

Discharging Capacitors Failed

Electric Balls occasionally materialized when large capacitors were suddenly discharged. Several ball lightning researchers experimented with huge submarine battery and capacitor banks. Capacitors discharge through thin exploding wires (EW) are often substituted for chemical explosions because they are safer, economical, controllable and a thousand times more explosive than chemicals and can reach temperatures over 45,000 degrees K. They are used to directly and quickly in microseconds ignite high explosives and also release explosive bolts on satellites or spent rocket stages.

Electroforming makes molds. EWs exploded in water blast a blank or metal sheet into a mold. EWs are used as an intense light source in flash bulbs and also help research metal spectra. They are used in welding. EWs ignite and strike very long arcs that are far beyond the breakdown distance of a spark. It is possible to accurately control the detonation and shock wave, which feeds energy into a gas shock tube ignition, including research into fusion energy. The rapid rate with which EW drops from highly conducting at 100,000 amps to nearly non-conducting at 100 amps is 20 nanoseconds.

In an attempt to generate electric balls, your author repeatedly charged a forty-pound, 114.7-microfarad capacitor to explode various wires inside various gases and beneath water, and within metal enclosures in his futile attempt to duplicate the Lawrence Leach fireball.

Batteries hold charge but, compared to capacitors, they cannot be charged or drained quickly. Capacitors are measured in elastance or D, rated in darafs, which is the reciprocal of capacitance, C, measured in farads, or charge over voltage.

Recent "super-capacitors"—porous carbon aerogel ("frozen air") and carbon-cloth types without chemical reactions or ceramics, which charge even faster than carbon—store over 5,400 times charge as conventional capacitors!—but even that is puny. These super-capacitors flatten surges and sags of power for semiconductor and pharmaceutical manufacturing plants, hospital defibrillators, and smart missiles, and might improve Tesla's Magnifier.

Magnifier Construction and Operation

In 1899, Nikola Tesla generated swarms of 2.5-inch balls that bubbled off his coils—a sensational feat that was not recreated until the end of the Twentieth Century. In 1899 at Colorado Springs, Tesla invented a giant tri-coil Magnifier system, and perfected it for the next several years to resonate Zenneck "surface" waves.

As is well known by the coilers today, but was not known in the 1970s, and contrary to early reports, there was no electromagnetic coupling between the 51-foot-diameter, 20-to-40-turn outer secondary and the 10-foot-diameter, 98-turn Extra (inner secondary) Coil. The outer secondary was fed electrically to the bottom of the Extra Coil, out of which arose a wooden mast and metal ball atop, but not electrically-connected with cable, as everyone believed, but instead apparently charged with a powerful single-electrode x-ray beam from a giant machine of unusual design, which DC-pulsed the Schumann Cavity with extra low frequencies (ELF) of about 8 hertz. Tesla guarded this aspect carefully, and even today, it remains a mystery.

The four coils include an 8-foot-diameter and 8-foot-high Extra Coil mounted 12 feet above the hangar's concrete floor. Upon this were wound 98 turns of number six wire. The 26-turn 51-foot diameter secondary generated 8 million volts at the top secondary turn, which was topped by a single-turn coil called the Reservoir Coil that prevents the secondary's top turn from arcing over to the floor. It glowed with a purple corona. A wire from the secondary's top turn ran to the bottom turn of the Extra Coil and it was at this point—if the Reservoir was not added, that an impedance mismatch would create arcing.

Beneath the 51-foot secondary was wound a two-turn primary that was loosely coupled magnetically with the secondary's bottom turn which went directly to several buried and well-grounded metal plates.

Tesla experimented with a fifth coil, a Conical one at 8 feet in diameter, placed atop the Extra Coil. It generated 12.5-million-volt continuous damped-sinusoid lightning bolts at 1,100 amps, and later an astounding 25 million volts at 2,400 amps.

In Utah, the 11.42-inch Discharge Sphere atop the Extra Coil possessed a capacitance of 38 picofarads. The heavy two-turn primary coil was 600 MCM wire of 64 microhenries inductance, and was loosely magnetically coupled to the bottom turn of the secondary coil through a low mutual inductance.

Both ends of the primary coil led to a capacitor bank where a number of 0.025-microfarad 40,000-volt WVDC oil-filled capacitors provided a total capacitance of 0.2 microfarad. These heavy and expensive capacitors were hung upside down after being purchased to prevent the oil from seeping to the bottom, thus destroying them when they were used.

The primary coil in isolation resonated at 44 kilohertz. A 10-kilowatt diesel generator output was stepped-up from its 60-Hz, 220-volt AC by a 100-kW, iron core transformer to 44 kV and drove the capacitor bank. The primary coil voltage usually was 50 kV.

Commutation is Critical

One secret lay in the gaps and in the commutation or making-and-breaking the primary current with a breakwheel at the correct rate. However, it was important for Golka to first understand the gaps.

On the second floor of his lab in 1970 and over the following years, your author visited Golka and saw him fabricate and test spark gaps. Golka perfected a "small" ten-foot-diameter Junior Magnfier that generated ten-foot lightning of five million volts. Before departing in late May of 1973 for Utah's Bonneville Salt Flats, he fabricated and tested two large spark gaps.

These two spark gaps were two metal cylinders with three electrodes. The main two were adjusted with the third, the trigger, to fire at any selected voltage. Each 25-cubic-foot cylindrical spark gap could operate as a pressure or vacuum gap. Tesla experimented at times with one gap in the primary; and at other times, with two installed.

The original gaps in Utah contained a heavy disk that resembled a giant piston and rod inserted axially into a large metal cylinder to form one electrode, and the tube or shell to form the other. This allowed discharges to radiate circumferentially to the shell, thus providing more cooling area. Without this, the gaps became very hot. To adjust firing voltage, a metal rod or trigger electrode was inserted axially from the other side so that it approached the piston-shaped electrode. By adjusting the separation between the two, which could be performed without disassembling the gap, firing voltage easily could be changed.

Although Tesla filled his gaps with pressurized coal gas, the Magnifier gaps in Utah used sulfur hexaflouride, a heavy viscous gas that liquefies easily under pressure. It has nearly ideal properties for spark gaps, except two. Its thickness prevents it from cooling fast enough and the metal gaps become hot. Worse, if water vapor enters the gap, high voltages decompose the gas into two toxic and corrosive chemicals. On one occasion, this leaking gas blinded Golka for two days.

Sulfur hexaflouride in 1973 cost sixty dollars per tank. Each of the two identical spark gap shells was a cylinder of 6.25-inch diameter, with a couple layers of round plastic blocks on each end, with two inner tubes that formed a seal. A tube could be inserted to force oil into the center and back out to extract heat and cool it. Many compression rods passing through the two plastic end cap blocks kept the cylinders and caps sealed tight under compression.

Inserted into one plastic end was a piston-shaped electrode, sometimes called a "mushroom electrode," Many other shapes were tried and rejected. From the opposite end of the cylinder entered another electrode, the trigger, which could be moved in or out to adjust the ignition voltage at which it would fire. If exposed to air at these high transient alternating voltages it can turn to a corrosive gas, a fact not mentioned in the manufacturer's specification sheets. The assembly was to be usually operated under oil or else it would gap over in air.

Instead of sulfur hexafluoride, Tesla used coal gas, a byproduct from manufacturing coke for the iron and steel industry. Coal gas was made by heating bituminous coal in vertical retorts. Tar, ammonia and sulfur compounds were removed by purifiers—several large boxes with shelves of a porous mixture of iron oxide or rust and wood shavings. Today, natural gas has replaced coal gas for household use, and thus a close substitute for what Tesla used. At that time in the early 1970s, no one knew anything about how to proceed, and there were no coilers or experimenters tinkering with giant coils.

Unfortunately, nowhere in his notes did Tesla describe how he generated electric balls. Possibly, the rotary breakwheel—Golka used a modified giant spinning saw blade—must interrupt the gaps at 4,200 breaks per second. Although this was important, it proved difficult to achieve. The slightest dynamic imbalance caused the breakwheel to wobble and break off a tooth. Once the breakwheel broke lose and shot through two walls. It travels at over 750 miles per hour, or at supersonic speeds, and was specially bowed and balanced. Even worse than dynamic imbalance was catching an electrode in the rotary breakwheel—very lethal. He tried many methods. In 1979, he elevated a car to spin a rotary spark saw blade with 30 large, triangular teeth, but discarded it as impractical.

More subtle damage came from radio frequency currents on the breakwheel. If these currents ever find their way through the pulley and belt back to the drive motor, the motor's winding insulation slowly burns out. For this reason, He replaced it with a 40-hosrsepower internal combustion engine in Utah.

Tesla at one time used an aluminum breakwheel, and even experimented with two specially machined counter-rotating breakwheels. Although he experimented with several breakwheels, here are some typical measurements—25-inch diameter, 40 to 20 teeth, and 1.22-to-3-inches between teeth. Driven at 1,000 RPM, they produced 1060 sparks per second. Golka used various diameter blades. Using the 40-hosrsepwere motor, pulley and belt, geared up the sawblade produces peripheral supersonic speeds, exceeding Mach One or over 750 miles per hour.

Tesla's breakwheel frequency was 4,200 breaks per second. Although Golka exceeded this, his initial break frequency was 1,800 breaks per second. He assumed that his gap operated as an ideal switch of unknown resistance in the primary. The secondary and Extra Coils were each 2600 feet of number 6 AWG insulated wire. The secondary had 26 turns and an inductance of 12 mh, but with an unknown capacitance. The 100-turn Extra Coil also provided a known inductance of 28 mh, but an unknown capacitance. Therefore, total inductance equaled the sum, or 40 mh. The spherical capacitor and the Extra Coil in its grounded state were replaced by a lumped-circuit equivalent.

The Extra Coil oscillated at 30 kHz. Each discharge pulse was a 24-joule damped wave train with a damped time constant. Simple network analysis yielded relations that allowed calculations of unknown parameters from measured frequencies. Many engineers fail to note that a capacitor need not have two plates—one will suffice, since every object by itself has capacitance. This was first described by James Clerk Maxwell in his two-volume *Treatise on Electricity and Magnetism*, but is rarely described in modern electromagnetic textbooks.

Optimize Capacitors and Windings

Adjust capacitors and windings to maximize voltage. Obtaining proper coil windings was a time-consuming headache—gained from an interactive iterative process, from rough calculations, and intuition gained from trial-and-error—tuning coils at low power, cranked up the power, and then retuned the coils again, and so on, iteratively converging on the correct value. To measure resistance, inductance, capacitance and voltages was not straightforward.

First, once the coils are tuned, all nearby objects cannot be moved during the experiments because the tuned coils are sensitive not just to humidic and barometric changes, but to all neighboring objects. Retuning the coils can take days.

Next, to adjust the primary capacitor bank, measurements were taken by placing neon tubes (sometimes several arranged in a cross-shape atop a rod) at given distances from the terminal. If the tube glows over the entire range of several primary capacitor bank values, then he moved the neon tube farther away. The range of capacity would diminish, until finally at a certain distance, the tube flickers briefly for good discharges. The mid-point of this narrow range is the optimum primary capacity.

Maximum lightning voltages were between 12 and 20 million volts at 1,200 to 2,400 amps. Obviously, when Tesla began, no one made instruments to measure these awesome repetitive-damped-sinusoidal currents and mega-voltages! Tesla measured his lightning voltages by attaching oversized metal balls to his Extra Coil's terminal.

By trial-and-error, Tesla selected the largest metal sphere from which small sparks were barely able to break free. This was the break-off (peak) voltage, and was like a rocket's planetary escape velocity. According to Tesla's Diary at Colorado Springs, the break-off voltage was 66,000 times the ball's radius in centimeters. Such larger balls are used only to determine voltage and then removed. With the small ball back in place, the coils produce lightning of that calculated voltage.

Of great importance to Tesla was adjusting the capacitor banks to get maximum lightning voltage. If the Magnifier was provided ample power from the power transformer (all it could use) then primary capacitance could always be adjusted from the coil to get the longest lightning bolts.

The alternate case was when energy was limited, that is, the Magnifier must be designed to obtain maximum performance from a small power transformer, as Golka did with two earlier, prototype 10-foot coils at his lab. The primary capacitance was kept constant at the greatest value that could be charged to the required lightning voltage with available power.

For many observers, a difficult point to grasp was how a two-turn primary produces greater sparks than a primary with many turns—say 30 turns. At first, in the early 1890s, when Tesla began working with conical coils, to him this loose coupling seemed remarkable, but there was no denying it. It is true that maximum voltages occur with low magnetic coupling. For example, with a coupling of 0.348, the capacitor must be almost three times the resonance value. This occurred because coupled coils possess actual inductance, and so their resonant frequencies must change.

Perhaps the secret was in commutation; it permitted shock-exciting the Extra Coil into air-to-ground discharges at a pulse rate, pulse energy and damping time constant identical to those Tesla used. Although four times Tesla's pulse energy, for some time the pulse rate was only half his; but at higher break-wheel frequencies, reduced pulse energy is used. Trouble existed in extinguishing the gaps, although this problem was solved later.

Constructing a Prototype

Before attempting the full-sized equipment, smaller prototypes were constructed. To avoid radio frequency interference problems, metal sheets covered the ceiling and walls inside the second floor of a former two-story school with metal sheets. His one-fifth-size ten-foot diameter prototype Magnifier was 7.5-feet high, with a 64-turn secondary coil in the center of this secondary coil, 67-turn Extra Coil measuring 38 inches high and 36.75-inches in diameter. The bottom of this Extra Coil was elevated about 15 inches above the wooden floor, resting on tree glass terminal insulators.

The secondary inductance (primary open) was 2.56 mh and 20 microhenries (primary shorted). Primary inductance with secondary open was 20 microhenries, and shorted was 2.25 mh. To calculate coil inductance, he used Wheeler's Formula. It gives inductance, which equals turns times radius squared divided by nine times radius plus ten times length. Although this famous formula proved useless on the 51-foot giant that he constructed in Utah, it did give him rough starting values with his 20-percent junior-sized Magnifier at his lab.

Coil frequency was 485 kHz, well above the 50 kHz that his 51-foot coil would use in Utah. Mounted atop and resting upon the Extra Coil, a 62.26-inch steel ball provided capacitance. Larger balls caused excessive current draw and diminished the spark size. The steel ball often drew 300 amps. This current entered through a giant Variac or variable transformer, which he varied by turning a large auto-steering wheel and column. This input was stepped up from 220 V to 40 kV by a second iron-core power transformer.

Next, this 40-kV, 330-mA current was interrupted by a high-speed saw blade and small pressure-spark-gap. The six primary capacitors were 40-pound, 0.025-microfarads, 40-kV units. Prior to this, it used old capacitors from industrial dielectric heaters. Unlike the giant coil later constructed at Utah, these two prototype 10-feet Magnifiers used two large lamps to cushion the transformer and act as regulators.

High-frequency lightning from the coil was hungry for capacitance, and the bolts would even be attracted to and hit glass or plastic. Golka pushed a fireproof plastic ball mounted on the end of a 15-foot fiberglass pole so that jagged bolts of lightning would repeatedly smash into the plastic ball. He twice pushed a three-inch thick glass plate eight feet from the coil. He place pliers behind the plate. The six-million-volt lightning went through the thick glass plate, continually striking the pliers.

When he brought a large 2-by-4 timber near the ball, keeping the lightning from striking it, and by using an extra line on the other side, the 2-by-4 instantly burst into flame from the frequency. However, if the wire touched the steel ball, then the Extra and secondary coils became detuned, the lamps glowed brighter, sparks would shot from the rotary (breakwheel) saw blade, and the lightning stopped.

Selecting a site for his full-sized Magnifier involved various factors. First, it could not be near any city. To determine where he would construct the large Magnifier, he examined FCC ground-conductivity maps of North America. High conductivity was his deciding factor; more ground current meant more efficient coil operation. As it happened, New Hampshire, western Massachusetts, western Connecticut and Tallahassee, all came in at a low one-ohm per meter receptivity. Runner-up California's Big Sur came in at a respectable two ohms. The Appalachian and Rocky Mountains proved good, and so did seawater, Utah's Bonneville Salt Flats, and wet loam. Swamps, oddly enough, were unpredictable and never guaranteed good ground conductivity. Cities were poor.

The bone-dry, rainless summer climate of Utah's Bonneville Salt Flats won. By constructing a full-scale coil under the open sky on the rock-hard salty sand bed, the open air also would hamper fires and dissipate poisonous ozone.

Next, he went on site to examine the Salt Flats desert and talk with U.S. Coastal and Geodetic Survey geologists. From the sites, he collected and tested pail-fulls of light-gray salty sand. Although when wet, the sand proved corrosive and quickly ate into iron nails, when dry it became as hard as stone and could be drilled. Its electrical resistance was low, and with higher voltages above 3 million volts, its resistance dropped lower.

Tesla had selected Colorado due to the rarified atmosphere in part because he felt it helped his unit establish an ionized path to the ionosphere. It turned out such elevation was unnecessary. Thus he later chose Shoreham on Long Island at sea level for his plant, although Niagara Falls would have been a better choice.

Golka relocated his Magnifier in 1986 to the temporarily idled six-story steel Molybdenum building at the Climax Mine operations near Leadville in Colorado. But too quickly, the conditional (based upon the market price of molybdenum) lease unexpectedly expired—unfortunately before his research could be completed

Zenneck Waves, Not Schumann Cavity

Wireless power was Tesla's objective. Wireless and lossless, power transmission via either Zenneck "surface" waves or the earth-ionosphere Schumann Cavity—there is controversy on the choice—could transmit massive electrical energy. If so, then it may be possible that the vast solar, hydro, geothermal and sea-thermal energy from the remote deserts, from distant waterfalls, from deep geothermal drillings or vents, and from the oceans could be transmitted—all at low loss without wires—to any location on earth, and hopefully be monitored without theft.

Tesla observed stationary waves, claimed that the ground and ionosphere behave like a spherical cavity resonator. His tuned circuits, capacitively coupled to earth, developed two wavelengths at resonance. The metal sphere atop the tower emanated electrostatic waves of length, f/c, while the earth terminal propagated longer waves of length: (pi/2)c/f. At the earth-ionosphere resonance, the elevated tower's electromagnetic emission became negligible.

The air above his tower provided a conductive path to the lower ionosphere via pulsed ionization. At these voltages and waveforms, air is more conductive than copper. The earth-ionosphere resonant cavity acts like a low-loss transmission line. Micro-pulsations in the earth's magnetic field, caused by such things as rotational vibration, cause immense circulating currents in the core.

From 1905 through 1988, all attempted explanations of Tesla's wireless electric power were fancifully paradoxical. The immoral paradox was this. Let us suppose that if Tesla was wrong, and then continued to bilk his investors, and attempted to defraud more investors—no, he was not that type of man, even his worst critics agree, and he possessed great integrity—then there is no logical alternative—they must conclude, *ipso facto*, that Tesla committed criminal deceit, gross misrepresentation, and massive criminal acts and also torts of both negligence and intent. (Your author took many law classes.)

If these critics are correct, then Tesla intentionally perpetrated major violations of civil and criminal law, all imprisonable offenses. Worse, if wrong, Tesla was massively stupid for life and short of quite a few bricks upstairs, and never repented of this major fraud and stupidity.

Your author wondered, "How could a genius with such integrity be dishonest for so long?" This mystery made no sense. Then, in the late 1980s, James and Kenneth Corum solved the mystery. Tesla was correct.

Tesla cited Arnold Sommerfeld in 1916 to explain his worldwide power system, where 95 percent of the energy appears as "current waves" and is propagated near the ground by conduction and the remainder through the tower, which did not sprout lighting in this mode.

A surface wave that became known as the Zenneck "surface" wave, that traveled along the interface between the air and ground, was described in 1909 by Johann Zenneck. The propagating energy was concentrated near a guiding surface and did not spread like radiation. It was Sommerfeld who proved that an electromagnetic wave could be guided along a single wire possessing finite continuity. It was Zenneck who proved that the surface of the earth performs similar to a single conducting wire.

Stating that both can be present in the wave complex but in varying proportion, Sommerfeld distinguished between the Hertzian component (the space wave) and the "electrodynamic" surface wave. Tesla asserted that the transmitter design determined the exact composition of these emissions.

Some questioned the existence of Zenneck "surface" waves, but later experiments proved that Zenneck "surface" waves propagate with less loss at lower frequencies and do not contribute significantly to fields generated by electric dipoles or quarter-wave radiations, but a quarter wave resonator can strongly excite Zenneck "surface" waves. Attenuating exponentially along the guide, without inverse-square spreading or diffraction (that occurs with Hertzian waves), the resulting surface waves are single-conductor, transmission-line mode. If parameters are optimized, the Zenneck "surface" waves exhibit minimal radiation fields and waste little energy. Zenneck "surface" waves propagate optimally in the ELF region, and below one MHz.

When Zenneck (1907) analyzed a solution of Maxwell's equations with a "surface ground wave" property, he analyzed the vertically polarized plane wave solution when Maxwell's equations are applied near a planar boundary, which separates a half space with finite conductivity from free space. Such a high-conductivity Zenneck "surface" wave is a function of the dielectric constant and frequency.

The Poynting vector, designated as **P** or sometimes **S**, is parallel to the planar boundary, and it decays exponentially, but with different decay constants, parallel and perpendicular to the boundary. Electromagnetic waves transport energy from source to destination at a rate of energy flow per unit area in a plane electromagnetic wave described by the Poynting vector or the reciprocal of permeability times the cross multiplication of vectors electric and magnetic field vectors.

The direction of the Poynting vector is the direction of electromagnetic energy flow, while **E x B** provide instantaneous values at each point. [pp. 894-7 of Halliday and Resnick and pp. 242-9 of Ramo, Whinnery and Van Duzer.] Even though we only know total energy flow through a region per time unit, and it is given by the total surface integral, we may prefer the convenience to think of the Poynting vector as providing direction and magnitude of energy flow at any point in space. Poynting vector or radiation density is watts per square meter.

A mistaken term, "surface (ground) wave," creates a misperception that energy flows only in a region near the surface. The boundary guides the wave, not localizes or confines it, so that most of the energy is not near the surface.

With horizontal polarization, the conducting boundary excludes the wave from near the surface. But with vertical polarization near a conducting boundary, wave energy reaches down to the boundary significantly. For a curved boundary—that is, propagating around the earth—the surface curvature creates diffraction effects and propagation beyond the horizontal.

Of academic interest, the plane Zenneck "surface" wave possesses an infinite source. This also applies with a wave with cylindrical symmetry. It also has an infinite line source. The propagation problem is two-dimensional. Aside from exponential decay factors, the field of a cylindrical Zenneck "surface" wave field is inversely proportional to the square root of distance, unlike free space emission.

Consider a vertical electrical dipole, a point source, positioned above a conducting plane. When far from this oscillating source, the field from this vertical dipole can behave like a cylindrical Zenneck "surface" wave. Although this depends on the direction we look at, near the boundary the wave approximates that of the cylindrical Zenneck "surface" wave. In 1909 Sommerfeld derived the equations—making a sign error that he soon corrected—that proved the dipole field morphs over to the cylindrical Zenneck "surface" wave close to the surface as the distance increases. There is an intermediate region, and also near the surface, where we can approximate the field with the cylindrical Zenneck "surface wave". But as we move farther away, the decay becomes the 1/r dipole form, and only slightly modified by its nearness to the surface. At large distances over a planar boundary, the direct wave overcomes the Zenneck mode.

Also, the wave lacks the Zenneck form, in general, and we must note the horizontal and vertical polarizations, the former from a horizontal and electrical dipole. With vertical polarizations, the field reaches down to the surface.

With a vertical electric dipole radiator over the globe, when we receive beyond the geometrical horizon, because there is no direct ray—ignoring ionospheric reflection and that from ionic meteor trails—only that part of the field that as receivable comes by diffraction.

We ignore anomalous propagations from atmospheric ducting. With lower frequencies, diffraction more effectively routes electromagnetic waves around the curves of the globe. The frequency preferred is in the HF range, where the field is a sum of modes from solving the cylindrical Zenneck "surface" wave over a conducting sphere. At that time, no one realized that ionospheric bounce between earth and ionosphere was the dominant mode for propagating early wireless radio waves around the globe.

Note that the term "surface wave" and "ground wave" are misnomers. Confusing matters, some authors assign different meanings to these terms. Further confusing matters, surface waves are sometimes confused with creeping and traveling waves of electromagnetic scattering theory, to which they are closely related.

John Wait, David Reiss, Gary L. Peterson, Toby Grotz, James and Kenneth Corum and others have written on Zenneck "surface" waves as they relate to Tesla. Although Poynting's Theorem is not debatable, its physical interpretation is open to criticism. [p. 131-7, Stratton.]

The Corums argue that efficient global wireless power is still possible. Your author agrees, not just on technical grounds, but also on common sense. Tesla would never proceed from Colorado Springs to the next stage—to Shoreham—and undertake the vast project if he had not already proven—beyond all doubt—that it was workable. Only bad fortune and inflation derailed Shoreham, and we may never know if our planet's history should have gone in a better direction—an accident of history.

Zenneck Power Transmission

Once oil reserves dwindle, and if the controlled fusion options fail—or because of future terrorism and oil shortages—then we might want to revisit Shoreham and rediscover what Tesla achieved.

Wireless power Zenneck "surface" wave and ionospheric resonators will not challenge existing energy markets and systems, but instead fill an unoccupied niche that will interlock and co-exist profitably with existing distribution. A global wireless power grid would link into the world's existing three-phase regional systems.

The earth has an over-abundance of pollution-free natural energy, particularly solar, but it is located where people cannot live—in major deserts. Powerful winds sweep uninhabited lone prairies and frozen steppes. The question is how to transmit energy with low loss from "there" to "here."

No matter how successful, the transmission of wireless power by the Zenneck "surface" wave resonator cannot threaten existing structures nor challenge or replace existing power grids. Instead, it would complement existing distribution exactly where the existing distribution systems are weakest—in long distance and transoceanic transmission. Placing utility plants at the fossil fuel origination sites—existing coal mines and oil and gas wells—would reduce (but not eliminate) the need for costly mega-tankers, coal trains, and fuel pipelines. These are not renewable fossil fuels, and they are becoming costlier.

Transmitted globally with low-loss Zenneck waves, low-cost electricity could affordably, through hydrolysis, provide hydrogen to future automobiles and to other internal combustion engines, including airliner prop and jet engines, and ship turbines. And more homes and skyscrapers would convert to low-cost electrical heating. Distance would be irrelevant, and harsh regions like Alaska, Antarctica and Siberia could be populated.

For example, Stalin unwisely populated Siberia, including the famous "Siberian Science City" But massive energy costs doomed the attempt, and Russia is now withdrawing its citizens from resource-rich Siberia.

Can Zenneck waves shift the balance to solar energy? An early advocate of migrating to solar energy, your author soon recognized massive obstacles—ones that now might be overcome with low-loss and wireless global Zenneck waves.

This would be far less destructive to Earth's interlocking habitats and natural environments, and also reduce "the human footprint" and its stresses upon regional and global ecosystems. And constructively harvesting the practically infinite, alternative, renewable, pollution-free energy from the pure wind and golden sun—direct from the world's deserts (some always face the sun), geothermal, wind-swept steppes and plains, and from remote dams—finally becomes profitable to supply the needs of all, including, most importantly, the poor pawns of despots—the starving peoples of our planet.

Before it is too late, to minimize the evil force of terrorism and oil, both our governments and energy firms must eplore Zenneck electromagnetic power transmission.

We just examined how Golka and Corum pioneered the rediscovery and recreation of Tesla's Zenneck Waves. However, unfortunately, Tesla Artificial Electric Balls are not true ball lightning. Not surprisingly, others also conducted experiments and created theories.

22

Failed Theories

Over the past two centuries, many experiments were performed, and even more theories created. Many theories are impressive but mathematically arcane. The history of experiments and theories would fill volumes. The following are favorites.

Submarines Generate Fireballs

Born in England, the late physicist James L. Tuck joined the atomic bomb Manhattan Project at Los Alamos during the last world war. He later went on in the early 1950s to help start Project Sherwood—the search for the world's first fusion reactor that "burns" deuterium and tritium—which was soon declassified. Tuck suspected that fireballs might contain a key to controlled fusion energy. He wrote about ball lightning and fusion in a paper that he delivered at MIT in 1971.

In the paper, Tuck described how he had submarine batteries delivered to Los Alamos and began a series of "bootleg" experiments on which his physicists worked during their spare time. After several months and under pressure to complete their research, they made a last-ditch attempt: They encircled the switch with a cellophane box and filled it with methane gas. Crouched behind sandbags, they watched while they activated the switch. A sheet of flame and thunderous roar frightened them.

They then took the 16-mm film from the two cameras for processing. On about 150 frames appeared a ten-centimeter glowing ball that went behind something and came out the other side. Tuck experimented with large low-voltage discharges and also used a 30-kV 75-kilojoule-condenser battery for high voltage arcs. The arcs left behind a yellow and red afterglow, but with the one exception, nothing that resembled ball lightning.

Golka repeatedly visited Tuck at Los Alamos, and Tuck later gave him the film. Later, after your author examined Tuck's film, he concluded that it was not ball lightning because after debris flew by, the pinpoint of light flew in a parabolic arc and was influenced by gravity. It did not float.

By chance, your author once saw something identical—lightning struck an old pear tree in 1974, and a similar glowing object (a glowing chunk of wood or pitch) performed the same flight pattern, but traveling farther—120 feet—glowing brightly along the entire parabolic arc. Your author also examined films from a failed Dutch attempt, which funded an experimenter with huge submarine batteries that generated tremendous currents. Finally, the Dutch government defunded this controversial project.

Then, there was the case of Dr. Paul Silverberg, formerly of Raytheon's Advanced Development Laboratory in Wayland, Massachusetts. Silverberg investigated a submarine where a green fireball floated off contacts in the engine room. The fireballs lasted one second. Silverberg calculated that 0.4 to 4 megajoules were needed for a one-second lifetime. He studied fireballs that shot from the reverse gear in diesel submarines. Silverberg theorized that an infinite number of frequencies created constructive interference on a nonlinear transmission line structure.

On old diesel-electric submarines, a large knife switch enabled the huge bank of lead-acid batteries—and each giant battery is as tall as a man—to accept charge from the diesel-powered electric generator. The constant-current generator supplied fixed current to the batteries, but the voltage would rise or fall. They could generate very high voltages.

To create fireballs, the machinist-mates would briefly open the knife switch for several seconds until the hair on his arm holding the switch would stand up. At this, he could slowly close the switch until it was one-eighth of an inch apart, and a tiny electric ball would pop out and drift for a few seconds to hit a steel bulkhead and explode, sounding like a small-caliber handgun. Some bulkheads had many black spots from this.

Golka used direct current from a huge bank of batteries, providing 200 volts at 156,000 amps, putting out 40MW peak power. By contrast, Tesla reached 12MV at 1,200 amps—much greater voltage but less current. A diesel submarine can briefly supply current exceeding lightning flashes.

Buoyed by the stories of fireballs appearing in conventional submarines, during the 1980s, he operated the Lionfish, a World War II submarine berthed at Fall River in Massachusetts, in an attempt to duplicate the phenomenon. At a veteran submariners' conference, he met retired diesel "boat" (submarine) sailors who described submarine fireballs.

Diesel subs are relatively inexpensive and very silent—more so than nuclear subs, despite acoustic suppression—and many nations today use diesel boats. In the late 1990s, Golka used massive submarine batteries to produce small quarter-inch ion-seeded ballettes that spun rapidly, but they did not float and were not true fireballs.

A single submarine battery is huge, and submarines carry many of them. Today's diesel boats contain 240 to 960 cells distributed in up to eight compartments to make the best use of space and ballast. They have come a long way since the 1940s, and possess six times greater energy densities than their predecessors and exhibit low gassing, and possess good low-and-high-current capacities. Automatic battery-health maintenance systems in the 1980s reduced the inadequacies of manual monitoring. Modern diesel subs have also adopted the blimp-shaped Albacore hull, so they move faster and get more speed and range from their batteries.

Just to give you some idea, such a battery can supply the main propulsion motor with 100kW to 200kW of continuous power for five days, and 7.5MW in short one-hour bursts. Voltages can exceed one kilovolt. Battery weight exceeds 500 tons, a quarter of total displacement. Diesel subs are deadly. For example, during the Falkland-Mulvanis War of the early 1980s, a British diesel submarine sank the General Belgrano, an Argentine battle cruiser. Designed for deep "blue water" service, nuclear submarines have encountered acoustical difficulties in the shallow water missions required in the post-Cold War era.

When your author was a systems sonar engineer designing attack class submarine sonar systems and their range and range-rate units, he was amazed at the monstrous size of the two vacuum tubes used for powering the thousand-plus 74-pound transducers lining the 15-foot-diameter, air-filled sphere hidden inside the round bow. The cable from each cathode, at the top of each giant tube, was covered with a thick insulator.

When going "active" to transmit, the current flow was so great that the voltage drop along the anode cable was so enormous that if that cable—which was in an inverted-U shape—was bent incorrectly, a one-foot arc would jump one foot across the cable to another part of the same cable! And that spark would penetrate through two inches of rubber on both its exit and entry.

With their rockets, cruise missiles and subrocs on board, could FBMs and attack class nuclear subs encounter explosions? Are surface ships occasionally at risk? Several years ago, a certain battleship's 16-inch round exploded, doing great damage. (Your author was inside the barbette of a sister battleship. After the Cold War ended, all were decommissioned.) This mysterious explosion was ruled a non-suicide, but never solved. Probably not ball lightnig, but your guess is as valid as any conjecture.

Theorists Stack Their Decks

There are hundreds of theories. Want one? Take your pick. But all share one quality—none work. Not one explains fireballs and detection skills, their navigation and gravity-defying flights.

All theories assume too much before their authors even attempt derivations. In other words, if armchair theoreticians dislike or cannot explain certain properties or boundary conditions, then it is easy to develop a mathematical model that does not wander too far beyond the boundary of respectable physics. They do so by ignoring embarrassing facts. Outside the technical fields, this is called "stacking the deck." In casinos, it is called "cheating." All fireball theories do this in spades.

They apply inappropriate boundary conditions to Maxwell's equations and propose solutions by requiring boundary conditions derived from physical realities that do not exist for the inappropriate solution. Of course, normally there is nothing underhanded about using boundary conditions if you select them realistically. For example, we reject some wrong solutions to Shrodinger's wave equations because they inaccurately model physics.

Like sick magicians, some fireball theorists pull valid formulas and derivations out of thin air, but annoyingly fail to list all their sources. Or the sources are obscure, or in a foreign language. This makes it hard for others to fight their way through these derivations. Tired of it, we reduce our rigor and jump ahead and miss things.

Some of these authors possess formidable skills and doctorates. They derive their theories for different reasons. But they are handcuffed to orthodoxy. They are not eager to risk the label of maverick by proposing anything too far outside the envelope of accepted and respectable physics—for good reason. It could adversely tarnish their careers. Most are sincere. Some of these authors write articles that are inferior but are intended for less-rigorous publications that lack specialized editors. Some do this for mathematical amusement, others because of "publish or perish" rules, and some others because it pads their resumes. Some do it for idealistic reasons. All these are respectable reasons.

One fireball theory listed a monotonically increasing integral bounded above a certain value: and, both the inevitable outcome in his derivations from his flawed initial assumptions, among other errors—and I suspect they were intentional—failed to get this integral to converge absolutely. But he conveniently ignored this. And unless we carefully checked all his derivations—which is a tedious process—we would never detect his errors.

Physicists make mistakes in derivations. So lengthy are their derivations, some use large artists note pads and check each others' derivations before submitting papers for publication. But not infrequently they make innocent errors in assumptions of, say boundary, and because others do not detect these errors, then everything after that probably will be wrong.

But there is another breed. Some authors—sick mathematical quacks?—are seriously mysterious. They are not joking. They think differently. They claim degrees they do not possess from imaginary universities that exist only in their minds. As you can guess, they take a different route in deriving their equations.

They invent their own strange mathematics and network analysis. If we cannot understand them, how can we disagree? They are incomprehensible mathematical magicians, pulling strange theories out of thin air. Their articles, published in fringe magazines—and superficially look impressive—but only to the uninitiated—are cluttered with mysterious mathematics and alternative physics that they alone grasp.

Theories Abound, None Work

Marching forward, ball lighting theories are like reincarnated cockroaches—they are older than moss and never die. In altered form and cloaked with increasing levels of advanced calculus to camouflage their inadequate foundations, they get resurrected and live on forever. Like cockroaches, they evolve and multiply. All these theories possess thunder, no lightning. None work.

One of the earlier theories, agglomeration, was proposed during the early-nineteenth century. Forest fires emit carbon particles and trees exude resin and pollen. Volcanoes fill the upper atmosphere with tiny volcanic dust particles. According to agglomeration, all these particles stick together or agglomerate into tiny clumps that stick to each other. Some masses grow heavier, sink downward as they combine with organic particles, and get ignited by electrical fields or lightning. Dressed up with advanced calculus, and in modified form, the agglomeration theory still is proposed.

A recent version of agglomeration is the burning of fine particles of carbon and silicon from lightning hitting and penetrating the wet soil. Pure silicon globules from carbon and silicon dioxide in the soil must have a 1:2 ratio. Vaporized chains of floating silicon slowly react with oxygen in the air. These nanometer-sized silicon chains form fractal aggregates that clump together in chains of nano-particles that slowly oxidize to expend the chemical energy stored in these fila-mentous chains as heat and light.

An early theory first proposed in the late eighteenth century compared ball lightning to a Leyden jar—an early capacitor or condenser that stored electrons on one of two metal plates. The Leyden jar was a glass jar enclosed in metal foil, filled with water, with a thin chain inserted. According to this theory, the light-ning ball also traps negative charge inside, where it is enclosed by a thin spherical membrane of dry insulating air. As charge leaks across this insulating shell of air, it gives off thousands of tiny sparks over the surface.

A Russian experimental physicist, a 1978 Nobelist, Pyotor Kapitza, while under house arrest by Joseph Stalin, created a famous theory of ball lightning that postulated microwaves generated during storms reflected off the ground and interacted with the incident waves to form cancellations and reinforcements that sometimes ionized air. Kapitza was famous for discovering superfluidity and absence of viscosity at low temperatures, and mentored Landau.

Kapitza compared a nuclear radioactive cloud's lifetime to that of a lightning ball. A 490-foot nuclear cloud radiates totally in ten seconds. By scaling, he knew a four-inch lightning ball should disappear in under one-hundredths of a second. To solve this discrepancy, he suggested that lightning emits certain frequency waves that are reflected from conductors, such as the earth, and interact with incoming waves and thus cancel or reinforce. This sets up an invisible grid of nodes and antinodes.

Energy at the antinodes ionizes the air in that region, which loses its electrons, glows, and slides into a node. As the node moves, it drags the ball along. The atoms inside the ionized ball oscillate back and forth in time (resonates) with the incoming electromagnetic waves, thus absorbing energy to stay ionized and remain glowing. He calculated that electromagnetic waves from a lightning storm must be 3.65 times the diameter of the ball.

Problems soon surfaced. Kapitza's model failed to explain how energy can reach a ball submerged in a water barrel or the enormous explosive power. It failed to explain how fireballs materialize inside or slip into metal or conductive containers, such as airliners. And, a large amount of radiation needed at 428MHs to 857MHz was undetectable by a Brazilian meteorologist. But although the nec-essary frequencies do not occur during storms, and Kapitza rejected his own the-ory, other theorists propose reincarnated versions of this discredited theory, albeit in modified or disguised forms. One example is the Maser-Soliton lightning ball.

Others suggest electromagnetic standing waves and transient nonlinear soliton waves that focus. If so, why have these waves never been detected experimentally?

Some theorize it is a nuclear phenomenon. Since the most abundant isotopes—atoms with an abnormal number of neutrons in their nucleus—come from water vapor in a storm and are from oxygen and hydrogen (protons), then one possible nuclear reaction yields beta decay reactions that produce positrons with energies around half a million electron volts. Energy is continuously generated inside the ball.

Physicist Shalger suggests that with a plasma toroid model, the toroid oscillates between a poloidal current mode or H-flux loop mode, where its structure displays electric dipole properties or the toroidal current modes or E-flux loop mode, where its structure will display magnetic dipole properties.

In one nuclear model, confinement is by a plasmoidal double boundary-layer containing a trapped-particle dipole-layer of the BKG wave. In the Poisson Model, four alternating and consecutive spheres of excess, trapped ions and electrons form the fireball. In an alternative version of the late 1970s, neutral hot plasma is supposedly surrounded by a thin sheath of negative excess of ten percent of charge and a positive excess of ten percent charge. For an eight-inch fireball, this plasma sheath should be one-quarter of an inch thick.

Alternative explanations involve chemical storage and oxidizing colloidal suspensions or vapor. When lightning strikes organic objects, it vaporizes carbon molecules, and a flaming gas of this vaporized carbon or pitch creates a fireball. But there is not too much carbon at 10,000 feet altitude where some balls form. And most fireballs do not form when lightning strikes nearby. Boris Smirnov resuscitated this theory in greatly modified form in the late 1980s by proposing that the energy is stored inside fractal structures.

One of the most innovative theories, the fractal electric ball model, was developed in the 1980s by Smirnov at the Russian Academy of Science's Institute of High Temperatures in Moscow and formerly of Thermophysics, Siberian Branch of the Academy of Sciences at Novosibirsk. Smirnov understood the failure of models that required energy from outside, such as electromagnetic waves. This meant that there was only one alternative—energy must be stored within.

With his excellent grasp of plasma chemistry and filamentary structures, of plasma chemistry and dust particle combustion, of aerosols and aerogels, not to mention nonlinear mathematics, Smirnov was able to incorporate fractals and diffusion limited aggregation (DLA) into an aerogel analytical model of an electrically charged structure of interwoven sub-micron filaments. This porous fractal cluster stores energy. Vacant pores fill this aerogel structure.

As a random motion, particle diffusion is like Brownian motion—the average walked distances of all particles is zero. But if a non-uniform distribution exists, a statistical net transport of material occurs. Although diffusion occurs in all gas, plasma, and liquid systems, even solids, the rates do vary greatly. Particles that attract or stick together form aggregates. For ions with strong electrical charges, they often form well-ordered aggregates or crystals. In these cases, the ion charge shape usually affects crystal shape. For example, NaCl forms cubes. On the other hand, without strong ordering electrical forces of charged particles, the aggregates may stick together only temporarily and lack distinctive shapes to form DLA clusters that are sometimes tree-like, never compact, and often fluffy.

The possibility of particles to reach a cluster controls the shape of that cluster. Starting with a uniform distribution, particles meet and aggregates grow. Arms of clusters catch particles, making it likely that successive particles will never reach inner older parts, and a fluffy shape evolves—one with many arms like trees or corals with unfilled gaps. Loosely aggregated clusters form fractal dimension numbers between two (areas) and three (solid volumes). In-between exponents include sponges and Mandelbrot structures, all DLA cluster fractals possessing self-similarity.

A chemically charged fractal cluster releases energy through a multiple combustion process. For instance, a charcoal dust cluster absorbs ozone and its combustion is intensive and slow, and soot and carbon particles control luminosity. In the Smirnov model, carbon and ozone slowly pass through four intermediary steps, where each rate constant depends on saturation temperature, with the final stage leaving carbon dioxide and monoxide.

The ball produces color and luminosity similar to flame compounds without fireworks pyrotechnics. The combustion of charcoal in absorbed ozone may satisfy ball lightning. Fractals explain better how colloids, condensed aerosols, soot and dust are formed and how soot and carbon particles affect the luminosity of a flame. Within several ten milliseconds after these materials thermally explode, ultrafine (80 Angstroms diameter) smoke particles clump together in chain-like aggregates with fractal structures. Results from experiments in fractal clusters published in 1979 led Smirnov to extend them to his theory of electric balls.

In the late 1970s, S.R. Forrest and T.A. Witten rapidly heated tungsten filaments electroplated with iron or zinc with a brief high-current pulse. The metal vaporized into a hot, dense metallic vapor of uniform spherical particles due to collisions, but did not diffuse far and created a spherical halo at one centimeter from the filament. They condensed and linked together in chains. These chain aggregates were collected on a slide and examined with an electron microscope. They exhibited fractal properties.

Turbulent and swirling motion inside some fireballs suggests "chaotic" internal movement governed by laws of chaos, fractals, and complexity theory. Boris Smirnov of Russia first proposed the fractal model for ball lightning. When plasma interacts with a solid surface, fractal structures that form the "skeleton" of a lightning ball are formed. Tiny, solid nanometer particles join fractal aggregates to create fractal fibers in the presence of an external electrical field, thus starting the growth of this fractal structure that was formed.

This aggregate or cluster of polymer molecules or macromolecules interknits together to create a polymer net structure or "skeleton" with rheologic deformation of matter and fractal qualities with significant dielectric properties that supposedly enable them to acquire and store sizeable charge.

The ball's skeleton is a collection of interwoven fractal fibers. The skeleton simultaneously posses the qualities of solids, gases and liquids. Decomposition proceeds as separated thermal waves moving simultaneously along the different fractal fibers, thus the ball glows on the front of each thermal wave, according to Smirnov. An individual fractal particle is between one and ten nanometers wide, and there are a thousand thermal waves in the fractal aerogel structure of the typical ball. Although innovative, the Smirnov model suffers from serious failings.

But the true fireball is not some weak electric fractal aerogel or burning soot. It has energy that can sometimes kill people and livestock. Fractal aerogels and burning soot cannot melt cables and pass through sheet metal. To counter this, advocates deny that fractal fireballs posses much energy, and that the cases of damage are really not from the fractal fireball, but from associated lightning. But most reported fireballs do not occur near a lightning stroke.

Although the Smirnov model is the most innovative theory since Tesla's version, like all other theories, it also fails to explain fireball characteristics that contradict physics, as we understand it, including their quasi-intelligence, navigation skills, detection abilities, sensing, fireball servomechanisms, and their "inertialess-negative-gravity" propulsion that "defy" the laws of gravity, at least, as we know them.

Some try to blend Smirnov's fractal aerogels with fiery foam structures. Some large fireballs contain interiors that resemble swirling foam. This theory proposes that lightning balls contain fiery aerofoam. But foams are nothing new. From the mundane of foods and beverages and industrial cleansers to the exotic of quantum gravitational bubbles embedded in the very fabric of space-time to the distribution of galactic structures, foams exist at all levels of scales of different matter.

Often containing gases but not itself a gas, and unable to flow like liquids, and not a solid, foams are soft matter. Its ramifications now even cast their shadow across quantum mechanics. Unlike those physicists who seek a better understanding of fundamental particles, the condensed-matter physicists in their quest for a Theory of Everything attempt to understand underlying quantum foam.

Shortly after bubbles of nucleated gas arise from a poured bottle of detergent or beer to form a frothing foam, the water drains down and leaves a closely-packed structure of polyhedral cells that changes as their gases inter-diffuse between the cells in sudden local rearrangements. As individual bubbles burst, the form rearranges its stresses and tensions to maintain static equilibrium of forces. Liquid viscosity controls the drainage rate and dynamic qualities.

On the surface of bubbles in everyday foams reside complex molecules called a "surface active agent" or surfacant that prevents the foam from collapsing under surface tensions by repelling water from the bubble surfaces and keeping the bubbles separated. Different surfactants exist.

Most liquid foams exceed 95 percent gas. Also less than five percent liquid, they are quite rigid because, where the bubbles are so tightly jammed together so that only with difficulty can they slip around each other, and the pressure within them grows and they become more rigid, taking on more qualities of solids.

Foam properties come from bubble shapes. Determining the collective and individual bubble shapes is a centuries-old mathematical pursuit. Resembling a demented soccer ball, the Kelvin cell with its six-square and eight-hexagonal faces efficiently packs foam. More superior and more efficient than packing bubbles, the recently discovered Weaire-Phelan structure combines two different shapes of equal volume.

Over time most foams age, deform and flow. If subject to forces such as gravity—smaller foam cells are less affected—and to dissipative factors such as evaporation, then the statics and dynamics of the forces alter and the bubbles constantly readjust themselves to maintain equilibrium of forces. If a fluid, gravity drains their liquid downward and evaporation depletes films. In a process named "coarsening," larger bubbles absorb smaller ones.

Most foams are fragile. But expanding liquids called foam sealants fill, seal, insulate and stop drafts are used in home construction. They fill molds, after which they harden and become a porous solid—changing states from liquid to foam to solid. Some foams have thick but less viscous walls and little gas, but the gas expands and the expanding liquid changes state into a foam.

In their most general sense, foams encompass a broad range, which is beyond our purpose here. Foams are so fragile that they must be confined to glass containers for study. But drops of aqueous foam can be suspended on 30-kHz sound waves and then squeezed to see how the sound propagates inside the foam. Other researchers use diffusing wave spectroscopy by shining a laser beam through the foam. The intensity fluctuates as the foam bubbles shift their internal stresses, reorient themselves, and abruptly snap from one configuration to another. With continuous pressure and forces, the foam continuously rearranges itself and flows continuously. The flow of system enfoldment in foams is governed by non-linearity combined with non-equilibrium constants, which allows multiple solutions and diversified behaviors.

If there is varying ambient pressure, if the liquid is not uniform and homogenous, then the foam constantly searches for stability equivalent to biological selection, and the same structure of a bifurcation diagram that describes how state variables are affected by varying the control parameters. The evolving forma states seek new stabilities. Bifurcation provides diversification by providing foams with new solutions that display broken symmetries. All this theory, postulate some physicists, translates to the fiery foam inside ball lighting—but how is the question.

Foams are a subset. More broadly, rheology mathematically examines materials—non-Newtonian "fluids"—that act in unusual and complex ways—for example, foams, mayonnaise, molten plastics, peanut butter, chocolate, Silly Putty and bread dough. Although heavier than mayo, molasses and honey, when disturbed, do flow and flatten—and are thus not rheological. But mayo retains its new shape for months. Rheology examines flowing elastic solutions and those with long-chain polymers, molds in industrial processes, and how macromolecules behave in microfluid devices of lab-on-chip designs. Over time and under the correct temperature and pressure, many traditional substances behave rheologically.

The Hill Vortex Theory, postulated by Dr. Edward Hill, formerly of the University of Minnesota, speculated that a lightning stroke separates positive and negative charges into bits of dust and molecules. This swirling ball resembles a miniature thundercloud, and the tiny discharges of recombining charges glow.

The magneto-vortex ring theory was pioneered by Dr. James Reed, formerly at MIT's Lincoln Laboratory in Lexington, Massachusetts. Reed frequently generated gases of sufficient energy density at atmospheric pressure by using a specially modified induction-plasma torch. Because it is a good conductor, plasma can be heated inductively without electrodes.

The three-centimeter plasma spheres appeared inside a quartz cylinder surrounded by an induction coil. Dr. Reed's neighbor saw a fireball during a storm.

This convinced Reed, who wondered if a plasma vortex created in the same way as smoke rings, which are generated by suddenly forcing air from a large chamber through a small hole.

The ring travels a few feet per second. Reed put a piston upstream from the plasma in an induction-plasma torch. He then tested it by blowing smoke rings from the tube that traveled across the room. Then he started the plasma. But when he pushed the piston many times, he extinguished the plasma, probably because of the low-energy density of plasmas.

Reed said that negative experiments are suggestive to others, and he hoped that his description would stimulate others to run a positive experiment. Although such rotating vortex rings were produced in the laboratory, their life spans are too brief. Unfortunately, no authentic ring-shaped lightning balls (lightning bagels?) occur in nature.

Spanish physicist Antonio Ranada suggested that linked magnetic loops explain fireball stability. He postulated that since lightning creates horizontal and sometimes vertical magnetic fields, that occasionally the two loops link to create a fireball of glowing ions trapped within the magnetic field lines. As the plasma cools, electrons re-combine with ions, thus increasing plasma resistance and weakening the surrounding magnetic fields.

Fireball bubble theories postulate that since liquids exhibit surface tension, and that since fireballs possess an outer shell with surface tension—or behave as if they do—that they will take a form with the smallest surface—the sphere. The surface will behave as though the surfaces were elastic membranes, which are everywhere in a state of uniform tension. Advocates in the 1970s of surface tension include the late James Tuck of Los Alamos.

Modern advocates of this theory postulate joined double bubbles, as occurs with soap bubbles. Such joined double bubbles possess the least surface area when their volumes are equal because this two-chambered geometry is more efficient than any other for enclosing and separating two equal volumes of space. If one of the bubbles is larger, the rounded surface of the rounded surface of the boundary, pushes into the larger bubble. In some configurations, one bubble encircles the other like a torus encircling a peanut bubble. Unfortunately for the theory, such intimately-joined fireballs are rarely reported and do not seem to exist.

Einstein's Ghost Screamed

There are enigmatic cases where twin fireballs, separated by a short distance, somehow sense how their twin is behaving. Some wonder how pairs communicate. One theory says that twins exploit quantum entanglement of the quantum states of their ions and charged atoms. Are there clues? Experimentalists have entangled two gas clouds of trillions of cesium atoms, with coordinated quantum states. They coordinate their quantum states, so if one particle of an entangled pair possesses an upwardly oriented magnetic field or spin, then the twin's spin points down.

Of course, which particle has which spin cannot be discovered until one of the two is measured. There is no way in quantum mechanics to communicate information by such "spooky action at a distance," as Einstein named it. In 1935, Bohr debated this with Einstein—Einstein lost.

But only in 1966 did John Bell prove it true. Then, in 1981, Alain Aspect separated the photons by several meters. But Einstein's ghost and critics maintained the "spooky effect" declines with distance. So in 1998, Nicolas Gisin repeated the experiment at ten kilometers, proving distance had no effect.

All respectable fireball energy theories fall into one of these categories. Energy comes from within—either generated or stored—or is fed in from outside the ball. Or, as your author suspects, is it something yet unknown? Or maybe a mix of these three categories.

The internally-powered models include those that propose radiation from long-lived metastable states of gas molecules, ionized air molecules returning to normal states, glowing tree pitch, electrochemical colloidals, electrified dust or soot, and the like. Uman and Lowke in 1968 calculated that a 20-decimeter fireball cooled at 100K/sec near 3,000 degrees Kelvin. Although such an imaginary fireball should maintain its constant size during cooling, the ball should rapidly grow dim and fade out—which is rarely observed.

Some fantasize that Casimir Effects cause fireballs to instantly appear and disappear. Not likely. It is quantum uncertainty that permits energy to appear from nothing, provided that it quickly disappears. The quantum uncertainty of both members of conjugate variable pairs—parameters—such as position and momentum, and energy and time—cannot be zero. It is a rapid see-saw: When we more accurately measure one parameter, we lose accuracy in the other. The uncertainty in position times the uncertainty in location always exceeds Planck's constant divided by two pi.

This relates to wave-particle duality, where a particle can be precisely located but a wave cannot. Since everything quantum is probabilistic, we can trade off one uncertainty against the other, as in the famous double-slit experiment. The momentum/position uncertainty occurs in quantum tunneling, used in tunneling diodes. Energy/time uncertainty permits virtual particles.

Wolfram's Cellular Automata and Fireballs

Is ball lightning a computing machine? No one understands the mechanisms of fireball containment, navigation and flight. It might involve exotic physics. Whatever it is, will the phenomenon be better understood not with differential equations but with cellular automata?

In his revolutionary book *A New Kind of Science* (*ankos*), Stephen Wolfram states that systems which change states—fireballs ostensibly included—follow computable rules, and their complexity possesses a fundamental upper bound (Alan Turing's halting problem), and all systems that reach it are equivalent. Most systems reach the bound, and this happens for most initial conditions.

Sophisticated (complex) behavior and patterns spring from simple rules (programs) of iterated repeatability, and they perform every possible calculation—even those difficult or impossible for traditional mathematics. Irrespective of the system or field, the behaviors and patterns are the same, pointing to simple and universal underlying laws.

Our traditional math lets us predict outcomes by "cheating," that is, by shortcutting the real computational work that systems perform. It works—usually. We are accustomed to modeling large-system behavior by making assumptions. And where it easily applies, our math proves precise. But systems are computationally irreducible.

As explained by Wolfram (pp. 435-457), in most cellular automata systems, reversibility is impossible. [See *The End of Certainty* by Nobelist Ilya Prigogine] But those few systems that are reversible can be just as complex. At a fundamental level, in classical and particle physics, laws are reversible. If we run time in the equations backwards or forwards, it makes no difference—the systems appear indistinguishable. For example, the traditional case of an overhead video of a billiards game run backwards is indistinguishable.

But, Wolfram disagrees (as do some physicists) and says the arrow of time is really not theoretically reversible. Although a few cellular automata systems can run backwards [*Wolfram* 441-457], for most cellular automata systems it makes a difference. Selecting initial states—deceptively straight forward, and something we learned from our earliest courses—is not without adverse consequences.

Frustrating physicists, traditional differential equations prove inadequate tools. Wolfram proves that simple programs that replicate turbulence and particle motions and collisions do follow simple rules of cellular automatons. Although on a micro scale, fluids, plasmas and gases are very different, on the macro-scale their distinctions blur. If particle number and momentum are conserved (and they probably are but may not be inside a fireball), and if there is microscopic randomness, then the same macro fluid or gas behavior will emerge.

Space-time diagrams demonstrate four patterns—of dull uniformity, periodic time-dependence, fractal behavior, and true complex non-repetitive patterns. From these, Wolfram introduces complexity and universal computations. Although CAs are more physical than differential equations, they have a history apart from Wolfram and emerged from John von Neumann's tedious transition rule and giant constructor with its arduous 29 states—definitely ugly. After Stanislaw Ulam, Arthur Burks and E.F. Codd, Edward Fredkin, Tommaso Toffoli, Charles Bennett and Norman Margolis refined CA, then John Horton Conway created his legendary *Game of Life*.

In the late 1980s, although chaos was well known, automaton research remained separate. At this time, Edward Fredkin at MIT wrote that the universe is a computer and everything a computation. But physicists associate computation with solving problems. So, what is the problem that our universe is solving?

In chapter nine, Wolfram innovatively speculates that automaton models can, through random network models and path independence, be incorporated into gravity and quantum theory. Although he conjectures that CAs are more fundamental than differential equations, he fails to develop this astounding idea. If true, then this is a great discovery.

A CA is a vast chessboard of cells that represent a finite-state automaton. Although more complicated, multi-dimensional geometric schemes are possible, that are more difficult, that only yield the same rules. Each cell can exhibit one of a finite number of distinct states or colors—Wolfram uses only black and white—governed by transition rules. Each cell is a simpleton computer that, at each clock pulse, examines certain (usually adjacent) cells, and according to specific rules, changes its own state accordingly.

Although Wolfram wisely uses the simplest CA, he arrives at the same rules that far more complicated systems also would. Instead of a linear one-dimensional array, checkered tessellation structures of any arbitrary form will do, without limit as to shapes or dimensionality (excluding fractals). And instead of discrete time-quanta, these complicated systems would flow. But for all that extra effort, that would only yield the same identical rules.

In most physics, time is bi-directional. Even in Feynman diagrams, interactions run just as well backwards. As for states, often simplicity is in the past and complexity in the future. Time enters Maxwell's eight equations and Schrodinger's equation in a symmetric way, so that we cannot distinguish waves that travel from future to past. But Wolframian science distinguishes between the arrows of time and proves that complex systems are often time-irreversible. But in the world of traditional quantum mechanics, distinctions between future and past blur.

Some physicists now ask if time is variable, or is quanta—or, if it will exist at all in a final Theory of Everything. Actually, no one knows, but there is no reason why time should flow at the same steady rate forever. In fact, time may not exist on many levels, such as sub-quantum levels. And, even on the most fundamental planes, the brain's ability to remember is not time, and the past all are memories that are only neuro-electrical signals of past events, that only occur in the present.

If everything in slow synchronism slows, instruments could not detect this slowdown. There is recent evidence that our fundamental constants might vary. [See Gabriele Veneziano.] Eight billion years ago, was the strength of the electro-magnetic force or alpha slightly less? And the Planck Constant may have been different. The gravitational constant or G may have been different. So, if so, then why alone should time be the only unvarying constant? And, if the rate of time varies, then could the cosmic background radiation, which proves the Big Bang occurred at 13.7 billion years ago (as determined by WMAP), be inaccurate? But does slowing the arrow of time fail to reverse it?

Researchers explore which CAs possess different symmetries and time-reversibility. For many CAs, the arrow of time flows in only one direction, and reversal of time sometimes undoes the prior steps—but often does not, and cannot determine past patterns. Then there is also the difficulty of accurately determining initial conditions, which is often impossible and not so simple as we think.

Cellular automaton cells are simple-minded minimalist computing elements, usually geometric arrays, with each individual cell an automaton able to exhibit a finite number of states. Each cell examines its present state and that of its neighbors, and from these inputs applies a fixed rule to re-compute its next state. Usually, all cells operate synchronously and apply the same rule.

In a simple one-dimensional system, lines of horizontal cells evolve down the screen—each new line being a successive system state—with time the vertical axis. Each line represents the next state. Each sequential clock cycle synchronizes the cells and creates a new line, In a simple system, each cell examines its two adjacent neighbors, yielding only two states. From this interaction, three binary bits yield eight possible configurations of each cell. There are 2^8 or 286 possible rules for this system's evolution.

In this minimalist computer, Wolfram discovered four categories of rules of behavior. There are simple repeating space-time diagram patterns where all the cells stay white or black, zero or one, or chessboard or alteration of stripes. This model generates dull uniformity. Then there is "nesting" or self-similar fractal patterns. Other rules generate apparent randomness, but their outputs seem to have some sort of pattern, although the sequences of state in each cell, when tested, are a statistical random series. Finally, there are rules yielding localized self-coherent, persistent structures that move in cellular space—structures that vaguely resemble subatomic particles created and annihilated in a quantum field.

Simple rules and programs produce complex behavior. Early researchers discovered that very complex—that is, complicated (not chaoplexity)—patterns were generated by simple rules, and that there is a low threshold or barrier of complexity, below which CA patterns are dull, even unchanging, and above which all four categories of behavior instantly appear. Once the rules become the slightest bit more complex, all four categories appear suddenly—there is no partial transition. But, after this threshold, adding more complexity to the rules, although it may add small details to patterns, does not change the four patterns.

Wolfram tested sophisticated systems—and there were no exceptions—so he concluded that although simple rules generate complex behaviors and patterns, that adding sophistication to these rules only added little new complexity, and all laws are accurately studied with simple one-dimensional binary arrays. In fact, Wolfram maintains that these rules apply everywhere. He writes that the cosmos itself may be one giant cellular automaton and the Theory of Everything will be reduced to several lines of code. It should, therefore, not be startling that ball lightning—whatever its physics—may not be an exception to the rules of cellular automata.

Of course, exactly how cells communicate varies. It could be by Van der Waal forces, by an exchange of bio-chemical signals or of subatomic particles, snowflake dendrite growth, onset of fluid turbulence, fracturing solids—in fact, anything.

Boldly, *ankos* claims to be a *Theory of Everything* in physics, claiming that the cosmos is a cellular automaton without continuum and with particles as only epiphenomenon, a secondary effect. Thus, motion and space are only illusions. Space-time is not a continuous flow, but is discrete, resembling signals communicated between cells. All fundamental particles exhibit discrete quantum states. So, if the universe performs computations, do the basic operations consist of particles and photons moving between them? No.

Although we may divide space into tiny cubical grids to make a CA model, this is a mistake for two reasons—there is no master clock to synchronize the cells updating and we must not impose on space a particular geometry. Instead, without a specific geometry, it is in a free-form network in which cells are nodes.

Although the node inter-connections are predetermined, their spatial coordinates remain unspecified. At this level, that shape and position lack meaning. So, rather artificially being built into this model, space geometry emerges naturally from it.

We perceive three spatial dimensions because each network node possesses three incoming and outgoing links. There is no synchronizing master clock. Updating signals or events propagate along network links. This affects the arrow of time and causality, making it flow one way, and thus is a *causal net*.

Are programs and algorithms better than differential equations to explain physics? We must not just examine how patterns are created, but how they are perceived. So, if a pattern looks random, then it is random. Randomness is not an inherent or intrinsic property of a pattern that is generated the way that pattern was created, because the mechanism that created that pattern was deterministic.

Wolfram's ankos has much to say about the inherent irreversibility and unre-producibility (and unduplicatability?) of initial conditions—and thus could it pose difficulties for the theory of a four-level Multiverse? Level I parallel universes experience our laws of physics, just as we do, but with different initial conditions. [For more, see Appendix on "Multiverse Fireballs."]

CA has weaknesses. The densely populated spaces of programs in CA systems recognize as valid every rule that relates a neighborhood configuration to a next state. But real life systems are much more selective about what they recognize as valid.

These days, it is fashionable to over-compare everything to digital comput-ers—including DNA, brains, consciousness, economics, politics, psychology, and religion. But some wonder if everything is made of more than sequential and combinational logic—and Wolframian science.

Although the older-and-faster analog computers fit into digital cellular autom-ata philosophy, today's advocates forget that quantum computers that were first explored by Feynman may one day replace digital computers. They operate dif-ferently—in parallel universes, explained by Multiverse Theory.

However, Roger Penrose brilliantly argues that something else is involved in the mind than mere *ankos*. What Penrose writes also applies to fireballs. If Pen-rose is correct, then maybe *ankos* cannot explain fireballs. Unfortunately, to dis-cuss this would require a full chapter and will lead us astray. So, instead, to examine Penrose's amazing views on this—the human mind, true consciousness, and robot intelligence—please study his excellent book, *Shadows of the Mind*.

As some ask, can *ankos* computational engines, running random programs on random data, spontaneously generate sophisticated systems—even human intelli-gence? No.

Wolfram's Principle of Computational Equivalence states that the computing power in most natural algorithms or procedures is the same because they are equivalent to a Universal Turing Machine. But can simple programs explain everything?

In the physical and life sciences, although simplistic computational paradigms can simulate things quite well on paper or inside a computer, out here in the real world there are constraints affecting growth patterns. From the surrounding media, there are developmental requirements (even energy transport), and chemical and physical properties. For example, this gives rise to Fibonacci patterns in leaf placement, but not because of Wolframic computations. Just because a computation resembles a natural process does not mean that it is due to a computation. For example, contrast Keplerian and Copernican explanations of planetary orbits.

Determining causality often takes much experimenting, observing and theorizing, and this also proves true for ball lightning theories. So far, all theories fail. One leading physicist, an expert on ball lighting, contemptuously stated, "The current state of ball lightning research is just so much mathematical masturbation." Towers of seductive mathematics disguise failed theories.

It is not our intent to spend much time on rehashing reincarnated and failed theories. Other books and conferences rehash quite well. Rather than reincarnate old theories with new mathematics, researchers must cross on stepping stones across a river to explore bold new directions—ones that will not defy quantum mechanics. There is one original, new hypothesis.

PART IV
Stepping Stone

23

Sagan-Hill Hypothesis

Propulsion theories were proposed by a brilliant NASA top rocket engineer, Paul R. Hill—but are herein modified and reworded from his text by your author for the first time and applied to fireballs. Since Hill is deceased, no one knows if he would have made these alterations. Although I was careful, any possible (but unlikely) misunderstandings in reworking and adaptation are mine, not his.

The proposed Sagan-Hill model is incomplete—maybe even in great error—but it is the first serious attempt to conservatively explore the "negative-gravity" behavior of fireballs and their control and manipulation of gravitational fields. This model "breaks the mold" in its honest admission of negative gravity (first suggested by Kip Thorne) and how fireballs "defy" gravity—or, more correctly, strongly control, manipulate and maneuver within gravitational fields. There is absolutely no other logical conclusion yet that correctly explains fireball navigation and propulsion.

Do not expect the answers to be immediately forthcoming from within the extension of existing physics. And, unfortunately, the supersymmetric extension inside the upcoming post-Standard Model [see Gordon Kane] will still fail to explain fireball propulsion and navigation—nor will it correctly include gravity. For that matter, neither can the present version of string theory correctly include gravity—for it also fails to explain fireball qualities, such as controlled manipulation of inertialess-negative-gravity for fireball propulsion.

Working in reverse, your author puzzled over fireball qualities, seeking possible explanations.

Explaining Fireball Qualities

First, color. With photon energies between 3.26 to 1.65 electron volts and resulting wavelengths between 3,800 to 7,500 angstroms (an angstrom is one tenth of a nanometer), fireballs generate corresponding colors between ultraviolet down through the spectrum to the infrared. Red and orange correspond to the lowest energies, and blue and green, to the highest energies. Yellow is midway between orange and green.

In most statistics, the average sum in frequency distributions list the colors of red, orange, blue and violet as roughly 15 percent; white at 25 percent; and mixtures at 25 percent. Stenhoff reports that a Spearman rank correlation test applied to form major surveys yield only loose correlation coefficients. But, unknown to many, the McNally results that Stenhoff lists are—and few know this—compiled from only a few select cases. Most of the original McNally 630 cases were, prior to my book, never published and even believed lost.

Stenhoff correctly referred to the poor correlations and uncertainties between the different studies and within each study as "perpetual uncertainties concerning color." The only color that all studies agreed upon was green—agreement within three percent. The internal and cross inconsistencies are caused by the inherent inadequacies in all statistics.

According to Hill (p. 62), the air radiates in any or all colors. Nitrogen provides Gayden green, atmospheric oxygen adds radiance to green, yellow, orange and especially red. If excited in a gross manner to simultaneously excite or emit, the color mixture appears as a blue-white. A mix of all the colors yields white. If the blues predominate, the blend resembles the blue-white of an electric arc. Some fireballs excite different spectral peaks and colors, or different combinations.

Whether cold or hot, ionized plasma looks like a flame because excited electrons fall back down to lower vacant energy levels and emit visible photons. As for colors, red and orange take the least energy to excite; blue and blue-white take the most energy. Fireball illumination comes from inside the ball and from its surface—whether the surface is clearly defined, fuzzy, or has a concentric halo or outer shell.

Light intensity is proportional to the photons per second. When a fireball makes sudden changes in speed, theoretically brightness should change with increasing fireball activation energy, but this is not reported. Any wavelength that a gas emits when excited, it will absorb; and through a short distance of plasma (if that is what a fireball consists of), the plasma becomes opaque to its own emission frequencies. Furthermore, an ion-sheath seems to surround some fireballs.

Some glowing fireballs seem cool or warm; but others, sometimes glowing dimly, exhibit great heat. Why? All plasmas are not alike. As for heat and light, cold and hot plasmas look alike because both are highly ionized with similar light emission. But hot plasmas are flames from combustion, and cold plasmas from energetic particles. Such particles include electron and x-ray photons with enough energy to knock electrons out of nitrogen and oxygen molecules.

A surface glow often obscures the edges of a ball, but not always and some are, in the words of one witness, "smooth and clear as butter." The reason is because plasmas have a critical thickness beyond which light will not penetrate. The absorption spectrum of an excited gas closely equals its emission spectrum. Any faint ray of light would be absorbed by excited molecules and re-radiated randomly in all directions.

It is impossible to determine colors for certain because witnesses who see a single color may see a narrow frequency band or several colors spread over a wide range, which will project the same appearance. And there is no easy way to get a tri-color camera with filters to determine the real fireball colors because of the unpredictability of fireball appearances. Spectrographic grating attachments on ordinary cameras could work, but there are no such reported cases.

How hot is ball lightning? Heat varies widely and its manifestation can quickly change. From heating and ionization data, and studying the best cases, we obtain conflicting indications that fireballs radiate intense energy at times, little at other appearances, and can switch from "lamb to lion" in an instant.

Second, propulsion. We must answer an important question—does ejecting elementary particles propel fireballs? If so, charged energetic particles will not work for propulsion. There are no apertures for such particles to come through a ball's shells (or equivalent envelope) and it is impossible for an isolated fireball to neutralize this current flow. Also, shooting down charged particles creates an air drag and a mild wind or breeze, which is never reported. If the charged particles are accelerated three million electron volts, the air will become radioactive. Radioactivity was measured in at least one Oak Ridge case and one Russian case (Dmitriev), but this is insufficient evidence. And any neutral particle beam could penetrate deeper than a charged particle beam. An intense photon beam would be visible and vaporize anything near it. All known particles and forces are inadequate.

Fireballs lack any visible means of propulsion. Let us examine the invisible propulsion force utilized by fireballs. If they retain mass, thrust forces propel them. We can eliminate several possibilities. Direct mechanical forces such as wind, floating, propellers and Bernoulli's lift are obviously absurd. Fireballs do not use buoyancy or aerodynamic lift, as do balloons or birds and aircraft, and they are not propelled by chemical action-reaction of rocketry because there is no visible exhaust and most balls fly silently, while those that make noise only sizzle electrically.

As for acceleration or force field propulsion, unlike the earlier inadequate candidates, this one is not inconsistent with fireball qualities. (The terms force field and acceleration field are synonomous.) The potential strength of the four known fields is enormous. Just as an example, assume God could separate all the charges totally from a penny—see *Physics* by Halliday and Resnick, volume 2, pp. 652-3—then at 124 miles above the earth, this separated charge could lift an incredible 4,500-ton weight with an amazing, high-speed—at a 100-g acceleration! With existing technology, it is impossible to separate these charges. Even our best capacitors make only a feeble separation. This thought experiment dramatically illustrates the enormous strength of fields. And, unlike particles, fields propagate at the speed of light.

Force that propels a fireball—which is made of light contents, gas—is modest. Thus the force field lines should not affect anything near it. There are four fundamental interactions or forces—electromagnetic (photons), gravitational (gravitons), strong force (gluons) between nuclear particles that hold the nucleus together, vector boson weak force that explains radiation that operates by the exchange of messenger particles. The fireball's propulsive force field is a negative gravity field or another field with similar properties, but is yet undiscovered. Could a force opposite to gravity, possibly with anti-gravitons, propel fireballs?

As for the deliberation exhibited by controlled flight, if unable to focus its inertialess-negative-gravity force field, the spherical symmetry in the balls' force field would be worthless for controlled flight. Fireballs use thrust vector control. Rarely do they emit visible jets—it is unknown if these jets are hot or if they are cold plasma—but some sizzle or hiss. These jets do not affect the balls' stability, and they are never used for thrust vector control. Some fireballs trail glowing tails.

One might suspect that fireballs—whether they are spheres, thin rods, fat sausages, oblate spheroids (pears), ovaloids (footballs), or other rare shapes—possess an axis of symmetry passing though their centroid or center of gravity, and through which points its vector thrust. Its inertial force, which should not be great, would be equal and opposite to the product of the ball's mass and acceleration—assuming mass does not change in which case the vector force would be the derivative of momentum, as with subatomic particles at near-relativistic speeds and de-massing rockets burning fuel.

Third, silent speed. Effortless in flight, fireballs dart, fall from clouds, circle witnesses, quasi (as if)-intelligently navigate, and closely pace fast-flying aircraft. But can fireballs fly at supersonic speeds? UFO skeptics such as Phillip J. Klass have long-held that ball lightning flies alongside aircraft at high speeds and has been mistaken for some UFO sightings. Several cases (described elsewhere) seem to indicate that supersonic fireballs do indeed fly silently. If true, it is because they eliminate energy loss from bow shock waves.

To eliminate air compression, their force fields control airflow. Because the force field velocity is the speed of light, a supersonic ball "instantly" signals air ahead that it is coming and notifies the air to slow and move aside. The force field ahead of the ball provides an equivalent force to retard air by adjusting its field energy gradients to replace aerodynamic forces. Pressure gradient signaling ahead fails at supersonic speeds, but the force field succeeds.

A ball is not streamlined. Because of a ball's blunt after-body, airflow separates from the round after-body to create broad turbulent wakes that provide a high drag at slow speeds and strong bow waves and strong trailing shock waves when supersonic.

To solve the flow problems of the ball and its inefficient aerodynamics, the acceleration fields before and aft of the ball must be mirror images. Controlling the force field around a ball provides airflow control. There will be no shockwave; and streamlines are speed-invariant, without separated-flow regions. There will be only a small turbulent wake and very low drag.

In flow separation, the rear of a ball has higher pressure that leaks forward to the lower-pressure within the subsonic boundary layer that is close to the ball. This causes flow in separation. These layers possess a steep pressure rise that stalls the low-speed boundary that then piles up to form a large region of dead air, forming a turbulent wake. Creating constant-pressure flow over its surface, the fireball solves the problem. In a conventional cannonball, air stalls in the boundary layer and there is flow separation. But by stalling air in the boundary layer and separation of flow lines, the fireball eliminates the undesirable rising pressure region. Constant-pressure control alters the fireball's spherical body into a streamlined shape.

This force-field control accelerates air around the fireball in a subsonic pattern, irrespective of speed, so that Mach-1 becomes irrelevant. This is not pressure-controlled flow as with a cannonball. Any insect in its path will not feel any difference in acceleration between different parts of its body and will move aside. The analogy is a satellite in orbit.

For silent supersonic flight, is there an alternative to the force-field control of airflow? If fireballs can de-mass or neutralize the mass of air near them, then yes. This would greatly raise the speed of sound. Deviations from atmospheric pressure due to air dynamics would be insignificant. The air would follow a subsonic and shock-free pattern. So there is a big question—is this from zero mass, that is, controlled mass, as your author suspects, or a counter field, as Hill suspects? Either will explain ball lightning propulsion.

Fourth, anti-gravitons. Hill proposes a virtual field exchange particle, hypothetically christened the Uon and yet undiscovered, that possesses a two-way exchange nature. It reflects from matter.

A hovering fireball beams small numbers of Uons downward that reflect back up, and many would re-reflect for a second round trip. Like other field virtual exchange particles, these reflections are absorptions and re-emissions. Like the photon-reflections from a mirror, Uons would penetrate the Earth and be concerned with the heavier particles of the nuclei and not the light electrons in their orbits. Hill suggests that it may be possible to convert electron-position pairs into anti-gravitational field energy.

The ball's field is not of static-electric/magnetic type and may be a negative-gravity quasi-static type that repels all mass. The ball is able to field focus, giving it control over its propulsion. If the anti-graviton exhibits a reflection property, then it will explain ball lighting fields. Exchange particles transport great quantities of energy without losing any. For example, gravitational fields are energy-conservative, and work done is independent of the path taken to move between two points.

Proposed Uon exchange-particles, like all exchange-particles, travel a bi-directional street. For example, the same number of gravitons reaches the Moon as travel to the Earth, which explains Newton's Third Law of equal action and reaction. But electromagnetic photons are absorbed, accelerating electrons to higher orbits, that then fall back and re-radiate another photon. Similarly, Hill postulates the Uon-reflection hypothesis. The Uon beam deeply penetrates the ground; and being concerned with mass, is reflected back by the nucleons. These reflected Uons return to the fireball and are re-reflected, thus balancing with the Earth's gravity and canceling g-loading on the ball's contents.

Fifth, energy. The energy source of fireballs is an enigma. All conjectured sources are external or internal (stored or self-generated). As critics readily admit, both models suffer from severe difficulties and cannot explain fireball behavior. Partly because of the Faraday cage and submerged-in-water fireball cases, the external model has fallen into disfavor. But advocates solve this by ignoring these cases or denying that they ever took place. The internal model, likewise, has serious flaws, notably the massive energy content. To solve this, advocates deny that fireballs ever exhibit much energy. Both advocates have a simple solution—If you do not like the facts, deny or ignore them.

As for the source of energy, no one knows. There were several cases where evidence of radioactivity was involved. But residual radiation is generally not detected because its radioactive rays may not be hard gamma rays. If a fireball emits a ray, because of its light mass, this weak ray is invisible, although it might still create some ionization. Radios near a fireball sometimes roar with static that drowns out the station.

Sixth, no shockwave. Eliminating supersonic shockwaves substantially reduces drag reduction. At subsonic velocity, the air ahead learns of the oncoming aircraft from the pressure increments transmitted forward from the leading edges, and combines to slow the air and move it aside so the craft can pass through. But at supersonic speeds, these pressure increments collect on the nose and leading edges.

But if the supersonic fireballs control the air ahead with a field that travels at the speed of light, then this control field will replace all the incremental pressures and absorb the kinetic energy. This makes the air smoothly stop before the leading edge without raising pressure and so there is no compressive heating or shockwave.

The fireball tailors the force field to replace the pressure increments within the air flow-field into an ideal constant-force and density flow around the ball at all speeds. This constant-pressure flow eliminates shockwaves. There are no rarefaction and compression waves. A person near a ball flying by him would not detect the passing wave of force because of its weakness. (But if the ball were massive, naturally the force would be felt). This saves the subsonic and supersonic fireball great savings in its propulsive field because the ball is converted to a shock-free, low-drag sphere with a constant pressure and density in the streamline paths.

In the Klass Hypothesis, noted authority, aviation editor, co-founder and UFO skeptic for CSICOP, Phillip J. Klass correctly maintained that fireballs were under-reported in the clouds and atmosphere, and were routinely and mistakenly identified as UFOs. Back in the 1960s, Martin D. Altschuller—who wrote chapter seven on atmospheric electricity for the 1968 USAF-sponsored, University of Colorado, Condon report—did then criticize this skyball hypothesis. But Klass' theories predicting supersonic skyballs eventually proved more likely to be accurate than they then believed. Balls of light and fireballs are reported near military interceptors flying at supersonic speeds. From examining the 630 Oak Ridge cases for this book, your author concluded that Klass was correct.

Seventh, horns and jets. Sometimes, one or two flaming horns or jets shoot out from holes that open in some fireballs—tongues of fire that reach out up to half the length of the ball, or longer. Although some jets hiss, most are oddly silent. Oddly, propane or oxy-acetylene torches (and your author used them for welding and cutting) roar always, and small jets of fire hiss or scream always—no exceptions. And when a jet opens up on a fireball, there is no evidence of action-reaction on the ball. The flame may be a cold zone of ionized air. Length of this flame differs and usually is steady, but may lengthen—although a few stab out and withdraw several times—before such "horned balls" disintegrate or explode. Dual-horned balls sprout two fiery horns that are not always identical: one may be long; the other, short.

Hidden Progenitors, Quantum Entanglement

As novel as it is—and probably closer to the truth than any other theory—the Sagan-Hill Hypothesis fails to explain the non-locality of Collateral Effects. Ball lightning is rarely caused by lightning, and often materializes out of thin air. Whatever causes it, also selectively collaterally affects nearby objects—sometimes on clear days. How, is to say the least, very puzzling. For this and other reasons, it is likely that a "Hidden Progenitor" causes fireballs and their—for want of a better hypothesis—nearby Collateral Effects.

Fireballs exhibit puzzling qualities that cause nearby unexplainable damage. An obvious question emerges—Does an underlying hidden shared cause create both fireballs and their unexplainable nearby effects? These Collateral Effects exhibit non-locality and—without reference to Einstein's EPR—exhibit an unexplainable "spooky action-at-a-local-distance." Collateral Effects fall into five unexplainable "spooky" categories.

- First, some fireballs instantly detect electrical changes near them—such as when a witness turns on a switch, and may react instantly.

- Second fireballs occasionally cause nearby—within fifty feet—wires and electronics to inexplicably burn out, even without a storm, but cannot be attributed to electromagnetic emissions or existing physics.

- Third, collateral electrical and electronic damage may be mysteriously selective, and melt one component but avoid the rest.

- Fourth, a few fireballs detect and mysteriously follow buried pipes.

- Fifth, other concurrent, unexplainable effects occur that hint at a deeper cause—a Hidden Progenitor.

Ball lightning is a result of something, just like a hidden magician who creates the fireball and its related Collateral Effects that interact with each other and thus seem to us related—but we never see the magician and are fooled by his illusion as to its real cause. Or fooled by a hidden operator who through invisible threads controls his marionettes. Likewise, we see fireballs that seem to cause their Collateral Effects, and we erroneously think fireballs caused them—but never guess the Hidden Progenitor as the real cause.

Non-locality rules the quantum world and affects the Hidden Progenitor, ball lightning, and Collateral Effects. In 1935, with his EPR thought experiment, Einstein failed to undermine quantum mechanic's Copenhagen interpretation. In 1952, David Bohm suggested a variation of the EPR thought experiment with photons. John Bell invented practical methods—Bell's Inequality—to test Bohm's version of the EPR paradox, and then Alain Aspect tested it to prove the quantum world violates local reality.

Bell's Inequality rests upon three assumptions of local reality—the universe is real and exists independent of us, that logic works, and no signal can exceed light. At first glance, non-locality seems to violate relativity—that is, nothing can exceed the speed of light—and we could instantly communicate with extraterrestrial astronomers across billions of light years (making SETI delirious). But non-locality cannot transmit information because the receiver only receives random numbers.

Bell proved that David Bohm's hidden variables exist. Hidden variables theories are non-local. A sub-quantum world, an underlying layer of reality, contains additional information as hidden variables that, if known, would predict precise outcomes of measurements. Unlike the Copenhagen interpretation, it says there already is a specific value, but the observer does not know what it is. Hidden variables control seemingly random quantum events. In Bohm's universe, the future, past and present co-exist; and the subatomic particles are secondary, process of movement, continuous unfolding enfolding from a seamless whole.

Unfortunately, progress was stifled for a half-century when John von Neumann made, in Bell's words, "a foolish error," when he wrote that hidden variables could not exist. In 1932 von Neumann incorrectly wrote that hidden variable theories could not explain quantum mechanics. Then, next year, Grete Hermann, correctly pointed out von Neumann's error, but no one listened to her. Von Neumann's formidable reputation frightened off everyone else—except Bohm—so the error stood unchallenged.

This error strangled progress. But in 1966, John Bell proved von Neumann's proof was fatally flawed—involving commutation Abelian groups, which include operations where order is unimportant. But, unfortunately, even after Bell exposed von Neumann's "foolish error," the error did its damage by altering the flow of progress and blocking research into Bohm's hidden variable theory—an accident of history.

Non-locality has great implications. The entire universe is a system with linkages that bind all participating particles into one cosmic quantum system. The cosmos resembles a single quantum system. (As asked in the appendix, do Multiverses exhibit non-locality and "inter-verse" linkages?)

A quantum entity is instantaneously (infinitely fast) influenced by what occurs at the nearby locality and other distant localities—even in distant galaxies. Non-locality is not debatable. In double-slit experiments, any electron moving through one slit depends on whether the other is closed or open. And the Aspect experiment used Bell's inequality to prove that we cannot avoid non-locality if we treat quantum things as real and existing independently of observation.

Discovered by Louis de Broglie and developed by David Bohm, and filling the universe, pilot waves guide real particles around. What occurs in any quantum part of the cosmos instantly and globally affects the entire pilot wave, irrespective of location. Bell's Theorem and the Aspect experiments prove reality does not consist of separate parts joined by local connections, but instead that all the universe's particles are interconnected, inter-dependent and inseparable.

Although no one disagrees with the equations of quantum mechanics, many disagree with how to interpret its paradoxes—notably the EPR Paradox, which has several forms. Bohm proposed the most important. For an example, consider two identical particles that leave and travel in opposite directions, with right and left spins superimposed.

Measured for spin, the wave function of particle A collapses or vanishes and that particle acquires a right or left spin. Angular momentum carries over into the quantum world, and it is conserved. So, when A is measured, it acquires a spin and B must acquire the opposite spin. When physicists on earth measure A, and discover it has a right spin, extraterrestrial physicists at that instant on the other side of the universe, measure B and detect a left spin. [Read Martin Gardner's. "The Guided Wave Theory of Louis de Broglie and David Bohm."]

By analogy, Bohm contrasted camera and hologram. When a hologram uses a laser and its coherent light to record a three-dimensional image on very-fine-grain film, it encodes intensity and direction in the troughs and crests of reflected waves or object beam that are in varying degrees of being in-or-out-of phase with the laser's reference beam. When light shines onto developed holographic film, its constructive and destructive interference patterns act like microscopic mirrors positioned at myriad angles, so that light reflects off the hologram exactly in the original directions.

Because intensity of each reflected wave (object beam) varies with direction, each eye sees a different view, thus creating a perception of depth. In a way, each hologram has billions of photographs, each taken from different perspectives, with a different depth focus.

To view a hologram, light shined onto the film from the direction of the original reference beam, creates interference patterns that diffract and reflect light to recreate the orientation and intensity, so that each eye sees a different reference pattern for each given point. Thus, the observer sees a 3-D virtual image floating behind the hologram. Unlike a simple photograph, which reflects the same intensity in all directions, a hologram registers both intensity and direction, and there are differing shadings for both ways. The reflection hologram just described is only one of many types, but the principle is the same.

By comparing lens and photograph with its point-to-point imagery, Bohm contrasted it to the static hologram, where waves from the whole object come from each part. The reflected light interferes with that from the direct path, and the resulting image is recorded. When the plate is illuminated with laser light, the interference pattern looks like the whole structure.

But in the universe, instead there is a constant dynamic wave pattern reflecting off an object and interfering with the original waveform, so that in the movement pattern, many objects are enfolded in each region of space-time. Particles do not exist continuously, but alternately come in and go out, so that the close condensations approximate a track. [For a description of Bohmian Mechanics, read pages 65-171 of Bohm's *Wholeness and the Implicate Order.*]

Developed by George Berkeley, Mach's Principle postulates that inertia depends on its relationship with all cosmic mass. Berkeley said that since all motion is relative, linear and curvilinear acceleration must be measured against something—that is, relative to the collective mass of all distant stars. Einstein later incorporated it into General Relativity.

From this, the Wheeler-Feynman absorber theory starts with the obvious—because of inertia, all matter resists being pushed. Likewise, forcing a charged particle to move requires extra energy above that of an uncharged particle—and thus it has extra inertia. For example, broadcast antennas that radiate electromagnetic radiation require extra energy because of this "radiation resistance." [See *Q is for Quantum*, pp. 435-8]

Born out of the radiation resistance puzzle, Wheeler-Feynman absorber theory states that every charged particle, such as every electron, knows instantly what occurs to all others at once in the entire universe. John Cramer's transactional interpretation is a legitimate alternative to the Copenhagen or many other interpretations.

Making the identical predictions as quantum theory, the transactional interpretation developed by John Cramer provides a different viewpoint than either the Copenhagen or "many (infinite parallel) worlds" interpretations of the four levels of Multiverses. No charged particle is isolated. For example, every electron "knows where it is" relative to all charged particles elsewhere, and vice versa. Each electron resists being pushed because of all distant charged particles throughout the universe. Push an electron and every other charged particle instantly knows it.

Because it ignored quantum effects, the original Wheeler-Feynman theory was classical. But to apply absorber theory to quantum mechanics, they used Maxwell's equations, which yielded two solutions—one describing a negative energy wave flowing into the past and the other, into the future. For pragmatic reasons, we usually ignore the full version of Schrodinger's wave equation, but which has another solution that allows relativistic effects. With its dualistic sets of solutions—the familiar one and a mirror-like image—Schrodinger's wave equation describes the flow of negative energy into the past.

When calculating quantum probabilities, duality manifests itself. A complex variable, state vectors describe quantum system properties.

Schrodinger's equations describe advanced and retarded waves. In a typical quantum "transaction," a particle "shakes hands" with another one somewhere in the universe. Without going into the details [see *Q is for Quantum,* pp. 409-413], there is difficulty in how to treat interactions that are bi-directional in time and simultaneous. The transactional interpretation semantically cleverly accomplishes this by standing outside of time, using pseudo-time.

A retarded wave propagates into the future; an advanced one, into the past. The advanced wave moves into the future, encounters electrons, which absorbs the wave's energy. The transactions simultaneously occur across space-time—they are atemporal. Transactional interpretation makes no predictions not already made by quantum mechanics, but is a different conceptual model that develops insights and intuitions. Remember, all quantum models are just that—only models, not reality. Quantum theory's probabilistic results come from underlying (deterministic) motions of smaller particles, that is, hidden variables.

Although greatly skeptical, your author cannot absolutely dismiss the following conjecture. Do fireballs enter and leave our Hubble universe from one of the four Multiverses? [Refer to "Multiverse Fireballs" in the Appendix.] Some physicists suggest that dimensions from other universes lie a millimeter away.

By way of analogy, assume a hypothetical two-dimensional flatland. Two-dimensional Hubble flatlander physicists live on this flat plane. They wonder if their flatland has a positive or negative curvature, is a sphere or saddleback, but discover it has zero curvature. Some conjecture that other undetectable flatland universes exist, possibly parallel to them and separated by a mere millimeter.

When unexplainable flat fireballs (or is it "fire circles"?) appear and disappear from and to nowhere, and "defy" gravity and "violate" other laws of their physics, some flatlanders wonder if these flat fireballs originate and return to another planar universe. Flatlander's Kip Thorne and Carl Sagan even propose a theory of travel by wormholes. But because Multiverse research is still in its early stages, to suggest any connection with fireball formation is interesting conjecture. [For more, read articles by Max Tegmark and Visser, Kar and Dadhich.]

How other theories might explain ball lightning is a mystery. I am no expert on quantum mechanics, so I cannot predict if Penrosian or Bohmian Mechanics, or even loop or string theory, or any other existing theory will prevail—perhaps in modified form. The leading contender, which your author favors, is string theory, but it has little empirical support and is valued because of its elegant mathematics. And, rather than emerge from conventional physics, it came almost entirely from advanced mathematics.

Although its math is beautiful, string theory fails to make real quantitative predictions or exhibit solid empirical validation, and postulates undetected particles. The equations of string theory are so difficult to solve that leading physicists only understand them in very special cases. This is not surprising, because even now quantum theory—which is well established—models only simplest molecules, and even the Standard Model fails ten ways. [See Gordon Kane.]

As original as it seems, is the aforementioned Sagan-Hill Hypothesis really correct? No, probably not. But compared to other fireball theories, it is honestly bolder. At the very least, it will expand our horizons into new directions. But the evolution of our present physics was not deterministic and might easily have gone very differently, down unexpected paths.

Paths Untaken Diverge

Accidents of history are common. Events that unfolded were never inevitable, and need not occur as they did—even in the basic sciences. The infinite paths untaken diverge—sometimes chaotically, sometimes dangerously and sometimes radically, going far from where we now arrived.

If you doubt me, then let me explain. There are precedents for my speculation. We might imagine how the paths not taken in physics might have unfolded differently—if only certain forks in the road were taken. For example, what if Riemann had not entered mathematics or if Grossman had not told Einstein about Riemann? Would we have to wait until 1960 for General Relativity? Or if after Heisenberg's allergy and recuperation on Heligoland Island, suppose his two friends had not informed him about matrix algebra? Students might still be saddled and be intuitively misled with Erwin Shroedinger's equivalent-but-different approach, using Hamiltonians and later Lagrangians—but with an unfortunate-but-alluring subconscious link to waves as having some sort of physical reality, which they do not. Without Shroedinger and Louis-Victor de Broglie to explain wave-particle duality in terms of "hidden variables" and pilot waves, it may not have filled David Bohm—the father of Bohmian Mechanics—with inspiration to find an alternative to the Copenhagen Interpretation.

As heretical as it first sounds, are there things that on a truly deep level we rely on—possibly like quantum mechanics, space and time—that maybe are not truly essential? But because of our mathematical consistency and only an accident of history, we chose a certain direction. In the exciting first three decades of the Twentieth Century, many physicists developed equal-but-alternative quantum ideas that philosophically lost out and are forgotten. Physics might have chosen different roads—all experimentally successful, but different. And extraterrestrial worlds might have taken these alternative roads and discovered the same laws differently, but that work just as well.

And, what if Riemann in the mid-nineteenth century had not developed higher-dimensional mathematics? Einstein would later have failed at General Relativity. Or what if Riemann had not died young of tuberculosis? He might have discovered electromagnetic theory and its eight equations before Maxwell and then realized that the fourth dimension was not spatial but of time, and then been onto some sort of theory of invariance and Special Relativity—but earlier than Einstein in 1905, and probably shortly after the American Civil War! Was this near-miss a path not taken—an accident of history? Classical physics would have fallen by 1870.

Or what if later Grossman had not informed Einstein about Riemann? I think it would be unlikely that Einstein (as he said) could have created his crowning achievement—General Relativity. And, unlike Special Relativity—which others already by 1905 were beginning to propose—then the much more difficult General Relativity would most likely not have been developed until recently; and even, as some string theorists dare suggest, perhaps "falling out" of the equations of string theory as a special case. Leading physicists suggest this. Brian Greene postulates in his book *The Elegant Universe* how, if Einstein had not discovered General Relativity—and there is no reason that others might have for some time—then physicists might have developed string theory first and seen General Relativity fall out of it!

But the reverse sometimes occurs. String theory was done backwards—something that accidentally fell back in time out of the twenty first century, back in time to1968, but only by two lucky accidental discoveries. The discovery of early string theory in that year by Gariele Veneziano at CERN that the Euler beta-function inexplicably describes properties of strongly interacting particles was an accident of history without explanation.

In 1970 the trio of Nambu, Nielson, and Susskind postulated an explanation for this curiosity—that particles were tiny vibrating strings. The theory was then swept aside by inadequacies and the success of Quantum Chromodynamics. Then, Schwartz and Scherk expanded the theory to include not just the strong force but also gravity. In 1984, Schwartz and Green tidied up some conflicts and incorporated all four forces. Over the next three years, physicists were able to derive equations of the Standard Model from it. In 1995, Ed Witten united the five string theories, to create second-generation, string theory. String theory walked before it was born. Because of this premature birth, the final steps in the evolution of string theory are still missing, and the field is fueled with conjecture. It was an accidental path taken prematurely—an accident of history.

Earlier, if Shroedinger had not succeeded, then Heinsenberg's equivalent—but different—theory would have done the job quite well. Shroedinger's wave theory and the transitional Bohr-Sommerfield atom—in my opinion, although they conceptually functioned as crutches, did damage—they conceptually held back our predecessors, and still leave anachronistic imprints of their misguided legacy. There is nothing conceptually intuitive about quantum mechanics.

Although quantum mechanics has proven, like relativity, to be incredibly accurate—Penrose and Hawking once debated which was measured most accurately (quantum mechanics won)—the endless controversy and debates about the meaning of quantum mechanics still rage. It is safe to say, as once did Richard Feynman, that no one yet understands it. If you try, as Feynman said, it will be intellectually and emotionally upsetting. And if you think you understand quantum mechanics, you are delusional.

Major paths that were never taken, we will never know. Stephen Wolfram in *A New Kind of Science* is amazed how many of his revolutionary discoveries—which may affect ball lightning research—were not made earlier. If his bold predictions come true, then the sciences will change radically in the next 20 years. It could have begun much earlier, as Wolfram writes, and we would not today be using advanced calculus so exclusively. If he is right, then perhaps we took a wrong turn at a fork in the road back there—another path not taken.

We might mentally re-run history, and wonder if only Morgan had adequately funded Tesla at Shoreham, could he have resonated the spherical cavity with Zenneck "surface" waves? If so, could he have pioneered low-loss, worldwide wireless power distribution? If possible, then making distant energy from remote deserts and jungle rivers available globally—burning coal and oil at their sources rather than transporting them and thus avoid the great cost and risk of ocean oil spills?—might have altered history and reduced the world's dependence upon oil.

But if Tesla had not done this research, or if his laboratory notes had been truly lost, then history shows us that no one would have re-discovered how to achieve this—for over a hundred years, if ever. Furthermore, if Golka had not become interested in ball lightning and Tesla, and had not pursued his vision, the Corums later would never know, and there would be no "coilers" or serious ball lightning research—and this book would not exist. The development of science is not a straight line. We can never know what never took place and might have been—never know the paths not taken.

To physicists, the enigma of ball lighting was a stumbling block. But the difference between such a stumbling block and a stepping-stone is what they will now make of this enigma.

However, the enigma of this stepping-stone is not the end, but the beginning. Our story continues, because there exists a mysterious phenomenon that, much like ball lighting, also involves inertialess-negative-inverse-gravity—another phenomenon that is even more enigmatic—but also very real.

PART V
Inverse Gravity

24

Clues

As we have already seen, all present candidates for the Theory of Everything in physics are inadequate—and, as pointed out by Brian Greene, string theory is a magnificent edifice still in construction. Pioneered by t'Hooft, Susskind and Maldacena, the holographic principle is a major stepping stone in string theory—or it might even illuminate the entire edifice of string theory—but we do not know yet—just as the equivalence principle, general covariance did so for relativity, and gauge invariance does the same for the Standard Model.

As Greene mentioned in a mostly-unpublished *Scientific American* article, we are still climbing this edifice. But by examining both true ball lightning and also the existing clues of aircraft characteristics and behavior, it is already possible to predict what—in vague terms—a true, final Theory of Everything in physics might look like. So, with this in mind, let us look at the conclusions first.

Inertialess Negative Gravity

Already, we can predict much of our future physics. But details are murky. For example, no one knows exactly what quantum computers will be like. Instead of transistors and Boolean logic, quantum computers will exploit quantum entanglement. Initial applications soon coming will boost accuracy of GPS radar and enhance telescope resolution—improvements that, with phased arrays, should see the oceans and continents of extraterrestrial planets.

But, even without quantum computers and by using conventional transistors, we will achieve astounding single atom transistors, although it will be difficult to interconnect and interface them at room temperatures. Today, although we are close to achieving transistor behavior in a single atom, it will be difficult to interconnect them and interface them with the external interface, all at room temperature.

We have studied electrons confined within quantum dots, semiconductor nanoclusters where we manipulate dots to create lasers, single-electronic transistors where current flow is one electron at a time, or even fluorescent markers for biomolecules. So small are quantum dots that quantum mechanics rules.

Smaller yet, future atomic-scale transistors will be small quantum dots that will pass currents of single-file electrons. Near absolute zero, if small magnetism is near, there is a sharp rise in resistance—the Kondo Effect. But in quantum dots, resistance drops. Although this is a curiosity, researchers are wiring together individual transistors with carbon nanotubes. Other researchers seek to incorporate all the needed circuit elements (Rs, Cs, transistors) in single nanomolecules—each invisible nanomolecule exceeding today's best supercomputer. Others achieved this long ago.

But quantum computers add vast improvements. Our successors will not be human—but machine intelligences superior to us humans in many ways. These developments are predictable.

However, other major discoveries unseen by present physics will radically affect our industry, our civilization, and our physics. For example—and it is the most important of all our examples—let us consider the propulsion. Make no mistake—this is the most important aspect.

After this book was written, your author (a rational skeptic) was reluctant to add this part on flatwoods. He added it only because of a connection to ball lightning, and their possible shared propulsion—for want of a better phrase—of "inertialess-negative-inverse-gravity."

Your author was present on the evening of September 12 of 1952 when an unknown aircraft of revolutionary configuration landed atop a small hill in Flatwoods of West Virginia and made national news. Trapped, your author and his family were witnesses. Before we, through taped testimony of the main witnesses, re-live that fateful night, let us examine aircraft characteristics.

From observations first collected by the late Major Donald E. Keyhoe, USMC and Richard Hall of NICAP, certain characteristics of aircraft emerged. They vary in size from half a foot in diameter, to a typical 30 feet, to a rare 3,000 feet long, but with the larger ones not disk-shaped but usually chevron-shaped. Surfaces are smooth without rivets or edges, and doors shut to leave no lines. Exterior surfaces lack structural features.

Internal combustion engines (cars) that approach too close will sputter, lose power, and stop running; and headlights grow dim and then go out; their radios spit static and then die; while portable radios die and flashlights fade out. Allegedly, diesel engines keep running. But after the craft departs, lights go on and engines restart, and there is no problem with batteries.

Consider this. Strong electromagnetic, electrical or magnetic fields simply cannot do this, and this defies electromagnetic theory (physics). Your author could not satisfactorily modify the eight equations of electromagnetics to account for this. In addition, intense magnetic fields from super magnets would permanently alter a car body's magnetic field, would do damage—and worse, such fields cannot penetrate the Faraday cage of the engine compartment, which is largely enclosed by sheet metal. And, note that vacuum tubes, transistor radios and integrated circuits also cease to function but are undamaged.

Television pictures dim, blur and distort, and some lose audio first—all indications that loss of power was not the cause. (Fireballs also have caused this.) Household power and streetlights fade and extinguish. But once the craft departs, power returns and there usually is no damage. Remember, there is nothing in physics, as we know it, or electrical engineering that can explain this. This is a valuable clue.

Inertial and gravitational mass are indistinguishable. Since craft nullify the effects of gravity upon their mass, they also nullify their inertial behavior. With a massless craft, exerting modest forces would produce high accelerations (over 20 g) so that it would seem to "vanish." In the Flatwoods case, the disk was heavy and left deep tracks when landing at dusk and lifting off at sunrise—evidence that it was no longer massless.

And consider this. If an advanced technology simply wished to, they could easily stop all electrical power over vast areas, or overload electrical power grids and cause permanent massive shutdowns of worldwide electrical power distribution, and blackout earth during winter—and destroy civilization. With their grasp of genetic engineering and nanotechnology—millions of years beyond us—they could easily destroy all human life on earth—without nuclear bombs.

Or they could easily divert a hail of smaller asteroids to destroy mankind. Or maybe, if they are feeling real nasty, divert a colossal asteroid to collide with earth. And, with their propulsion systems, make no mistake—it would be a trivial task.

Most nearby witnesses seem unaffected. But a few witnesses are temporarily paralyzed, although involuntary bodily functions are unimpaired, and these witnesses instantly return to normal when the craft leaves. Teeth fillings and metal implants do not heat, as with induction, but it is uncertain if pacemakers cease functioning. Half the paralyzed witnesses lost consciousness, but first felt or heard a humming, and some felt a prickling. Other medical effects occur.

There are occasional exceptions. Jacque Vallee points out the alleged contradictory (nice and nasty) behavior—perhaps part of an elaborate smokescreen of confusion to bewilder and ridicule our researchers, and throw them off the track.

Animals hate such craft and are spooked in some unknown way. Dogs are terrified and cower. At Flatwoods, the two terrified dogs refused to accompany the witnesses up the hill, even several days later. Dogs and witnesses have both been simultaneously paralyzed. Insects also sense such craft and go silent. In some cases, dogs brought near a landing site panic, even after several days, unless there is rain.

Propulsion

As a young boy, Einstein was fascinated by a compass and magnet. Although history is silent, it is also likely that when young Einstein picked up iron nails, he wondered why the force of his small magnet could overpower all the gravity of our huge earth at short ranges. Why is gravity so weak? And later Einstein certainly wondered why gravity lacks an inverse (repulsion).

Gravity is dramatically weaker than the other forces—in scientific notation, it is 1×10^{-43} that of the electrical force—but equals it at the incredibly tiny Planck length, or 1×10^{-35} meter, probably forever beyond our reach. For comparison, the Large Hadron Collider at CERN will probe down to "only" 1×10^{-19} meter.

Unlike the other three forces, gravity cannot be stopped by any shield, and has no inverse. Physicists who pioneered quantum theory justifiably ignored such a weak force. But as weak as it is, gravity is our enemy.

Since fireballs and craft defy inertia and gravity, we suspect that both exploit an unknown aspect of the yet-undeveloped Theory of Everything in physics—and can provide physicists with clues. When we understand the inertialess-negative-gravity propulsion—which for light fireballs, is much weaker because it is smaller and lighter—we will also understand that of such craft.

NASA's former in-house expert and eminent rocket engineer, the late Paul R. Hill attempted to reverse-engineer such revolutionary propulsion. Hill made many brilliant observations in his book, *Unconventional Flying Objects,* which your author does not completely agree with, but innovatively adapted to explain how fireballs also defy gravity.

Controlled and directed, in magnitude and direction, inertialess-negative-gravity simultaneously affects every bit of matter and energy aboard such craft. This is radically different from the familiar action-reaction propulsion of Newton's Third Law in jets and rockets, where the pilot cannot pull too many g's—that is, take tight turns or accelerate too suddenly—as everything in the rocket or jet also is affected. If any rocket attempted to take the right-turns and instantaneous reversals extraterrestrial craft do—it is just like hitting a cement wall—they could never do it, and in their futile attempt would be destroyed.

Because their inertialess-negative-gravity vectors (for each individual particle within a craft) can be directed not just against gravity, but instantly pointed in any direction, the craft can motionlessly hover silently or instantly move rapidly in any direction—sometimes so fast that they "disappear" off the radar screen. Some reports indicate that they work undersea—implications that could obsolete not just rockets and jets, and navy ships, but also our ballistic missile and attack class submarine fleet.

Such craft affect generators and wires, and stop the flow of current. Nothing else known in physics can do this. So why don't they seriously harm human bodies? This is odd, because we are full of electrolytes, potential differences (voltages) and current flows of ions and charged particles. And they do not affect gold fillings and metal surgical implants. However, some witnesses who came too close beneath hovering disks reported heat and other unpleasant effects.

The disks affect magnetic compasses and cause radio hum before disabling them. They do not boil battery electrolytes or melt wire insulation. They disable ignition-type internal combustion engines, but not diesels. But after the craft leaves, all these effects instantly cease and there is no damage. Objects near craft are not stirred up, and the craft are usually eerily silent.

So rapid is the acceleration (sideways and upwards) that craft seem to vanish from radar and sight. Radar detects some, but others are stealthy (invisible to radar). They instantly accelerate to incredible speeds—over 9,000 miles per hour—in the atmosphere, without sonic booms. They streak across the sky in violent maneuvers, diving and climbing at astonishing speeds.

The metal hood and body of a car should act like a Faraday cage—a metal shield—that acts as a barrier to strongly attenuate (weaken) or stop penetration through steel by electromagnetic (radio) waves. Paradoxically, steel cannot shield the electrical ignition of a car from an inertialess-negative-gravity field.

In contrast, the floating Flatwoods "Monster" had a small field confined to a small volume, and thus it failed to dim a flashlight only 15 feet away. The floating giant "monster" was really a small robot—a "Halloween scarecrow"—just a small sphere, perhaps only four feet wide (if inflated, even smaller), floating silently ten feet above the ground that let down a skirt just to seem huge and frightening. It was classic. If the those aboard the disk had a sense of humor, they were laughing.

Although craft shapes vary, most are metal disks that spin, a few are cigar-shaped, often with portholes. Sometimes, triangular, boomerang, isosceles trapezoid and square shapes also appear. Usually they are bright white during night-flight, but in daylight are metallic and reflect the setting sun, as with the disk carrying the Flatwoods "Monster". However, recently, some craft are reported over a thousand feet long! For you readers who are mechanical engineers, just imagine the disastrous static and dynamic structural stresses placed on such colossal craft.

Finally, extraterrestrials knew hundreds of millions of years ago that we on earth would one day attempt to reverse-engineer their craft. Did they carefully contrive "red herrings"—deceptions to confuse and throw our physicists off the track? Probably. But, on the other hand, the characteristics we described above cannot be concealed, as we will see.

Encounter

Halloween came early to Flatwoods in 1952—complete with an alien craft and a floating monster. It has been over fifty years since a mysterious machine landed on a remote hillside of Flatwoods in central West Virginia. What exactly happened on that night has been debated. Did an alien disk land on the hillside of the Bailey Fisher farm just in back of Kathy May's house? And, did an alien monster try to kill them? Whatever it was remains unclear. What is certain is that it captivated the nation.

In 1952, as a boy your author and his family were witnesses to the Flatwoods "Monster"—so I know it took place. I know for a fact that a craft of revolutionary design visited Flatwoods on that night. As for other skeptics, you were not there. I was there. And one skeptic (not Klass) who recently did visit Flatwoods, never climbed the hill nor talked to many of the important witnesses.

Another skeptic (I will not mention his name and embarrass him), while speaking at Bristol Community College in Fall River, admitted (published by a reporter in the New Bedford *Standard-Times*) that he saw such a craft, and that he disagreed with his own skeptic's organization on this issue!

It was early twilight when my dad drove us through Charleston and then drove northeast on Route 4, which parallels Elk River for 65 miles into Sutton and then north five miles to Flatwoods.

Today, dammed at Sutton, Elk River forms Sutton Lake, which feeds Elk River meandering southwest for 60 miles into Kanawka River. Headwaters originate east in the Appalachian Mountain range. Flatwoods lies in the middle of

West Virginia, and Clarksburg lies 57 miles northeast of Sutton, a 65-minute drive.

On the early evening of September 12, 1952, my mom, brother, sister and your author (then ten) were riding through Flatwoods in a green Hudson, driven by my dad. He drove east along Elk River to Braxton on Route 4, through Duck and Frametown, Sutton and north to Flatwoods. (Route 79 did not yet exist back then.) After my dad fixed the failed brakes, we drove north through Heaters and (I assume) north on Route 19. Dad had hoped to reach Delaware to visit his brother. We were returning from Denver, bringing back my mom and sister, who earlier had taken the train west. Well, the next day on the 13th (if I remember, it was sunny), dad changed his mind, because I was already late for school opening, and we drove north. We were returning to Massachusetts, from the Spears Chiropractic Hospital in Denver, where my sister and mom had traveled earlier by train.

The world-famous Spears Chiropractic Hospital, the largest ever constructed, founded in 1943, closed March 1984, and later was demolished. Earlier, we drove west to pick up mom and my sister, and were driving back east. Dad wanted to visit his brother in Delaware, but later changed his mind and drove north. At any rate, that is how we ended up in Flatwoods.

But our brakes failed (bad coincidence) at the foot of that hill, and after driving on the edge of the road to slow and stop the car, my dad parked on the right side. Without brakes, dad could not drive away and we would have been killed going over a cliff.

I remember that the very unpleasant odor worried my parents—there was something frightening about it—and I cannot compare it to anything else. We all had a creepy sensation that something eerie was going on—and we wanted to get out of there. My mom and dad even openly wondered if an alien saucer was nearby—it was that spooky.

After 25 minutes, my dad poured mineral oil into the master cylinder, located just under the floorboard below the driver's feet, and we safely drove north. Next morning, on the car radio, we heard the news of the saucer and "Monster".

Back home, at school, when I heard the other kids talking about the television show, *We The People,* and the Sutton "Monster" (which they called it back then), I told them I was there, but they didn't believe me, so I kept quiet. That show and other coverage set off a near national panic.

The encounter dramatically impressed me. So much so, that two decades later, in 1975, I returned to Flatwoods and drove back down. I met and taped interviews with several witnesses, including Kathy May and Dale Leavitt. At any rate, after that, I lost enthusiasm, until this book. And recently, I came across researcher Frank Feschino's web site and was pleasantly surprised that he will not let Flatwoods fade into history.

Then I read one mistaken skeptics's "hatchet job" article [Investigative Files: "The Flatwoods UFO Monster," by Senior Research Fellow Joe Nickell, November/December 2000] on Flatwoods in the *Skeptical Inquirer*, published earlier by the *Committee for Scientific Investigation of Claims of the Paranormal*. To you dogmatic skeptics, I will say this—I was there, CSICOP was not.

On many things, I do respect CSICOP (Phillip J. Klass)—well, at least on 95 percent of their debunking—but they are insanely, rabidly anti-UFO, as also is the *Skeptics Society* (Michael Shermer). In fairness to other skeptics, I can understand why they justifiably are so skeptical, as was explained so eloquently by Carl Sagan in his excellent book, *The Demon Haunted World.* And a skeptic, the brilliant Martin Gardner said the subject is odious—and legitimate scientists who get too close get tainted with its stench. (Yes, unfortunately, Gardner is correct on that.)

But skeptics cannot find a way to discredit these seven witnesses (and me and my family) on this case. So they allege the seven (twelve) witnesses saw a hissing, real stinky, giant owl. But on this case these skeptics—and they are brilliant—are mistaken.

Since 1952, I often wondered if Project Blue Book's first Director, the late Captain Edward J. Ruppelt, USAF knew more than he pretended about the Flatwoods "Monster" and extraterrestrial craft, and was playing coy with the late Major Donald E. Keyhoe, USMC—or did the orders to Colonel Dale Leavitt come from elsewhere? Or, as with the Washington Sightings, did the Pentagon again keep Ruppelt and Bluebook in the dark?

At any rate, I doubt all those sinister conspiracy theories, and instead I propose—**The Stupidity Theory**. I suspect that the Air Force investigators were and are under-funded and mostly incompetent, as are those at the DSA and CIA. From what I hear, they were snickered at and not considered serious heavyweights.

To survive, ufologists need crazy conspiracy theories; and I suspect that the extraterrestrials planned it this way to discredit serious scientists from researching legitimate sightings, like Flatwoods. Do extraterrestrials foster crazy red herrings (Crop circles? Alien abductions? Valle Hypothesis?) that encourages weird contradictory theories? Like a magician's misdirection, to keep our attention off the target? To get our attention off the real hard information?

Extraterrestrials throw up more smoke and deception, to sow confusion, to discredit this field, and to make UFOs an "odious subject." Michael Shermer of the Skeptics Society wrote that psychologists should study why people have a need to believe in weird theories, and why some people earn their livings from weird theories. [Shermer's question answers itself.] While we watch their weird stage show, the extraterrestrials are having a good laugh.

This confusion technique would encourage certain people to propose weird theories, but much harder for our leading physicists (if they value their careers) to "draw a bead" on extraterrestrial propulsion technology. If I were an extraterrestrial, I would do exactly this. And it takes no genius to create such mischief. If there are any competent government groups, they are small and insular. But most are under funded and incompetent. An exception, NASA's in-house UFO expert, eminent rocket engineer Paul R. Hill, was highly competent.

And now, before we hear two key witnesses, tell their encounter in their own words, let me first introduce you to the witnesses.

Witnesses

Five-year-old Peggy Harvey Clise saw the glowing saucer move fast across the sky and settle down on the hill behind her cousin Mary Fisher's farm. She told her granddad, but her family swore her to secrecy that was only recently broken, when she disclosed this to researcher Bob Teets. On his porch (in wheelchair?), Cecil Gumm saw it land across the road up the hill. Henry Marple at Heaters and Jimmy Fisher (now in Sutton) saw it.

Also, on the schoolyard playing football, five boys also saw the disk—trailing red fire. Neil Numley (age 14), Ronnie Shaver (10), Tommy Hyer (10), and Eddie (Edison) May (13) and Fred May (12) walked up to Mrs. May's house. Her two sons, Eddie and Fred went inside to tell their mom that they were going up the hill to look for it. Mrs. May's relative, National Guardsman Gene Lemon (age 17) joined the six.

Investigator Ivan Sanderson later said it took the boys about 20 minutes to reach the hilltop from the football ground in the valley. It took your author, walking uphill from the May house to the tree, less than ten minutes to reach the oak. The time Eddy and Fred took inside their house and tell Kathy May, their mom, probably took two minutes.

Kathy May

While still back in New England, I got Kathy May's phone number from the sheriff's office. After talking to Kathy for twenty minutes, she agreed to an interview and gave me directions on how to get there.

I drove down—two days and quite a long drive, too. Finally, arriving at my motel at 6:30 that evening in late April 1975, I phoned Kathy May, and asked if we could meet in the next day afternoon. Since she had an appointment later, this was not possible, but if I came in earlier, she would not mind. Since there were things for me to do later—investigations and pictures—I felt that the morning would be well spent. As it turned out, it was time spent well, and I learned more than I expected.

When I drove up into her driveway, I immediately noticed that her house was a small farmhouse, with slightly peeling white paint, with a classic T-shape with an enclosed porch out front. The entry door was to the rear, and all the windows had awnings. I went inside and met her mom (if I recall), Kathy May and Harry E. Stewart, who were engaged to marry and relocate to Parkersburg.

I found Kathy (Lemon) May to be articulate, sincere, sociable and pleasant—an attractive lady, and quite youthful-looking in 1975, and also quite intelligent and able to communicate her thoughts forcefully. Kathy laughed easily, spoke cheerfully in a gracious southern accent, and was personable, friendly and people-oriented, expressive, and often exhibited flashes of her keen sense of humor. She was a woman who was sincere, direct and pleasant—someone you would like as a friend.

I rarely listened to these tapes since 1975, and it is now 2003 when I transcribed hours of taped conversations, then re-grouped and re-edited them. Because of the passage of time, I am sure that I misspelled names, for which I apologize.

Kathy's voice was lively and very interesting—but something I cannot convey to you. You will notice disjointedness because—for logical consistency—your author regrouped and condensed her conversation (parts selected from several hours of tapes) for a logical flow. Certain observations are repeated—but in different words.

And I noticed minor inconsistencies between Kathy and Dale Leavitt's testimonies. As any experienced police officer or trial lawyer will tell you, this often occurs in many good-faith eyewitness accounts. For logical consistency, I regrouped her paragraphs to give a chronological time line. Because of long tapes, although I omitted much conversation, nothing important was added or deleted from what Kathy May told me, as follows…

The boys said it looked like a great big silver mirror that had this red flame behind it, that was, after it. And when it started coming down, they said it looked just like a big silver door. It seemed that it changed shape. They heard this noise. It was just like a big silver mirror with a red trail. I don't know exactly how long the tail was because the boys saw it, and I didn't, but it was not long. Jerry Markel also saw it coming in. It leveled off to land.

As for the trail behind the saucer, its length, I do not know exactly because the boys saw it and gave a description of it. They said it was like red flames trailing out the back of the disk, but it was not long. There was no smoke.

Jerry Marples, not Howard Heater, saw it land on the hill. He saw it coming in from behind. He was the one who called the sheriff. The *Webster County Echo* Editor was there. He came over at three o'clock next morning. Well, I am sure it was next morning because I did not recall anyone. Jerry Marples was down in Heaters right down below here [on my map], and saw it come in. Cecil Gomes was sitting on his porch actually saw it land. Howard Heaters—who you asked about—and who founded Heaters, did not see it.

[*Author*—Also, a local businessman, Jack Davis of Flatwoods on that evening saw a disk the size of a garage flying 50 feet over the ground near three hills, and landed on the hill. He saw a disk with a small "tail" trailing it, with a pink-red-dull orange glow, but with a topside like a mercury vapor light—a white luminescent glow on top. Davis later said, "It did not resemble a meteor to me. To me that object that I saw was a craft…It looked like it was something controlled." A Birch River resident reported a bright orange craft circling in Flatwoods.]

I worked as a beautician in Sutton. I got home and had been there less than half an hour and was still in my uniform. The flashlight set right there [on the coffee table], where I am now pointing. It was almost dusk when my boys rushed in and said, "There's a flying saucer that just landed up there on the hill in back! We are going up there!" And I said, "Deed [sic], you are not going up there by yourself!" And they said, "Yes we are!" and took off. And I yelled, "Wait a minute!" and picked up that flashlight, and I said, "What happened?" I didn't know the other boys were out in the road.

Then when I stepped down in the road [down several side steps into a dirt driveway] and looked [rightward], I saw a purple red flare up there. So I told the boys we are not going right up to it, just far enough so we can see where it landed, you know. Then we will come back and call the law and have them come and investigate.

Gene Lemon came over. He was out on the road. [A National Guardsman, and her cousin, he lived the next house downhill to the left.] We all walked along together, sometimes one in the lead, and sometimes another It took us about five minutes to get there. Neil Nunlee was a tiny fellow. We had two dogs, a collie and a spitz [barks a lot, a good watchdog, and resembles a small husky] with us, but they just whined and refused to go beyond—and not even the next day, because of the smell.

I remember turning around to see if I could see the lights, the streetlights of Flatwoods. We do have streetlights here, you know. [laughs] I guess at first I thought it was "fog" further up, but then I realized it was not a real fog. But the odor was there before we got to the fog. Once we got into the fog, it got thick and we could not see that well in it. It was really misty, and we ran into that before we ever got there. I remember just when we got up there, I turned and everything got real misty—like a real warm mist—and I remember turning around looking to see if I could see the street lights, and there was fog.

The "monster" hid in the dark right on the dirt roadway [path] near the tall oak. The tree was mostly taken out since, but some of it is still standing. Gene's flashlight hit the "monster" right in its face. It instantly lit up! Like a Christmas tree! It was orange, black, green and red. I must have turned green when I saw it! They were just like drapes here [on your sketch]. The skirt, where it folds, was flared out and was darker. And when it shook back and forth, the drapes did not move. The shoulders were real round and [its hood] was an ace-of-spades, stretched up, and the head [inside] was round. But there was no collar.

Truthfully, I did not stick around, to look to observe it! It looked like it was encased in a fog. Gene is the one who screamed and fell back. Neil jumped over atop of him and ran. If it was going to scare us, it did its job!

So we came upon it before we knew it. There is a gate there, unless someone has taken it down. [Today, only one post remains.] Somebody told me they took the fence down, and on the other side of it, that gate, was the tree, and on the other side of the gate that thing was there hiding in the dark. Gene took the flashlight. Yes, it was dark. Before the thing lit up, it looked just like a dark hulk. It had a shape, uh huh.

The "monster" did not glide or move toward us. It was only hovering, just floating, and had no feet. The whole thing was shaking back and forth. The "Monster" [lit and] acted up instantly when the flashlight beam hit it. It looked like it was metallic cloth. Protruding from its front "neck," one on each side, two antennae were straight and thicker than a pencil, and pointed slightly down at a 20-degree angle.

It was already dark. When Gene's flashlight beam hit it, it lit up instantly! Like a big Christmas tree! We all noticed it, all at the same time. It was not in any bushes. It was right out there under the limb out in the open, right in the roadway. The whole thing suddenly lit up—the whole thing—instantly! It was lit up from within—from its inside, from its interior. We could not see through it, but we sure could see its outline. No, I am glad I did not see through it. It was black, though up here on your drawing, around here. And the two antennas were black, and this [on my drawing] glowed red, and this, black.

It was a hideous sight, and I wish I had not seen it. But I said that if I had a camera there instead of a flashlight, I would never have to work another day. [laughs]

Even though the "monster" shook back and forth, its drapes did not move. They just fell down and folded like drapes. No, they did not move at all. The drapes were not moving or anything. The two antennas stuck out like this. About 25 degrees and was two or three feet long. It had no arms, no waist.

Later, the artist drew it wrong for TV. I looked at it and said that it did not look like that, but he ignored me and refused to change it. We stayed in New York for three or four days altogether. Of course, they had rehearsal for one whole day, and then we went on the next day. In New York, the artist drew it wrong, and he drew it wrong on purpose, I think to make it look scarier. But it had no waist. It looked just like your drawing here does.

I never was good at distances, I will tell you that. We were close to it—from here to that door—and we were onto it before we knew it. The round head was inside an ace of spades. The head was definitely round. It had no collar. Truthfully, I did not stick around very long to examine it very much! The eyes were the size of dinner plates, you know, like two pie plates. They were about a foot apart. They were really two portholes more towards, closer to the top.

I was less than twenty feet from the "monster". It made a hissing noise. It sounded like frying bacon and like if you had taken a silver dollar and dipped it into [indistinguishable]. Gene dropped the flashlight, screamed—he screamed so loud—and fell back, and stumbled over one of my boys, and took off running. I found it [flashlight] two or three days later. I cleared the fence, and one crawled through the bars going through, and the only thing that concerned me was the kids crossing that fence. I turned around to see if they all made it. But Lord, two of them were in front of me. The others were gone, long gone

As for how fast the "monster" moved, he seemed to bob along, but floating over a foot above the ground, just hovering, without swift steady motions. He was not really moving, just kind of quivering. I am glad he did not move towards us! It floated [over a foot up]. It had no feet. With those long drapes of its dress, it came down in folds, just like a skirt.

It lit up from inside. It looked like maybe metal, like it was made of possibly metal. That is what it looked like. If it wanted to scare us off, it accomplished what it set out to do. [laughs] It did not have any material [cloth] about it. It seemed to be metal. That is what it looked like. It was made of some material. It looked like metal; it did not have any material about it. If the "monster" intended to frighten us, it did its job!

Back and forth the whole thing was shaking back and forth, and hovering, but was not moving toward me. It lit up instantly, just like that, when the flashlight beam hit it. The two antennas were straight, with nothing on them. They were thicker than a pencil and went straight out, and came down at a twenty-degree angle from the "neck," right here on your drawing.

[*Author*—From Kathy May's directions, I drew sketches in 1975, with some major differences from, but still rather similar to Frank Feschino's recent drawings. There were three differences. I drew dual, two-foot neck antennae sloped down 20 degrees, and drapes that were not symmetric vertical ridges, and that went higher. And there were no jets shooting down at the bottom, nor rays shooting out of its two porthole "eyes." Otherwise, our drawings are nearly identical, but not as dramatic as the incorrect one drawn back in 1952.

At the house, Mrs. May's grandmother phoned sheriff Carr and deputy Burnell Long, but they were out answering another call from Woodrow Eagle of Duck—a miner who served in the Second World War and worked around planes—who saw a "small piper cub" crash in flames into a wooded hillside at Sugar Creek on the south side of the Elk River below Gassaway and phoned the sheriff, who came to investigate. But the officers could find no evidence of any accident. Upon returning to Sutton, the sheriff immediately drove north to Flatwoods to investigate May's phone call.]

What it was, truthfully I have no idea. I was afraid that that thing would follow us home! Gene led a party of ten or twelve back up the hill that same night around 12:30 or one o'clock in the early morning. I do not know if any were the National Guard. Then the sheriff came out and just looked at two haystacks, and then came back off the hill. And I do not remember, but maybe they came out later with the National Guard.

The editor Bill Trekle was at the court house, and he heard the call come in and he came out, and came on this road here [on map] and saw it. He was the one who saw it take off. In fact, he did not come out to talk to us until the next morning. This road going down here on this [my] map…and this down by the railroad bridge. Bill saw it come around and around before taking off. It spun around and around, he said first. It did not take off straight up.

He went down the wrong road, and that is why he was up on the trestle to see it take off. It [the disk] just went around and around before it took off. There were two sets of skid marks, both landing and take-off skid marks that looked like two sled runners about a foot wide, and they were there, and you could see that they were fresh skid marks about a half deep into the ground. You could see the big skid marks going right down the hill. The machine landed on a flat area before moving down into a gully area. It went down a grade, spun around and around before it lifted off. It went around in a circle and was hovering after it took off.

[*Author*—At sunset, just after seven, there was still light, and it did not darken until 7:30, and at 7:50 it was difficult to see the unlit "monster".]

They took no dogs up the hill. But our own two dogs stayed at the house and howled until we came back in the house. And the next day, when my daddy would start out or anybody would start up the hill, they would start howling. But they would not howl continuously, only when somebody would start up the hill.

When that creature lit up, the "oil" hit me! There was that much "oil" over there, oh yes. Oh my Lord, I had that big streak right down my uniform Did it have color to it? It sure did.

At 12:30 to 1:00, Gene Lemon led a party back up the hill. The National Guard came out that night. The sheriff had no dogs. It was our dogs that howled. The National Guard wanted to know what that thing was. They took samples and were here every day. They took pictures of everything, even of my uniform. The "oil" looked like it was sprayed on. It was not spongy, but like dark greenish gray "oil", but did not look like motor oil. The National Guard took the plastic samples. One was large. I should have been given one of the 1-inch-by-1.5-inch samples. The next morning many people came up the hill like ants, all day until dark. A lot of people probably got samples. I do not doubt it. I stopped going up there, but my daddy would take people up there.

Ivan Sanderson mentioned five reported landing sites, including Sugar Creek, Holly, and Elk River. This man and his wife were from New York, were coming up the river, driving along it, and he thought it was a plane crash. Her husband was driving. But she went into shock, and they had to hospitalize her.

Burnell Long, he is not dead. I think he is alive. And Charlie Whitaker was desk deputy and is still alive.

I did not talk to Ivan Sanderson, my daddy did. I did not because I was being interviewed in New York at that time. We spent four days in New York. We spent one day there, and went on TV the next day. But Sanderson and his assistants talked to everybody else.

Two "newspaper" men came here and scraped off the matter from my uniform. They came back the next day and said they were really Air Force Intelligence officers, and apologized. The agents were in the car. [*Author*—Despite Al Chop's denials to Major Donald E. Keyhoe and later denials by Captain Edward J. Ruppelt, two plainclothes USAF intelligence officers who impersonated newsmen did visit Flatwoods and interview Kathy May. Were Chop and Ruppelt honest?]

There is a little house. It is a cabin now [later it burned down]. They used to call it the *Braxton County Monster Shack*. That was Fisher, a farmer, who is no longer there. He sold his farm and moved [to Sutton].

The Jaycees made a mold of the "monster" and made copies, and gave me one—but I loaned it and never got it back. I have seen copies on some peoples' desks. I did not know who made the Jaycees mold but they said they would make only so many and then destroy the mold. [Copies in 2003 are still being sold.]

The Jaycees are not in the phone book, so the Mayor [Craig A. Smith] can get one of the models for you. There was a TV ad out of Charlestown four months ago, and the man had one sitting on his table and said you can order it for five dollars. I do not know who made the Jaycees model, but they said they would make only so many and then destroy the mold. I signed them permission, to the Braxton County Jaycees, to have this made. They helped the county out a lot, you know. Their [Jaycees] model resembles the "monster"—but only a little. [laughs] I forgot how many that people ordered—maybe 300.

[*Author*—One man who owns a gas station is still selling these models as of late 2003. Actually, the quaint Jaycee sculpture that Kathy showed me looked more like a novelty gift store item. The real "monster" looked "boring"—like a giant inverted green coffee cup with a red golf ball (or beech ball) balanced on top of it! But the Jaycees sculpture looked exciting—like a red-and-green, giant salt-shaker, complete with two arms and claws! Lunch anybody?]

But they [Jaycees] never did give me a model, but a friend did give me one later. I feel their word should be as good as gold—at least that is how I feel about it. I gave my written permission that they could make these models and sell them to help our county Jaycees. They made so many that I thought that would break the mold! [laughs] I have got one. Would you like to see it? I will bring it right down.

Some of the vegetation up there in the area died and looked like it was scorched…The other two spots that came down on were secluded, oh my, yes. One of them was Holly…It looked like fire on the end of it. So that a woman saw it come down and thought it was a plane crash…Up on a hill, it was a mountaintop. So later they went up there and searched the area but never did find that ship. No plane, no plane crash, nothing. And they thought they were secluded.

I saw two or three of the black plastic metal pieces, but the National Guard had them and took them. One of them was pretty good sized, about a half an inch by about an inch and a half. And I did not remember how many exactly that they did get. It was smooth and cracked on the edges. I did not remember if anybody tried to scratch it or not. I should have at least been given one of the samples anyways.

They estimate that a thousand people came up there. I mean, they were up there like red ants! All over! All day long until dark! And some of them after dark! I bet a lot of them got samples. I would not doubt it.

My oldest son Eddie [Edison] is now employed with the CIA and lives in Woodbridge, Virginia. He has the brains of our family. [laughs] I cannot tell what he does nor can he tell me, and I cannot and do not have his phone number. I never questioned him because I knew he could never tell.

They thought we may be contaminated and die in six days. Our eyes and nose were burning, oh yes. The boys vomited all night, starting 30 minutes after. All the boys took penicillin, but I did not take any. I did not know who their doctor was, but my boys' doctor was Doctor Hutchinson, but he is now dead. Some went to Doctor Eackle, but I am not completely sure. He is still alive in Sutton. My mother was in the hospital. My grandmother was here at the house, and she was the one that washed the "oil" off our faces and necks, and shirts.

That is how penetrating that "oil" was, I am not kidding. Clear through. And in our mouths and nostrils. That is why the National Guard and even the doctors were so concerned about me. There was no telling what it was and no one knew. They notified us.

The boys were out of school for a whole week. It was on a Friday, so it went right through the next week. The boys' mouths were raw, and they had diarrhea, but each day I could tell it was getting better. By morning, Gene Lemon was bad. There was one of the other boys, I do not know which one—I think it was Neil Numley—who I think went back to school on the third or fourth day, but he still was given shots at school. Odor still clung to the hillside the following day.

Grandma was there [at the house], but mom was in the hospital after gall bladder surgery. Mom could not believe it. She did not know about it until the next day. Mom came home several days later. She had gall bladder surgery, and they took samples. She could not believe it—that it had landed back here; it was all over the hospital. And of course, she did not know until the next day that me and my two boys were involved. [laughs]

Personally, I have not been up there [on the hill] for a while. But my daddy would take people up in his car. I just could not go. I think I made one too many trips up over that wall. [laughs] To tell the truth, I went to sleep. I would have drunk that night—except I don't drink!

Arnie Stewart of the *Braxton Democrat* wrote the National Guard for what the object was. Later, he called me and said he got an envelope back from the government, but could not open it until 7:30 that night. He asked me if I could get the boys at my house around seven.

So, he opened the letter and read it to us, and in there was a five-by-seven picture of the "monster", which looked like it. And they said that Mrs. May had given the best description of it, they did, even of those that had helped build the missile. That helped make me feel better. We had to accept what the government said. They told me to keep quiet and not to give any more information to anybody.

They said that these were ships the government was building to send to the moon. All four they were testing that night came down, but they could not locate this one. The one that landed out back was having oil trouble, and two men were inside. That is what they said, it is in that letter. Jimmy Heart [Hearst?] of the *Charleston Gazette* had it for six years. Call Jimmy Hearst of the *Charleston Gazette*.

They warned us to keep quiet and not tell anyone anything about it. They said it in that letter to me. Truthfully, we believed it, what the government said then. But now I am not so sure, and I now think they didn't know what it was and they wanted to keep us quiet.

This ended Kathy May's taped interviews. After this, I thanked her for her kindness in allowing me to interview her, and permission granted to publish it in this book.

Re-creation

Before leaving her house, I gave Mrs. May a paperback copy of Major Donald E. Keyhoe's *Aliens From Space*, and then with her permission I recreated that fateful night of 1952. The following was excerpted and adapted from my photographs and part of my taped description, as I walked up the hill, trying to get a personal sense of the events on this hill on that night so long ago. Into my tape recorder, I said…

Exiting Kathy May's house [which was maybe 55 feet from the road], I now descend down the steps onto a dirt driveway and turn right and now walk up a gravel path on a gradual hill. To my left down a slope I see a house and two dogs. I guess that's Kathy's cousin Gene Lemon's house. To my right is a staircase.

From the house, I walk up this gravel driveway, up a gradual slope for about 165 feet—to a rusty iron fence and gate, which is closed but easy to get around. I walk upslope on a gravel path for 80 feet—no difficulty at all. Up to now, the slope goes downhill to my left. At this point, this gravel path or road forks right to a small "house" with a tire hanging from a bar on two poles near it, with some trees near it. Kathy May says folks call this shed "the monster house." [Some years after my visit, it burned down.]

Instead of following this gravel path to the right to visit the "monster house," I will do it later. Instead, I will now walk straight along a high "ridge." The right slope slopes down at a thirty-degree angle. After 180 feet, I am stopping at the stump of an oak tree, to my right, which is about two feet in diameter. Just before the oak, a "ridge" runs right, with a thirty-degree slope on either side.

I am walking straight ahead. On my left is a cornfield, about 160 feet square. To my right, the down slope is 35 to 40 degrees. On my right is a thorn tree. Past the cornfield on my left are woods, and to my right, but down the slope, are more woods. [Deleted conversation]

What I see now before me is all that is left of this old oak—just a tall trunk—a ghostly wreck pointing upward to an empty sky from which once came here a revolutionary aircraft, perhaps from a distant alien world.

Mountains

Earlier, while driving down along the new superhighway Route 79 in my car in late April of 1975, to get a sense of this landscape, I taped my observations of this rugged terrain that these craft flew over and landed in. If they hoped to avoid civilization, well, they made a wise choice landing in this strangely beautiful, wild and mountainous land. Edited only slightly, here are my taped impressions. They give you a "snapshot" view of this strangely beautiful land. My sentences, as follows, were short but descriptive, as follows...

I am now driving south at 60 miles per hour. Frequent hillsides slope up at 30 to 45 degrees. Wooded farms. Wild mountain streams and lush green valleys and majestic mountain slopes—God's country. But not much crops, but lots of grazing hills. Soils give way and small streams sprout out of mountainsides and tumble down. Fog over hilltop is in the rain. Friday April 25, 1975. Overcast skies. Occasionally terraced sides to stop erosion. Low-grade iron ore, red soil.

The land only recently was cut through by this superhighway. Rugged landscapes, wooded mountain tops. Occasional farmhouses here and there. Countryside is pretty barren. Rivers swollen with rain, in the valleys become swollen with brownish red sediment. Springs pop out from hillsides and little rivulets tumble down the rocky hillsides, swollen. Green with grass or spots that are red or brown from iron ore. Some of the scenes in the distant villages down there resemble beautiful toy landscapes.

These roads are hell on cars. A 4,000 foot mountain range in horizontal sedimentary rock sandwiched. Tiny hilltops are flushed with soil. But when the highway cuts through hills, sides rise up about a hundred feet or so. With giant steps on their sides to halt erosion.

Visibility in rain; mist floats up off some hilltops. Some faults and...Entering Braxton County, with mountain tops that can see tufts of fog that look like smoke, suspended. Hills around. Not good crop soil, but good for grazing. Layers of green in the soil. Small rambling buildings of farms. 79 is a two-lane highway with a breakdown lane. Few trees, which are up the hillsides and tops, with the valleys cleared out for farming and grazing. Mix of pines and broadleaf trees. Can see ridges along top of hills, but some are double ridges. Rarely did I see a cemetery.

Exit four, Flatwoods is one mile ahead. North on route 19, Flatwoods left, Sutton right. On 19, Flatwoods, one mile. Lumberyard on right side. North 19, is same as north 4. After the Mobil station is a railroad crossing and parallel to the road. A few horses and cows, just like back home. Driving though Flatwoods, I pass through Heaters, Inc. and stop into an antique gas station with a cracked toilet bowl, but it's water tight, with the crack filled with bright green moss, so it's okay. There is a deserted barn on my left, and I turn around and return to check in to my small motel.

Rather wide, Elk River flows rapidly in late April. The stretch between Sutton and Flatwoods has a lonely stretch of road with few houses. Topography above Sutton is visibly higher…The Sutton Reservoir and its dam, which did not exist back then, are impressive.

Radar

The famed investigator Ivan Sanderson determined that radar tracked the flight paths, which made a right-angle turn. Sanderson determined that two craft landed, a third seemed to crash, and a fourth blew up in the air. And southwest near Charleston, an object disintegrated in the air. Sheriff Robert Carr and Deputy Burnell Long rushed out to investigate a fiery crash below Gassaway on Elk River.

Since on September 12th, the real darkness does not set in until about 7:45 p.m.—long after the 6:58 p.m. sunset—and the near-full-moon does not rise until 8:10 p.m.—then the unlucky seven could not see the "monster" until they almost walked into it. On September 12 of 2003, your author simulated their encounter.

The individuals include in 1975, County Sheriff John Stalnaker, Charlie Humphreys who is office deputy. But the sheriff in 1952 was Robert Carr, dead in 1975, and his Deputy was Burnell Long, who was still alive in 1975.

What was the purpose of landing? Was it a reconnoitering mission, maybe malfunction problems, or a psychological behavior test of us, or were they testing government and societal responses for possible First Contact? Or sending out conflicting misdirection to confuse our physicists and discredit curious researchers? Or just a quirky sense of alien humor? For now, your guess is as good as mine.

Phone calls and telegrams came in from all over the country. *Fate Magazine*, which covers strange stories, sent a native of Braxton County, Gray Barker, to cover the story. But the most qualified investigator was Ivan Sanderson of New York City with assistant Eddie Shoenenberger, who arrived several days after the event. An experienced investigator with scientific knowledge, Sanderson had appeared before as a guest on television and radio.

Sanderson obtained aerial maps of the area and interviewed the witnesses. Quickly realizing that it was a big story, Sanderson contacted Bert Ash and Earl Walter, members of the Charleston chapter of the National Speleological Society, who quickly drove to Sutton to assist. Next morning, Sanderson and his assistants thoroughly examined the site. They questioned the witnesses separately, in pairs, and as a group, then interviewed residents.

Next morning, the moon set at 8:08 a.m., and the sun rose at about 6:24 a.m. Since half an hour before sunrise, the eastern sky lightens, the machine was easily seen rising. Several years later, author John Keel investigated, and a couple told him they saw the "monster" on the next night and an alleged craft that rose from woods.

Dale Leavitt

After Kathy phoned Dale Leavitt and I also talked to him, I drove to his house set above a gentle sloped hill. He had a pleasant voice with a decent manner—definitely a leader. He added new facts and, as with Kathy May, I openly taped him, with permission to publish asked and granted. He added new information from his perspective, partially transcribed as follows…

Stalnaker lived up at Little Birch. I believe he died three or four years ago, unless I am mistaken. Well, I was commander [Colonel] of the National Guard Company and I got a phone call from someone near Frametown, just west of Sutton, and they said that an airplane just crashed down there, across the river in the woods, and they only saw the thing as it entered the woods, so they did not apparently get a description of the plane.

So I got a truckload of men out of the National guard and went down there, and we went across the river and up a real steep hill and searched the whole big woods area, and there was no sign of anything. It was a ways just this side on Lyndon [indistinguishable]. It would be dark soon; and. anyways we made a search of the woods as best we could just before dark. And after dark, we came back and I got a call from Flatwoods, and then I later heard many stories, you know, about how the Air Force had called me and asked if I could…and so on.

I don't remember the name of the commanding officer offhand, but he was out of Air Force Intelligence. So I gathered up soil samples, surface and subsurface, and the grass and weeds, and there was an oily substance on them, mostly around where they saw the "monster". And so we sawed some of the leaves and limbs off the tree, and sent all this into them, as requested. And finally, and this was a short time after the happening, this was two or three days later.

People trampled things quite a bit by then, and we were out there on two or three occasions to talk to different people about this. A lot of people, as a matter of fact. I saw people playing on a grade school ground down there, in a ball game or something. We talked to them. We talked to people who went on to the hill. And people who lived in the area generally, and some people who had sighted it from the school ground.

So, a little bit later, I heard about one fellow who was driving up along Elk River, and he came up with his wife and child. This is the actual story, at least at the time it was reported. This fellow was afraid to report this encounter at first. He felt others might say he was crazy. But he came up the road down there, and his car stopped and drifted off the side of the road. And there was an object setting on the side of this hill, and so, after that he smelled an odor, and there were little people all around it—apparently working on it.

So he went back to his car to start it, but it still would not start, so he decided it was not going to start, and he rolled up his windows and locked his doors, and stayed inside his car. And then one of these came up and scratched his hood and tested his windows, and all that sort of thing. He reported this something like a year later. He said that this thing left and then his car then started up—just like that. His wife went to the hospital, she is the one. As far as I have heard, it was just a case of fright, of nerves. I do not remember her name, off hand. I have all this in a big folder, all the clippings out of the papers and other material.

I recovered some metal out there, it looked like solder, and I sent them to Washington, straight to the Pentagon, as requested. But before I did, I took some of the solder that went down into the grass. Anyway, I cut the solder in two. The pieces were about the size of the end of a matchstick, and just about like the shape of it.

Anyways, I kept part of it. I sent the samples into the intelligence agency address, as requested. So I put it and the samples in this box and sent it to them. It went to Washington straight to the Pentagon. I used to have the names and addresses, but I am not sure.

It looked like it melted and then hardened when it hit the ground. That is just how it looked, just like a teardrop. It lay on the surface of the ground and did not enter. We thoroughly searched the ground. As for color, it looked very much like solder. It was easy to cut, just like solder, but I would say not easier than solder. Not like lead. I could cut through it. There was a service station up there in town that I went to, and there I took a knife and cut one of the specimens in two.

We really should have had a metal detector out there. We did not check the radiation because we had no Geiger counter. [Author—Did the two incognito Air Force Intelligence officers (disguised as newspapermen) use a Geiger counter?] We took soil samples from the surface and subsurface. We went down about six inches, and took samples from that depth, and then some from three inches down, and from the surface.

We walked up there, right up this path on your map you drew, past this iron gate, and then there was a fork right here that splits off, and there is a grade here, with a house or shack here [the "monster shack" which later burned down], with a tree and tire-swing now here, and a cornfield down here. And then there was a fence here, or whatever remains of it. There was a big 60-foot oak tree here on your map that may have been a white oak with many branches on it. And that is where the "monster" was, so they never got beyond that.

Now, the saucer apparently landed just on this side of the tree, and slid down a little ways leaving a track, right there on your map. Kathy did not see the disk, although others did, including children on the playground, which is located—say, you came out here on route 19, here; I will draw it for you this way. The schoolhouse sets here, and the playground is there.

The dark gray material that was on Kathy's apron [She still has the apron.], she said was not slimy. Well, what was on the grass was similar to an oil, perhaps when a motor burned something half way and then deposited out on the ground. We could tell that it was kind of oily-like, that was more than an oil. But you could take it and smear it. There was not a heavy coating, and most of this "oil" was confined to the "Monster"'s area.

We had to go outside the area where people had milled around on, so there some of it might be reduced. [Undecipherable. Something about being on the road.] I did not keep it around that long because I packaged it up. And the "oil" that was on the grass. I picked that up with the grass and mailed it in too. And I never did keep any of it...I walked from here, went to the road, like fifty or sixty feet.

I had letters from Ivan Sanderson. The boys on the football field saw something on it that shined and it looked like either the setting sun reflected off it or it lit from its own light—I do not know which. They said a short red tail trailed behind the disk, but I did not recall anyone describe its exact length. But it came from the edge of the disk, not from its center. [Author—Did anyone take pictures?] Not a soul. And no one could get a real good view for long because it was largely over the hillcrest from where they were at. One of the boys that was up on the playground, maybe it was more than this one.

The landing down the river was a pretty significant one. People saw it come down, and they thought it was an airplane that crashed at Sugar Creek.

At any rate, the skid tracks came in here [on your drawing], stopped here, then skidded down the hill [four feet apart; undecipherable]. There were reporters there from many newspapers. As for the tracks, where it started to lift off, in this area on your map, we did not get to examine it until the next day. And by that time many people trampled around the site.

We were out there that night, but there was no point in looking too closely, even though there was a full moon, as I recall, although I am not sure if it was full. At least, it was not really too dark. We walked up there, but we could not tell where it was. I am not sure we wanted to meet it in the dark, but we looked. So we went back up there the next morning, but it wasn't early enough! [laughs] Boy, I am telling you! And there were 1,000 curious people up there, and a total of 2,000 over the next four days.

There is a possibility that I have some printed material someplace, but offhand I do not know where. I have not looked for it or seen it in years. I had letters and documents that came from Washington thanking me. Of course, I had to pay the freight for shipping everything to them. It's possible the Pentagon did not want to acknowledge their interest. Or it could mean that they wanted to cover their trail, but it could also have meant anything. I spent some twenty dollars for postage altogether and they never reimbursed me. [laughs]. I had the letters and things in my file. And I also got letters from people, and kept those.

Stewart, the old man, died. [*Author*—Mark said that Lee Stewart left the state because of something oblique.] I am sure that he is out at Somersville. I do not know of any [alleged] photographs of the disk that were taken. I think they printed that letter in the paper.

I moved since then. I don't know if Mark Washman let files go to a radio station up north. There was a photo of me holding the droplets. There were six, from were the "monster" floated. The people had trampled about, but we were careful, and these were on top of the surface. As for how long the black film lasted, I don't know.

You say that Kathy noticed two antennas, thicker than a pencil and three feet long that came, sloping downward, from its neck. We were curious if the thing had any antennas, and we asked the others, but most did not notice it, but Kathy did notice it. I believe that would be right. There were wrinkles on its glowing green "gown," but with no belt, no arms. From the picture drawn in New York, it sure did not resemble it. Well, some people say it was just an owl. That sure was a real big owl, wasn't it? [laughs]

They tracked it with Air Force radar. Not Sutton, but from other radar sites, but I do not know the operators' names. [Was radar disclosure purposeful? Or did their craft lack stealth technology?] They conjectured that several saucers flew in from the Eastern seaboard at 7 p.m., passed over Baltimore traveling northwest and then veered around and headed back over central West Virginia fifteen minutes later from the northwest to southeast. It was visible over Maryland, Pennsylvania and our state—and seen at [indistinct] on the grade school ground by the boys playing ball. It flew right up that river, and meteors do not do that.

After another half hour of taped interview, and making drawings, I asked Dale Leavitt if I could use his conversation in my interview, and he agreed. Thanking him, and saying goodbye, I drove north to my motel.

Odor

I had been getting busy signals each time I dialed Sutton's local newspaper, the *Braxton Democrat*. After five calls, I decided it was close to quitting time, and since it was Friday night, there was no way to reach Arnie [sic] Stewart on the weekend. He lived out of town. At 4:30, I decided I better get down town to Sutton—fast. Since it was a four-mile drive, this was no problem. At fifteen minutes before five, I parked my car, walked over to the *Braxton Democrat* office.

Well, Arnie Stewart was long gone, so I talked to the present editor, Mark Washman, who said that he did not have the letter—Mr. Stewart took some materials with him when he left for parts unknown. Stewart lived out of the county before, but now he lived out-of-state, and any hope of tracking him down seemed remote.

Initially, Washman sounded understandably cautious, the same reaction I also got at first from Sheriff John Stalnaker. Washman told me that Jim Hearst of the *Charleston Gazette* was also a former editor there. The *Braxton Democrat* had articles—"the same old stuff, nothing new"—that I was welcome to see. I asked him about the awful odor, but he was unsure.

In 2003, I checked UFO odors at one website—www.nidsci.org/articles/pdf/rullam.pdf. Odors are rarely reported with extraterrestrials and their craft. As you can already guess, most odors are quite obnoxious. This website said the Flatwoods odor was a sickly smell resembling hot, grease metal of very strong intensity, and with a hedonic tone: This foul smelling mist was a nauseating odor—atrocious. It made witnesses ill and caused eyes to water, burned and hurt nostrils and throat. As one of the witnesses, I say that sounds accurate—and add that it was unlike anything that I ever smelled since then.

All the other odors listed on that website were also unpleasant, some described as a strong burning smell, an awful odor that made a mother ill, and described by witnesses as mixed with sulfurous smoke, nauseous, felt prickling throughout body, had to stop, lost balance several times…like sulfurous and brimstone, noxious, foul, nauseating, witnesses sick, lost sense of taste and smell, would not swallow properly, of bad garbage and chemicals. It seems that if aliens are trying to discourage wildlife, dogs and humans from getting curious, they are doing a good job.

Lee Stewart, co-publisher of the *Braxton County Democrat*, in a videotaped interview in December of 1996, said he and several men armed with shotguns returned with Lemon an hour later and smelled the "sickening odor." Stewart talked to witnesses, who were trembling in fear, and one was in near collapse. The strong sickening, burned-metallic odor remained, but there was no sign of the "monster".

"It was sheer turmoil," said Stewart. "The three boys that were there were very, very sick to their stomachs, and all of them were wheezing and coughing…We just spotlighted around because none of us were [foolishly] inclined to hunt for something we didn't know what it was [12-foot 'monster'] in the dark."

Garry Harris, Publisher/Editor of the *Mid State Star* newspaper accompanied A. Lee Stewart (reporter) for the *Braxton Democrat* to the encounter site on the morning of Saturday 13th, and saw a sizable number of National Guardsmen investigating the site.

Washington

Two months before Flatwoods, did aliens test our responses with the famous Washington Sightings that began on the late night of July 19th? It began on that 1952 Saturday night, at 11:40 p.m. Seven mysterious blips appeared on the air traffic radar screen at the Washington National Airport. They flew leisurely at 100 mph from south-southwest to Washington, DC. Air Controller Edward Nugent called supervisor Harry G. Barnes to look. Air Traffic Control also had the same objects on their screen—slow-moving solid, visible radar targets—and even saw a bright orange craft from their tower window.

Seven white disks slowly flew over the White House and the Capitol Building. One week later at 8:15 p.m. dozens of multi-colored craft flew leisurely over the capital, and when F-94 jet interceptors arrived, one alien craft playfully circled an F-94.

With an eerie feeling that the aliens were listening, Barnes later said that they understood his radio traffic and instantly reacted. The alien craft resembled small kids out playing, directed by curiosity. Radar targets of unidentified targets continued into the evening of July 20th. For the rest of the week, strange lights appeared over our capital. Diameters were 100 to 200 feet, and speeds, at times exceeded 900 mph, made with right-angle turns and instantaneous reversals.

But, shockingly, the Pentagon kept both Project Bluebook and Captain Edward Ruppelt totally in the dark! So, Ruppelt and Bluebook did not even know until July 22nd in the morning—but only by accident because they read about it in the newspaper!

On Tuesday July 29th, General Samford held a kangaroo press conference to brazenly debunk the spectacular lights as mere temperature inversion. Samford and the USAF did not believe their own red herring explanations, as explained later by Captain Edward J. Ruppelt, and so the case is still listed as explanation unknown. Sardonically, Ruppelt wrote that temperature inversions occurred almost every night that summer, yet there were no sightings.

The warmer air trapped under colder air, with a difference exceeding six degrees, bends (refracts) ground lights to allegedly create moving skylights. It is mostly hogwash, and these spectacular sightings were never explained. Today, after the airliner attacks of September 11 in 2001, no aircraft are permitted near the White House. All intruders are challenged and may be shot down. On November 20 of 2003, a "flight of birds" (so they say) was mistaken for an intruding aircraft. Probably, but what if extraterrestrial craft once again fly too close and even over the White House—as in 1952?

Time passed, questions remain. At the Bicentennial Monster Festival held September 12-14 of 2002, Mayor Margaret Clise and researcher Frank Feschino, Jr. attended, with Stanton Friedman lecturing. At 7:30 on the 12th, people gathered at the school playground to commemorate the exact time fifty years ago when an extraterrestrial craft flew over and landed. Today, Kathy May is elderly and her sons Fred and Ed (Edison) and Ron Shaver all have silver hair.

Although time passed, questions remain. What was the Flatwoods "Monster"? Did a floating robot spook them? Did it land on the hill for repairs? Were they testing local, public and government responses? Did their test dovetail with the earlier Washington Sightings? Did they test our responses? To see how ready we were for disclosure and First Contact? Or to let our Pentagon know?

Were Washington and Flatwoods the beginning of clever alien misdirections? Red herrings that now have become Vaudeville freak shows to distract and discredit serious investigators? Obfuscating smokescreens to mask future sightings? To sow confusion—confusion that hides aliens—like scientists burying signals in noise to mask signal existence? Or was Flatwoods just an amusing prank—alien humor?

Boston

On the evening of April 23 in 1966, an estimated thousand citizens saw strange machines flying over greater Boston—in fact, as far outward as fifty miles away.

Our sighting began just past eight, when hundreds of witnesses in the Taunton, Easton Brockton, Bridgewater and other areas, far south of Boston, saw a dazzling light radically misbehaving in the northwesterly segment of the sky, above the horizon. It was so impressive that I wrote down the details as follows.

At 8:10 p.m., dad walked inside and said there was a strange light in the northwestern sky. We ran out the front breezeway door, left past the front of our garage, and looked backward (northwest) across the field out back.

There, above the skyline of dark northeastern white pines, in the distance appeared a white light of great intensity—dazzling—absolutely brilliant. It moved back and forth almost continuously, sometimes smoothly and other times erratically. Occasionally it stopped. Upon reversing itself, often it did so abruptly—making instantaneous reversals, like a ball bouncing off a wall. But at times it flew jerkily and erratic—not like any aircraft I had seen before.

Sometimes it dropped below the tree line and then swung over and gradually rose as it did so. Several times, this brilliant light diminished in intensity and immediately a green and red light came on—seeming to revolve around the craft—for twenty seconds. Then the bright white light would instantly come on, and the small green and red rotating lights would cease.

The light maneuvered within a rectangular area above the horizon with compass bearings between approximately 340 and 355 degrees, and elevations between 15 and 25 degrees, as measured by rods and brick supports atop a level cement platform, quickly set up by your author, and later measured and with trigonometry calculated the next day. At times, it was tighter. Your author looked at the light through 16-power binoculars.

The motion often was uneven and instantaneous in its reversals that sometimes occurred, and the motion near the center between reversals was often slower than the motion occurring near the two sides. But mostly, it was constantly moving, usually back and forth. The dazzling light was clear cut and not fuzzy. It was a solid object—a machine.

At times a pinpoint of red light appeared left of it and moved about. We heard no sound from any of the lights, nor did the witnesses (near the dog track) under it later report any noise. There was no connected motion between the intense white light and the two red lights. At times the light would be moving slowly and then suddenly speed up, reach the endpoint and quickly shoot up. The period of the faster "swing" from side-to-side took one second. After eight minutes the white light moved slowly to the right, drifted down to the horizon, disappeared behind the distant tree line. (Later, other witnesses said the light seemed to suddenly disappear).

At 8:45 an airliner flew from south to almost overhead, at 50 degrees above the horizon, flying to Logan Airport in Boston, as they did then every night. Since I spent several hours on some nights outside, in all weather, the night sky was familiar. Halfway across the sky, the airliner turned around and flew back in the original direction it came from. This was abnormal, especially since it was several minutes after the alien craft had departed. Your author saw a few more planes in the next two hours—mostly, airliners traveling from south to north.

There was no fog or mist at our location. The sky was overcast. From the estimated cloud ceiling, under which the airplane flew, I later calculated the maximum distance and from my measuring its elevation, or bearing and azimuth, that would put the alien craft over a trailer park. I later calculated that at times it flew 110 miles per hour and (assuming a circle) making a complete revolution with a radius of 55 feet. At 11:10 p.m. light rain began, and I went inside to listen to the radio.

Later that night, when I listened to Boston's WBZ-AM-1030 news, it reported dramatic sightings all over Boston. In a brief taped interview, Ray Machetti [spelling unknown], an Air Force veteran and airline employee, stated that he saw a cigar-shaped craft, 35-feet long, with porthole lights flashing every ten seconds "like car headlights." The strange alien craft disappeared over Boston's Mystic Bridge.

Mrs. Sawyer from Salem saw a saucer with red-blue blinking lights before four and five a.m. on this night. Then on Columbus Avenue, a South Boston man reported. Mr. N in his car with his wife on Hamilton Street was going left onto Adams, and saw a strange aerial craft.

WEEI reported that on April 20th, in Foxboro on the Sharon line—about 10 miles west of Taunton and 30 miles south of Boston—a saucer hovered there the previous night, and that many observers went there the following night—and the saucer hovered again.

The next day, Leonard Lawrence on WEEI of Boston hosted his usual Sunday afternoon, phone-in talk show. But when nine out of ten callers talked about their sightings of the prior night, this eventually annoyed Lawrence. Then, on Monday morning, WPEP carried a program on local sightings. Linda of Norwood around 11 p.m. saw a spectacular light maneuver for almost an hour.

A Bridgewater woman driving west on Pleasant Street, crossing over Route 24, saw a the mysterious light so dazzling that she stopped her car, then got out to watch it. A woman (Leonard), who then lived in Lakeville, further to the south of Bridgewater, also saw the light [same one?] from a second floor window, about ten miles south.

The light was dazzling, clear cut, and not fuzzy. But in the Taunton-Bridgewater area, far south of Boston, we were only a small part of the total massive Boston sightings going on up north, which were never explained. And all NICAP (Keyhoe) ever gave these spectacular Boston sightings was a tiny write up.

To be perfectly honest, if I had not been a witness to the Flatwoods "Monster"—no, absolutely not, I would never ever believe it—never—and I would doubt all sightings of such craft. (To hide, do extraterrestrials foster sensational claims to sow confusion and misdirection?) As it is, I doubt most sensational claims. So, I am not a dogmatic (conservative) skeptic, but a rational (liberal) skeptic.

The *Committee for Scientific Investigation of Claims of the Paranormal* has an important unspoken beneficial purpose—to protect legitimate scientists from kooks, and to debunk anything that attracts kooks. CSICOP "runs interference" to protect legitimate scientists from having their good reputations dragged down into the mud, as one saucer kook tried to do to Carl Sagan. That said in CSICOP's favor, I must honestly accuse them of sometimes being too ubiquitously iconoclastic.

25

Among Asteroids and Comets

Some skeptics (Gell-Mann, *The Quark and the Jaguar*) doubt Kip Thorne and Carl Sagan's fanciful proposals to sidestep the speed of light by passing rapidly through wormholes to distant galaxies. But there is no need for Thorne's wormholes or even for fast relativistic velocities. Slow velocities will do quite well. And by cautiously predicting our own technology, we can predict what others long ago also accomplished.

Interstellar travel is for robot machines—not for flesh and bone—even inside spinning wheels that simulate gravity. In fact, interstellar travel is dangerous even for machines. Although radiation shielding is possible, it is inadequate at practical interstellar speeds—at a "slow" (non-relativistic) five percent the speed of light. Even flying that slowly, there will be serious radiation damage to craft structure and contents.

No one can repair and replace damaged cells. Aliens could try nanoparticles smaller that bacterium, and only a few 100 nanometers wide, that seek radiation-damaged cells, enter them, and release DNA-repair enzymes or enzymes to initiate auto-destruct (apoptosist)—but this solution is doubtful.

Skeptics justifiably doubt that star ships travel frequently between stars. Fortunately, there is little need for frequent interstellar travel. There are better solutions. For example, lasers could communicate between established star civilizations, to form a galactic-wide federation, using highly directional, "laser network party lines" that quarantine earth.

Perhaps for billions of years, nearby star-civilizations have carefully, through phased array, "ganged" telescopes that carefully watched early earth and detected the emergence of an atmosphere—the gases of early life in the early Cambrian Period (and even possibly 3.5 billion years ago at the dawn of earth life)—that indicated life had begun.

On exo-planets, bright blue suggests an earth-like atmosphere of gases, and life-supporting gases like oxygen and nitrogen, which absorb certain infrared wavelengths, while green plants reflect red light. If alien biology is convergent, then we already know which gases to search for—planets with atmospheres changed by life.

And if convergent evolution (homoplasy) is universal, as championed by Simon Conway Morris, then aliens long ago predicted our earth's evolutionary path through our early Cambrian explosion (sudden proliferation of new life forms at 570-500 millions years ago). As NASA indicates on its Web sites, we too will soon be able to detect extra-solar alien earth-sized planets—and later, with better resolution, even see their continents and oceans.

Origins

Long ago, as alien robots spread across our galaxy, they launched swarms of individual intelligent robot capsules toward stars, including early earth, perhaps at only five percent of the speed of light, but still too dangerously fast to avoid fatal collisions with interstellar subatomic particles and specks of dirt, and occasional rocks, boulders, even brown dwarfs or orphaned interstellar planets, and debris ejected from other solar systems—not to mention dangerous comets in their narrow elliptical orbits with apogees that reach out beyond the interstellar boundary out to two light years. Perhaps ninety-five percent of each fleet was destroyed.

Once the surviving robot capsules ("seeds") entered our earth's cometary and asteroid belts, they "blossomed" and mined the belts and constructed larger vessels. From studying earth's DNA, these intelligent robots grew various hominids capable of existing on earth. Some were exactly human, others rather different—but all designed to later confuse and discredit ufologists, and to camouflage alien existence through clever misdirections, confusion and ridicule. And used for First Contact with our civilization. Unlike the intelligent alien machines, hominid aliens will have things in common with humans.

Why do alien robots living in our asteroid and comet belts refuse to reveal themselves? For good reason, although they may have considered disclosing themselves openly back in the early 1950s. Perhaps the Washington Sightings of 1952 and other sightings, such as the Flatwoods "Monster", were testing our responses—then, judging from our responses, they wisely changed their minds.

As for the most likely reason for non-disclosure, it is likely we would have misunderstood, leading to catastrophe, for several reasons. First, it would be disastrous for us to suddenly discover that long ago many intelligent machines took up residence in our solar system and were watching us, then bio-engineered modified clones of humans, and that these alien robots had superior spacecraft, and could (if they wanted) bioengineer the total collapse of humanity or redirect hundreds of asteroids and comets earthward.

Well, you see how the many chances for misunderstandings could lead to anarchy, collapse of our civilization, and a search to discover the secrets of alien propulsion, and possibly try to destroy the aliens, forcing them to destroy us.

Or maybe it is worse—they will prove that our religions are false. Or even worse, that there is no hell to restrain us. With that knowledge, care to guess what might happen to our civilization?

Evolution

If most aliens residing within our solar system are highly-intelligent robots, then we have little in common with their religion—if they have a religion. Billions of years ago, most bioorganic aliens ("meat machines") replaced themselves with far superior intelligent robots. If convergent evolution (Simon Conway Morris) is accurate, then early bioorganic alien evolutions everywhere probably paralleled our own evolution, as follows.

On earth, of the many parallel evolving hominid species, only Homo sapiens survived. Reproductively incompatible with us, and though rather intelligent, and although he almost beat us, Neanderthal also vanished. Paleoanthropologists do not yet know who was the first hominid—perhaps *sahelanthropus tchadensis*—but now know human evolution did not resemble a tree or straight line, but a bush, with many parallel hominid species.

Although the early hominids walked upright, looks were deceiving, because with their new bipedalism they still had small brains. Why they evolved to stand upright is unknown—perhaps to reach fruit on higher branches. (It could have been a sudden spontaneous mutation, as with "Oliver the Bipedal Chimpanzee," discussed below.) But what caused their subsequent explosive seven-fold increase in brain size? Such grotesquely huge brains of *homo sapiens* require incredible energy, so the evolutionary forces were powerful.

Perhaps it began 2.4 million years ago, when the MYH16 protein isoform, crucial for powerful jaw muscles, mutated out. Thereafter, the skull was unconstrained by a bulky chewing apparatus, freeing brains to grow. Since powerful masticatory muscle sculpts bone, hominids lost their large skull crests, from which earlier powerful jaw muscles were attached and their brain cage grew explosively.

Of course, added intelligence means the usual advantages like tool making, but also permit powerful social groups to work together, to construct shelters, to hunt together, and to cooperate and coordinate their actions—a powerful evolutionary advantage. There were side effects: as hominid brains grew in size, skulls grew thinner to dissipate the increased heat.

The loss of the MYH16 protein isoform also led to the evolution of the larynx and thus complex speech. But evolution made a compromise that led to medical choking crises not faced by other animals. Behind the tongue lies the human larynx. Poking up and resembling a partly buried trumpet-shaped pitcher plant, the larynx contains nine cartilages, with vertical vocal cords inside. Its dual function is most unusual.

At every swallow, the epiglottis instantly drops like a trapdoor over the opening to close the entrance to the larynx. At this, the laryngeal apparatus rises up and forward under the tongue, which instantly kicks the liquid, food or saliva over the voice box back into the esophagus and gullet. While this instantaneous coordination between nerve, muscles and cartilage occur, minor food particles and secretions may be trapped in the small trough that runs between the epiglottis and tongue, and several pits behind the laryngeal opening.

Vocal cords shut at the slightest irritation or cough reflex—coughs that exhale air at 50 miles per hour. But with age, the weaker muscles and slower reflexes create more accidents, with liquids and hastily-chewed foods, even saliva, being "mis-swallowed, to "go down the wrong way," and thus leading to unpleasant coughing to expel it. Also, half of all humans inhale stomach contents while asleep. Some cough up and inhale. Among older humans, this can lead to pneumonia. Animals do not have the human vocal apparatus (a Rube Goldberg contraption), and so do not encounter such human difficulties—nor share in its advantage of speech.

Chimpanzees lack a human larynx, and cannot speak. The uniformity in small inheritable genetic differences is so uniform for H. sapiens, who descended from a small group of ancestors (perhaps under 2,000). Chimps are ten times more genetically diverse, and the gap between chimps and humans is 20 times greater than the gap between any two humans.

With greater genetic diversity, evolutionary anomalies occur. For example, "Oliver the Bipedal Chimpanzee" is well known, and many films exist. In the early 1970s, famed trainers Frank and Jane Burger purchased a mysterious chimpanzee, named Oliver, who had a very small human-like head, with a short jaw and strange ears—not quite human. He had short hair and unusual eye coloring. With his bird-like voice, many un-chimp-like mannerisms and with unusual anatomy, he resembled a five-foot-tall hairy humanoid. He walked with a human gait.

His hips were shaped differently, more humanlike, and facially, he looked more human. He stood tall, with shoulders back and knees locked, and walked bipedal like a human—all the time. Watching Oliver walk upright like a human gave most people a creepy feeling.

Oliver did not act like other chimps. He did not have to be trained to act human—it came natural. Other chimps shunned Oliver, who preferred socializing with humans. His intelligence far exceeded other chimps, and he was quite likable and cooperative, unlike other male chimps, who get "grabby" and aggressive, and run off to climb trees.

Oliver had awareness and understood, which was shocking. [See Penrose, *Shadows of the Mind.* pp. 406-411. (animal consciousness)] He learned quickly and solved problems beyond other chimps. He preferred human mannerisms. He would mix his evening drink by pouring a shot of whiskey and some Seven-Up, stir it, walk over and sit down in his chair to relax and watch TV. He would smoke a cigar and drink his nightcap.

Next morning, without training or prompting, he would spontaneously get up and walk over to pour himself a second cup of coffee and get a refill for the Burgers. They would send him on chores, and he would take a wheelbarrow and empty hay and straw from stalls, feed the dogs (get pans and mix dog food), and so on. Oscar acquired human habits. Oscar displayed un-chimp emotions, including tears of remorse and sadness at separations.

By 1976, reaching sexual maturity (but ignoring female chimps)—and possessing the strength of three men—he tried mating with Mrs. Burger and other women, so at her extreme insistence they sold Oliver to a New York lawyer, who some say constructively exploited Oliver (taking him on a tour of Japan) and then sold him to California animal exhibiters, who used him in shows. They then sold Oliver for lab experiments and he languished for seven years in a 5 by 7 cage. Although they claim Oliver was not used in lab experiments, he showed signs of rough treatment. He had been dragged around and exploited for 20 years.

Gratefully, Oliver was rescued and now happily resides in Boerne, Texas, at the Primarily Primates sanctuary. Today, Oliver has arthritis and is partially blind.

Cytogenic analysis and mtDNA sequencing and hoology comparisons of Oliver to African chimpanzees of known geographical locations was published in the AJPA [Ely, Leland, Martino, Swett, Moore]. Oliver is a member of the Pan troglodytes troglodytes subspecies from Central Africa (likely trapped in Gabon) and has 48 chimp chromosomes, but of different sequencing from other chimps. Interestingly, there are other rare reports of such bipedal apes and chimps in Indonesia and Africa. In May of 2004, the *Discovery* channel aired a one-hour investigation, *Humanzee,* of Oliver.

However, even without the advantage of speech, "Oliver the Bipedal Chimp" (hominid?) possessed an advanced understanding, intelligence and preferences far exceeding other chimps, indicating that early voiceless hominids were probably far smarter than we think. Oliver is evidence that if voiceless hominids were shown how to make bows and arrows, how to attach obsidian points to spears and axes, how to make fire, teepees and huts, and so on—then they would have learned and taught others. Perhaps they could not easily discover it themselves, but they could easily learn it and copy it—and then even innovate. Oliver throws a new perspective on hominid evolution and on other new issues, including a re-examination of the evolution of religion and ethics.

Does Oliver suggest Darwinian evolution took Paleozoic "quantum jumps"? Or is non-Darwinian Periannan Senapathy's brilliant *Independent Birth of Organisms* (Genome Press, 1994—and highly recommended), more accurate than Darwin? Well, if not on earth, could IBO occur on alien worlds? Is IBO possible? Highly complex, IBO better explains the Cambrian explosion (Biology's mysterious Big Bang) of organisms with grab-bags of genomes assembled from DNA soups to explain missing links, eyes, complex organs and RNA/DNA enigmas. IBO is parallel, like a bush, while Darwin is linear, a tree. Can nature use both (maybe combine) Senapathic and Darwinian evolution?

Aside from Oliver's shocking mistreatment, what is equally amazing is the lack of recognition (until recently) of the implications for ethics. If we recreate (bioengineer) earlier hominids, should they be protected by our Bill of Rights?

Although modern chimpanzees and baboons do create simple tools, the early hominids developed brains to mentally envision tools within untouched rocks that could kill animals, cut their tough flesh, and crack the bones for their rich marrow, and cook with fire—a powerful evolutionary advantage. Although the evolutionary history of the human hand is murky, our shorter fingers and palms with flexible thumbs provided hook, power and precision grips for better use of these early tools.

Hairless, *H. erectus* (sometimes called *H. ergaster*) could breath better, so survived in the hot sun. *H. ergaster* could associate things, helpful in tracking and hunting. By shaping stones for hand axes, and able to coordinate plans, he leaped from prey to nomadic predator. He began rudimentary value exchanges, primitive currency. Each troop was held together by male leaders with harems. *Ergaaster* had expressive face muscles that communicated social messages, and that refined displays to discourage injurious conflict.

Ergaster followed the Nile and migrated into Asia to become a new species. But over time, *ergaster* failed to evolve further, and continued to hand-hold chipped rocks and failed to put them onto sticks for spears and axes.

Still, early hominids like *ergaster* that emerged 1.9 million years ago still could not discover hidden connections, deeper meanings, and think and communicate with symbols—and it never occurred to them to attach their sharp stones on a spear and create stone-hurling slings. Hominid characteristics did not evolve separately, but in a parallel synchronicity. And oddly, and this is controversial, although *H. sapiens* existed 200,000 years ago, it was only 60,000 years ago that some rewiring or change occurred inside their brains that led to a sudden explosion of sophisticated tools, elaborate graves and jewelry.

Exactly why did *H. neanderthalensis* go extinct? They had brain sizes slightly greater than ours, probably spoke well, improved their projectile points, buried their dead with ceremonies, certainly were physically powerful, and definitely proved formidable hunters. But, unlike *H. sapiens*, the Neanderthals (for reasons still unknown) preferred their caves lower in the valleys and also formed smaller social groups. Perhaps preference for small groups was now a social handicap, although earlier being loners had survival advantages during earlier glaciations (and limited food sources).

Neanderthal could not adapt socially to larger groups and greater interactions. Extinction was not from failure of individual adaptation, but failure to evolve into more social hominids. Their brains were hardwired differently through 260,000 years of harsh evolution, and could not re-adapt quickly enough to new social changes. Still, they almost won.

Why did hominids become hairless? Because when they killed animals, cut and scraped the skins, they discovered they could use the dried furs for clothes and protection against the chill. Those with more natural body hair encouraged more parasites and diseases that weakened them, so they could not raise their young, which died, thus killing off the gene of longhaired hominids. But tiny hairs retained an ability to detect crawling ticks, fleas and mites—a distinctive survival value for the newly hairless hominids. But why several odd anatomical areas retained hair is a mystery. So, note this law—at times, hominids by their behavior directly and significantly altered their shapes and evolution. For hominids, evolution alters behavior, and individual and group behavior alters evolution.

Neanderthal could shrug of pain, suffer broken bones that re-healed, and survive endless minus thirty-degree days! These powerful supermen were the Klignons of the Ice Age. With superior senses and awesome hunting skills, they proved tough, cunning and formidable adversaries. In fact, they almost prevailed over H. sapiens. They loved meat. Challenged by ice, Neanderthal thrived in small bands.

Facing a parched Africa, H. sapiens dwindled until only 2,000 survived. Both moved into each other's territories, back and forth. Neanderthal almost won. Then something pulled H. sapiens from the edge of extinction. At the edge of extinction and driven by harsh natural selection, imagination soared and only the most innovative survived. Then, about 25,000 years ago, the last Neanderthal departed in Spain, somewhere near the Gates of Gibraltar.

Neanderthal certainly had different DNA and thus could not interbreed with us. Did we possess some slight evolutionary advantage that led to better communication, superior tools, more travel, and greater trade—and thus more children? And it does not take much of a difference. For example, although 99.4 percent of human and chimpanzee genes are identical, it is that tiny 0.6 percent that involves bipedalism, speaking and understanding grammar.

We will evolve no more, and in our future, in a sudden quantum leap, our successors will be super-intelligent machines—but of what type? And of what ethical nature?

Orientations

A question of evolution here—and not to be irreverent—but when did early hominids acquire immortal souls (if such exist) that survive after death? Did those with half our brain size have souls? Did H. erectus? H. ergaster? H. habilis? Or maybe H. georgicus? Did "Oliver the Bipedal Chimp"? Maybe half a soul?

And when during their gradual transitions did early hominids—which all stood erect, but still were dumb, but not as dumb as apes—exactly acquire immortal souls? Did one early hominid have a soul but his parents were soulless? (That is awful!) Or did some have partial souls? Through future genetic engineering, mankind will create early hominids. (Is that ethical?) Although earliest hominids were really stupid, will it be murder to kill one but not another?

Even more important to us, do alien machines have immortal souls? And, even more important—do aliens have ethical scruples? For the answer, we will need to examine our own ethics and world religions.

Humans worship God in one of four ways. First, their monistic world view (all is one) includes Hinduism and Buddhism, Confucianism, Shintoism and Taoism—all emphasizing the oneness of our universe. Reincarnation presents a chilling possibility—a premise that if souls reincarnate (returning as humans or insects or animals—why not as plants?), then their number must be finite. Thus, will overpopulation yield children without souls—the soulless? The Hindu idea of heaven is a soul dissolving like raindrops into an ocean, losing all identity.

Second, the polytheistic worldview of many gods includes animism—the belief that everything has a spirit independent of its physical being. Polytheism includes some Roman pagan religions, including emperors who claimed to be gods—but were cruel tyrants.

Third, the monotheistic world of one god includes Zoroastrianism and Judaism, which created the concept of a god personally involved in our history, plus its successors—Catholicism, Orthodoxy and Protestantism. They believe god is composed of three equal entities—Father, Son, Holy Spirit (Creator, Redeemer, Sanctifier), and believe that Jesus came to save mankind from original sin.

Christians believe Jesus was born without original sin and is the third entity in God—who has three equal but distinct entities (Trinity) and took human form to suffer and be crucified to redeem mankind from the sin of our origin or nature, original sin, and that he arose from the dead, and that there is life eternal after our deaths—in heaven or hell, depending upon our earthly faith and deeds.

Catholics and Orthodox believe Christ becomes present through transfiguration in their Eucharistic host (wafer), of unleavened bread and wine, which through transubstantiation becomes the blood and body and soul and essence of Jesus Christ. They believe that after death, that most saved souls endure atonement and purification for their repented sins in purgatory, all before they can enter eternal paradise. The living must pray for those souls in purgatory to hasten their transition into heaven. Grateful, they will pray and help us. One tombstone reads—Pray for me and I will pray for you.

Mohammed founded the Moslem faith and accepts Jesus as a prophet, but nothing more, since Jesus was not crucified—therefore he never rose from the dead. Mohamed's teachings were later collected into the Koran. In India, Islam gave birth to Sikhism. Moslems also believe in angels.

Fourth, atheists claim that it is possible to prove there is no god, while agnostics also claim there is no god but differ—claiming god cannot be disproved. Some Deists believe in God, but not in an afterlife, and prefer to not think about hell—which makes everyone uncomfortable. Freethinkers are unsure of anything.

How did these four religious worldviews—and especially our beliefs in an alleged afterlife—all evolve? Through our Bible's brilliant Jungian myth of Adam and Eve, its Genesis symbolically depicts the opposite nature of man and woman—how woman tempts man to fail her so she can judge him. But only God can safely judge (that is, with emotion of resentment), and when we do, guilt begins and pursues us—the so-called "hound of heaven"—but we often never know why.

Evil tempts humans to hate it, and through this mechanism, it gets into new souls (if humans have souls), their hearts and intellects. It does this first to children, to separate them from their common sense that they are born with. Injustice (cruelty) tempts humans to hate (resent) injustice—to fall—and to hate those who tempt us through injustice to resent them. Evil seeks to clone itself in new victims, like a parasite. Extraterrestrials understand this better than humans.

Although the story of Biblical Genesis (Adam and Eve) is a fable and not literally true, on the other hand, this Jungian Myth of Adam and Eve is not just a silly story of good and evil. It cleverly encodes deep information about the human man-woman-God relationship, and consequent guilt from deep resentment, doing wrong (and suppression), and "the hound of justice" of our conscience (the silent inner voice of God) that pursues us—which humans suppress and ignore at their peril—when they deny doing wrong, when instead they should feel sorry for doing wrong and be emotionally healed. This means that there is a correct way (without resentment) to resent evil (wrongness) and be righteously "angry."

This Jungian Myth portrayed in the fable of Adam and Eve (that is, all *H. sapiens*) cleverly embeds our essence and the carnal knowledge of good and evil, and the role of food. For brilliant and insightful Jungian analyses of human nature, please read psychologist Roy Master's *The Adam and Eve Sindrome* [sic].

But things for us are about to change—big time. Alien super-machines (and also our descendents) may not share these psychological qualities with us—simply because they are not human. Few clergy comprehend this—laity, even less.

Almost certainly, extraterrestrials comprehend this. They are advanced robots, machine super-intelligences—without need for intimacy or delineated sexes, without gluttony, without need for sleep or food or recreational substances, perhaps without greed, perhaps without true individual natures. They are machine civilizations, long ago disconnected from things that define their own bioorganic predecessors—and from us humans. No, not evil, and probably better than us—perhaps indifferent. But very different. Perhaps without the sin of our nature, our origin (original sin), that was created in us by our evolution.

Evolution was inevitable here on earth, as elsewhere. As we are, aliens were. As aliens are, we will be.

Whether there is life for you after death is nice to know, but more frightening is the answer to two questions—Are you going to hell? And what will hell be like? People who doubt hell have very good reasons (often personal secrets) for doubting and secretly fearing eternal hell's existence.

But what if aliens prove there is no hell—I mean absolutely, totally prove it? Well, guess what—after that few will be lawful, good or obey any laws—a catastrophe that could destroy civilization.

Do aliens know if Jesus Christ rose from the dead? If he did not, was he a vile criminal and disgusting fraud? Or were his followers all dupes and cons? Are charismatic healings real (they are), but "something else" and not proof of life after death? And are Christians dangerous fools for spreading such vile lies? If aliens can answer this question—and your author is certain that they can—then just this one answer will be earth shattering. Many people will not like either answer, and will express their anger—a no-win for the aliens.

All religions, many of them so fundamentally contradictory and divided into nine or so major groups, certainly cannot all be correct. So, all religions will be angry at alien revelations. But only one faithful group—Moslems—may call for a jihad (holy war) against all aliens—satanic infidels.

Machine civilizations evolved above their early "territorial killer imperative" (Konrad Lorenz and Robert Audrey). Like us, do alien robots sleep and dream? Need psychiatrists? Do they "party"? Smoke pot? Drink beer? Watch football? I don't think so. Unless alien machines have artificially morphed themselves to simulate our personalities, they will differ from us.

Unlike their own early biological ancestors, alien robots do not need food, do not secrete feces and urine, and thus have no need for mouths, teeth, noses, hair, lungs, genitalia, lymph, blood, colons, intestines, and so on—you get the picture—most of the things that make us human. And they may have less need to verbally communicate, and will do so to each other by wireless data streams—and comprehend faster than we do. Most will possess excellent vision and odor detection.

Eating is pleasurable, as Sigmund Freud recognized, and who described oral and anal fixations—all things alien to alien robots. They may not possess our emotions. Aliens may not even need to feel enjoyment nor satisfaction, pleasure nor relief from eating meals or sexual rituals and the joys and trials of mates and raising children.

The man-woman "original sin" connected with eating—described so eloquently in the classic book *Eat No Evil*—which was so brilliantly recognized by the psychologist Roy Masters—will not exist for such machines. They will lack many things—no families, schools, marriage, sports, vacations, travel, sleep, exercise, television, alcohol or drugs, youth, growing up, and no birth and death as we know it.

Disclosure

Some think we need no God, no heaven, no hell. Some hate God. But make no mistake—this is deception, for without God, there are no guardrails. Religion is the moral glue of ethics that holds civilization together.

If from alien contact, we learn for a proven fact—beyond any doubt—that there is no hell, then this will create permanent anarchy and destroy our earth's civilization. As for Christians, most will meekly accept the aliens' ironclad proof that Jesus did not rise from the dead and that there is no God.

But some religions will not enjoy being told Allah and Mohammed are frauds, and will dislike desecrating alien infidels—maybe calling for a holy war (jihad) to slay all aliens. In the ensuing spiritual and military vacuum, there also will arise Hitlers and Stalins, for they also know that without a hell, there is nothing to fear but terror.

The aliens understand this. So, is this what deters them from disclosing their existence? Obviously, they know that First Contact must be handled delicately, or there could be disastrous consequences. Disclosure that benevolent intelligent robots resided in our solar system's asteroid and comet belts for possibly billions of years—far longer than we have been here—and have DNA-cloned humans and humanoids—creates serious potential for major misunderstandings on our side.

It is bad enough that they violated international law and that witnesses were injured, which certainly have legal consequences in criminal and tort law—as all lawyers will tell you. International law is toothless and vague, as it is, but imagine the nightmare of crafting and drafting joint national, international and alien laws!

Aliens know that in earth's history, among humans, not a single inferior civilization that encountered a superior one, no matter how well intentioned, ever survived intact—and most were decimated from contact.

When will earth's First Contact occur? Certainly, it will occur when we construct inertialess-negative-inverse-gravity propulsion craft and rapidly explore our asteroid and comet belts, and then encounter them. Aliens know this. They will not wait for this and may act before then.

Compactification

Machine intelligences do not need earth-like planets. They do not eat food, so need no mouths, no stomachs, no lungs, no bodies as we do, nor can they be harmed by weightlessness or vacuum, nor be confined to humanoid body shapes or environments that we require. They resist radiation and sustained high acceler-ations. With endless raw materials from comets and asteroids, the orbits about our many stars (like our sun) would be perfect—superior to planets.

In spreading from one star to the next, spreading across our galaxy, alien robot machine intelligence civilizations could benignly coexist with earths like ours. Far advanced beyond us, such intelligent robot civilizations would be less invasive and have smaller "footprints"—use fewer resources, use less energy, pollute less, go dormant, stay hidden, and be friendlier to the environment. They would not age.

Interstellar space is full of dangerous things—subatomic particles and radia-tion, and coronal mass ejections, and other things we can never anticipate until our ships hit them at lethal speeds. Since interstellar travel is dangerous to all fast-moving craft, thus star civilizations launch swarms of slow-moving truck and house-sized craft, using Compactification—compressing everything that then "unfolds" or "expands" when it reaches its target stars and begins by mining that star's comets and asteroids.

For interstellar intelligent robots, they are ageless Methuselahs, so time and aging is no problem for them. But for us humans, living only ninety years, time is precious. Alien machines may live for thousands of years. So a voyage to earth taking 2,000 years is not that long—to them. And since they are machines, very little that makes us human will give us anything in common with them. And since they are probably some type of robot benign Borg collective, loosely net-worked together, then we may never understand them ever—at least until our descendents evolve to become like them.

At any rate, when the intelligent alien fleet's robot "seeds" unfold when they reach a target star, they use asteroid and cometary raw materials to bootstrap upward, to unfold, to grow—just as tiny seeds contain the blueprint for mighty sequoias and draw their resources from the ground.

Alien robot machine intelligences are stellar "Rip van Twinkles" that sleep for a thousand years on alien fleets, until they reach distant stars. They then created more fleets that fanned across our entire galaxy.

Man measures himself by himself. He sees science fiction starships like those on *Star Trek*, populated with hominids and occasional human-like androids. The truth will be more mundane. But there will be surprises.

Aliens put out red herrings. They cloned different-shaped aliens, frequently conduct strange landings in remote locations, and may conduct stupid abductions and stupider medical experiments. Their real sightings are submerged in this clutter of statistical nonsensical "noise." They camouflage their existence to discredit and discourage all serious scientists from watching them. Yes, we are being watched—perhaps benignly, but probably indifferently. If nuclear war had begun, or a mega-asteroid loomed, they would not get involved.

Maverick Hypothesis

Aliens never interfered in our history before. So, why are there so many sightings today? In the maverick Vallee Hypothesis, astrophysicist Jacque Vallee speculates that aliens and saucers do not exist, but that they come from some "deeper parent reality." It occurred through history to pagans, Jews, Moslems and Christians, as visions and miracles, elves and leprechauns, and more recently as flying saucers—and most recently as alien abductions.

His Hypothesis says that none of these are aliens—just a sundry expression from this "deeper parent entity" that we do not yet understand. This Hypothesis of a hidden cause, so fuzzy and indistinct, is misunderstood by every ufologist, and none comprehend its magnitude. Harvard's Dr. John Mack, MD, is exploring this real-but-unexplored psychological phenomenon. So, Vallee's phenomenon may be different from true alien craft.

Tagamets

Creative ufologists suggest alien astronauts left legacies of their passing visits—even seeded Earth with life. So, did alien picnic poop on our landscape accidentally put alien DNA here? From which humans evolved? And did alien astronauts flying godly chariots purposely embed secret messages into our DNA? Not very likely. Well, maybe in other ways?

For example, in the days of the western open ranges, ranchers branded longhorns to identify them. Over a hundred years later, large UPC bar codes of the

1980s spawned large families of tiny codes of the new millennium that identify and carry information. Although invisible and too tiny to see, except only under the microscope, brand new microscopic barcodes from fluorescing particles ("quantum dots"), made of tiny metallic bars—overlaid with series of silver and gold metal stripes—can be injected or mixed into anything to uniquely identify it.

Alternatively, microscopic glass ribbons are only 20 micrometers wide and 100 micrometers long. These tagamets can tag larger biomolecules, including DNA. Glass segments contain ions of rare metals (including cerium, thulium and dysprosium) that fluoresce different colors and exceed 100 billion patterns. Such tagging mixes tagamets into explosives, paints, and "badges of authenticity" for designers, and to authenticate currency, for ID authentication and sorting.

But problems loom: What if these microscopic ribbons and particles get loose and contaminate the environment? Are there safety concerns? Confusion worries? Like our dumps and landfills of decades past, will future generations also curse our irresponsible legacy? Are these tagamets biodegradable? If not, what are the environmental and health dangers? Mixed in oil, grain, explosives and fungible substances, these dispersibles could spread throughout the environment, land and sea, and then the wildlife, rains, winds and oceans will disperse them.

As for imaginary aliens purposely leaving behind imaginary tagamets, nanoparticles or even evidence in altered Earth DNA—as proposed by UFO zealots—there is absolutely no evidence of this.

DNA

Aside from robots, are alien life forms similar or different from ours? Evolution is probably convergent (Morris) and not divergent (Gould), and always converges on similar biological behavior and forms, because form follows function.

If earth's evolution could be re-run many times in different ways, it would give an evolutionary trajectory that would produce biologies much like the ones we recognize—because similar anatomies and behaviors exploit similar niches, and natural selection drives all life to evolve to solve these challenges in similar ways However, alien DNA may be sequenced slightly differently, but on larger functional scales, they will mirror our forms. There are limited evolutionary problems, niches and solutions—and they reappear endlessly here as everywhere else.

Around deep-sea, caustic thermal vents, strange life grows. Only recently discovered, these strange organisms form a third kingdom, alongside plants and animals. And below 30 feet, deep in the solid earth down to two miles and embedded within solid rock, slowly grows microscopic life—only recently discovered—which in total outweighs surface life. Subsurface life inside rock may have predated surface life, is commonplace and hardier than predicted. And life can even withstand tens of millions of years of travel through space, shielded from radiation, encased within rock. Or did it survive the two mega-asteroid collisions 3.7 and 3.5 billions years ago—hidden deep inside the earth? Massive rocks were hurled into space. Some may have contained life. Did Earth seed Mars with life? Or re-seed Earth?

In the mid-Cambrian 540 million years ago, the evolutionary Big Bang blossomed when massive cosmic debris dumpred vast amounts of organic compounds into our early oceans. The violent solar system seeded Earth with water and organic debris that sparked the Cambrian explosion, stimulating the early sea life to evolve quickly, and provided it with rich raw materials. Although 3.5 million years ago, our solar system matured and gravity from Jupiter and our sun swept up much debris and "herded" them into stable orbits and clusterings, our sun's suspected dark twin, *Nemesis* (death star) periodically deflects comets and asteroids inwards every 26 million years, ostensibly explaining Earth's periodic mass extinctions.

At 225 millions years ago, if the vast Eurasian lava basalt flows had not ended the Paleozoic Era, even then, according to Morris (convergent evolution) then Cambrian sea life nevertheless would still have evolved into life similar in form, structure, biochemistry and behavior to today's life forms. And we would recognize them as familiar.

Dinosaur physiology had an edge over early mammals in lower oxygen, and thus gained an early lead. But if the KT asteroid had missed Earth, then dinosaurs (which were birdlike and warm-blooded) would evolve into modern hominids, and resemble us. Independent lineages faced with the same challenges evolve similar complex processes to exploit available niches—even when they are separated by a hundred million years.

Evolution is plastic. When one door shuts, another opens. Creatures that lose a feature, even complex structures, when faced with re-adaptive evolution, will re-evolve it—or go extinct and be replaced by some who can evolve. Examples include stick insects [*Nature*, January 16, 2003], where winged species evolved from wingless ancestors, whose own ancestors were winged—crawling flightless for 50 million years, only to re-evolve wings to fit a new niche. Other examples exist.

Life evolves both in gradually adaptive steps and in sudden jumps of punctuated equilibrium. Life survives. It is not static. It evolves, is plastic, and adapts. It changes endlessly to solve old problems faced ever anew, and grows—even re-evolving complex structures.

Signatures

When life is widespread on extraterrestrial planets, their atmospheres will show us a unique signature. The combined "breathing" of all life will alter their atmospheres in ways that we will soon be able to detect.

New telescopes will see images of the first galaxies and Earth-sized planets circling other stars—and, according to a NASA website, possibly see their oceans and continents. The old 5-meter (200-inch) Mount Palomar telescope was exceeded by the 10-meter (33-foot) Keck in Hawaii, designed by Jerry Nelson. With its 36-hexagonal mirror segments, the Keck was revolutionary.

Next, a TMT or Thirty-Meter Telescope is scheduled to open in 2012. Occupying an area the size of baseball diamond, it will see objects one tenth as faint as the Keck. Its improved adaptive optics, which compensate for atmospheric turbulence (shimmer that makes stars twinkle), will have a focus sharper than the Hubble Space Telescope. Already, Europeans are planning the Overwhelmingly Large Telescope or OWL. It will have over 1,000 mirror segments and be 100 meters across!

If Simon Conway Morris and the theory of evolutionary convergence are correct, then alien worlds will harbor Earth-centric life forms recognizable to us—although perhaps not with the same exact DNA sequences—but with an organic biochemistry based on carbon, oxygen, nitrogen and hydrogen.

Although silicon originally looked attractive as a substitute for carbon, biochemists are unable to offer much hope, although if it exists, it also would put detectable oxygen into their atmosphere.

Most likely, we will look for atmospheric "signatures" of oxygen, ozone, carbon dioxide, methane and water vapor in their atmospheres. Now, since geological processes remove oxygen from the atmosphere, they work against the accumulation of oxygen. Therefore, an atmosphere rich in oxygen has an equilibrium that is out-of-balance chemically and suggests an active life agent, such as green-plant photosynthesis. Or, even earlier in its evolution, an infant planet's microorganisms would feed off the hot gases and rich water vents, and some such life forms will generate methane.

Since it takes sunlight 500 years to break down methane in a planet's atmosphere, the discovery of methane would prove that it is continuously replenished by either many comets striking it, or seeping up from underground reservoirs leftover from that planet's formation, or a byproduct of bacterial life on and under its surface. Carbon-based life emits an elevated ratio of carbon-12 to 13 isotopes—sufficient for future space-based infrared spectrometers to determine—but still not solid evidence.

And if we detect methane levels exceeding 100 parts per million, we could still not categorically exclude a geological phenomena. But detecting excess methane together with oxygen would be strong evidence for life. Of course, machines and intelligent robots do not produce these signatures of biological life, so they could be orbiting stars that lack planets harboring such organic life. If such machine civilizations wish to remain hidden, they will be almost impossible to detect—until we develop spacecraft with revolutionary propulsion.

Enrico Fermi

Enrico Fermi (Fermi Paradox) wondered if there was extraterrestrial life, why are they not rushing to send signals to us? (Would you?) Thus, if they are not willing to talk to us, they do not exist? This is arrogance and conceit—not proof.

SETI fails because aliens quarantine earth. On the other hand, skeptic Michael Shermer claims there are only two other galactic alien civilizations (if that)—or bad odds for SETI. Stung by setbacks, SETI asks, "Where the hell is everybody?" Is Shermer correct when he writes that there are only two other civilizations in our entire galaxy, maybe—and thus SETI should terminate itself?

Or, maybe aliens use directional lasers or other means and refuse to talk to us. Therefore, SETI and Optical SETI (OSETI) may never detect any signals. And SETI may suspect this, but dare not admit it, for fear of losing their jobs. There are other explanations.

But lasers may not be the alien choice. Walter Simmons and Sandip Pakvasa suggest that aliens disguise their locations and hide their messages. Alien signalers could split their message, transmit them in opposite directions to mirrors located far from their planet, and the signals are redirected to the intended receiver, who recombines the photons to reconstruct the sent message. Rather than encode the pattern as a time sequence, they might do so by spatial positions.

If the image is very small, then Heisenberg's uncertainty principle, which limits how much information can be discovered about a particle, makes it impossible for SETI to ever determine the direction in which the alien electromagnetic photons are traveling. Even if SETI discovers how to detect and read the message, their act of reading the message will introduce such uncertainty that they cannot determine its origin.

In fact, it will require extremely sophisticated technology just to detect the message. To recombine the two beams and recreate the transmitted message requires SETI to detect the arrival times extremely accurately. Photons arrive extremely close in time and mimic noise. (Come to think of it, for our military to disguise a transmitter's location would be useful on earth itself!) Aliens are smarter than SETI.

We detect no alien signals. None. It is downright damn creepy. The absence of alien signals is telling SETI that something is awfully wrong—that we are quarantined. There are two good reasons—hiding is beneficial for them and us. Consider this—if we learn that aliens exist and do fly such fantastic saucers, we will be hell-bent-for-leather to develop these amazing craft—which will put us up into face-to-face collision with aliens in our solar system.

And when we learn that aliens—who are intelligent machines—lived for millions of years inside our comet and asteroid belts, and are a super-machine civilization, well, things could mistakenly go from good to ugly—and fast.

Furthermore, aliens may—the operative word is "may"—have a devastating impact upon our religion, which—and make no mistake about this—underpins our civilization, and thus could destroy our world forever. If aliens prove beyond all debate that hell does not exist, then only fools and stoics will obey any laws. Why should they? Make no mistake: our civilization will collapse.

SETI might better look closer for alien intelligent robots hiding amongst our billions of asteroids and comets. But unless aliens want us to detect them (or they get careless), they will be extremely hard to discover.

Comets and Asteroids

It is getting hard to find stars without planets or rocks circling them. But aliens do not need planets because they are machines—intelligent robots. Traveling at slow interstellar velocities, they long ago spread across our galaxy and colonized most stellar orbits in our galaxy. They are optimized for and operate within the resource-rich comet and asteroid belts of stars.

Long ago they bioengineered a few hominids, no doubt from our own DNA, so their biological astronauts or "bionauts" are at home on their occasional visits to earth and thus seem to have something in common with us.

A billion comets circle our sun. Comets are dirty snowballs of rich ices that include carbon dioxide and water ices, ammonia, organics, chemical and sand-like substances, and ions in the gas tail. Comets originate in the outer Ort Cloud and the closer Kuiper Belt. Long-periodic comets possess periods that exceed 200 years—others much longer. The short-period comets usually exhibit narrow cigar-shaped orbits. Long-period comets exhibit elliptical, parabolic or hyperbolic shapes and approach the sun from every direction.

A comet's coma or head is a large cloud of gas and dust, which can be many times the size of the Earth. Gas of the coma is made of hydrogen, carbon, nitrogen and oxygen atoms of both ions and molecules. Within the nucleus, water molecules break down to form a large—larger than the sun—cloud that surrounds the coma or nucleus. The nucleus can be tens of miles across. The nucleus is dark and irregular in shape. The coma shrouds the nucleus, making it almost impossible to see.

When a comet nears the sun, its nucleus heats and outgases, releasing gas and dust, and by the solar wind—fast ions and atoms out-moving from the sun—that sweeps the outgas from the comet away from the sun to form a tail that leaves a trail of dust-like material. This dust-like material, along with meteoric debris, creates annual meteor showers or the random and sporadic meteors.

"Plutinos" or Kuiper Belt Objects (KBOs) are 900 to 1,400 kilometers across, and we discovered over 800 icy bodies, including Quaoar. A member of the inner Oort Cloud, our tenth planet and ¾ the size of Pluto, reddish Sedna, lies 13.5 billion kilometers out, but its elliptical distance varies between 900 and 76 AU from our Sun. Sedna's odd orbit results from perturbations by a massive body, yet undiscovered.

Formed in the Kuiper Belt, gravitational interactions hurled most debris free of our solar system; but the remainder migrated into the Oort Cloud, which is more dense and closer than once thought. The orbital dance of planetary objects is both chaotic and complex, and thus often unpredictable—even with perfect data.

Micron-sized-micrometeoroids are constantly swept out of solar orbit by the solar wind of photon pressure and planets that act like giant vacuum cleaners. In the Mars-Jupiter asteroid belt, millions of rocks' orbits re-cross, and some re-colliding asteroids continually re-fragment finally into micron-sized dust that creates forward scattering and causes the glow of earth's zodiacal light when it is between us and the post-sunset and pre-dawn sun. Fainter than zodiacal light, another ephemeral light called gegenschein in the eastern sky is due to backscattering of sunlight. The best latitudes are between 35 and 25 degrees. Interplanetary particles are not too hazardous to spacecraft in our solar system—moving at relatively slow speeds.

But even then, a microscopic impact from an interplanetary particle can deactivate a slow-moving "snail" (16,000 mph) communications satellite—if it hit in the wrong place. In the early 1960s, satellites with detectors and microphones picked up pings when particles hit them. The Space Shuttle had triple-layer windows and its outer component was often replaced due to pitting. Left in orbit from 1984 through 1990, NASA's Long-Duration Exposure Facility (LDEF) suffered impacts from orbital particles and cosmic rays, and particles punctured its large solar cells with micro-craters.

In reality, these "slow" orbital and interplanetary encounters are just trivial. But for interstellar spacecraft traveling at far greater velocities, perhaps at "only" five percent the speed of light, they will make catastrophic encounters. Interstellar space is full of subatomic particles and even debris—even asteroids, comets and errant wandering planets ejected from star systems. And a billion comets circle our sun. Some comets apogee out two light years from our sun, and probably do so for other stars. Hit any of these and your fast interstellar craft is toast.

Even ignoring comets and their debris, although dangerous to fast interstellar travel, there is a worse danger. Contrary to popular belief, interstellar space is not a perfect vacuum. Although fewer subatomic particles and radiation pervade interstellar space, these particles will be hit at catastrophic velocities. And, subatomic particles and radiation in such encounters with any craft traveling at ten percent light-speed velocities would be catastrophic to semiconductors.

Alien robots that live amongst our asteroids would mine and exploit the rich contents and mine carbonaceous chondrites, which are dark gray because of their high carbon content. Many contain carbon as graphite or amorphous carbon, carbides, and occasionally diamonds. They contain up to five percent carbon, some as carbonates that were once formed with water. Since the rest are organic compounds, this hints at a possible origin that was biological. Many of the organic compounds inside these meteorites are long-chain saturated hydrocarbons, which are similar to paraffins and waxes, while others were fatty acids possessing large molecular weights.

Although these hydrocarbons resembled the end-byproducts of once-living organisms, these compounds formed in the solar nebula during early planet formation at 4.72 billion years ago. In 1969, two fresh carbonaceous chondrites were analyzed and discovered to contain amino acids and long-chain hydrocarbons—non-biological, to be sure, but contained within and essential to Earth's life forms. Although 55 amino acids contained in the two chondrites are not found on Earth, eight were associated with protein synthesis, and eleven less-common amino acids are important to Earth's life forms.

Even though most exobiologists agreed that nonbiological processes formed these amino acids within meteorites, they were amazed that such complex molecules could survive for billions of years in the harshness of planetary orbits. Did these amino-containing meteorites seed the Earth's early oceans with organic molecules that then provided stepping-stones to early life?

And if biological amino acids do form on asteroids in the vacuum of space, as seems to be the case, then it meant one thing—they form and do survive for a long time under harsh conditions. It becomes more likely that life evolved everywhere—with precursor rich molecules carried in by a hail of comets and asteroids into the early pre-Cambrian seas. And thus extraterrestrial alien life may be rather commonplace.

Achondrites

Discovered, in 1984, in Antarctica, a greenish, four-pound meteorite—ALH 84001—an achondrite stony meteorite, was identified ten years later as a rock, similar to a Martian shergottite volcanic rock that was once hurled by an asteroid impact from Mars. It was 4.5 billion years old. Although it seemed to contain fossil evidence of early life on Mars, most scientists remain skeptical. Since then, there are over a dozen Martian meteorites that were identified. ALH 84001 crystallized out of magma four billion years ago on Mars. The meteorite contained oxidized iron, reduced iron, and carbonate—they do not naturally chemically coexist—and only certain anaerobic bacteria can easily manufacture these iron minerals.

The mineral assemblages resembled those that were produced on Earth by our own early bacteria. Under an electron microscope, the scientists saw that these carbonate globules contained rod-shaped fossil bacteria which resemble those that existed on Earth four billion years ago. But there was one problem—they were one-percent of the size—100-nm-long—and possibly too small to contain genetic material. But did they? Unlikely. While the debate continues, from elsewhere come new discoveries—and new life forms.

Inside rocks taken from volcanically hot seabeds north of Iceland in 2002, biologists identified a new tiny group within Archaea, the other recently discovered domain of life on earth, alongside but separate from eukarayotes, which include bacteria, animals and plants. (We exclude hardy unalive prions, infecting the neurological tissue of "Mad Cows." Viruses are also excluded because they may not be life because they do not reproduce outside a host cell.)

Attached to the surface of a larger related Archaea, these tiny spherical microbes measured only 400 nanometers across. Critics earlier correctly claimed that the minimalist fossil genomes of the Martian meteorite ALH 84001 were too small to reproduce. However, these new discoveries may provide a different perspective.

When the biologists tried to identify the microbe with polymerase chain reaction tests to identify its DNA, they could not, and only with difficulty did they sequence a piece of DNA that had a sequence different from other archaea. They concluded that these PCR tests prove inadequate to reveal life's diversity and that life may be far more diverse than we anticipated. According to its discoverers, this symbiotic microbial hitchhiker possesses significantly fewer than the existing 470 genes of the previous record-holder of the previously smallest genome.

But no one understands how a microorganism with so few genes can survive the great pressure and intense heat in the harsh sub-oceanic environment. One thing appears certain—this discovery seems to reduce the theoretical minimum to sustain a cell. But where did it come from? This minimalist genome may either be a highly evolved cell that streamlined its genome to adapt to a harsh environment or is a descendent of a primitive life form on earth that evolved here—or came from a meteorite.

At least 85 percent of earth-striking meteoroids are chondrites. The remainders are irons, stony-irons and achondrites. These were melted and their condritic structures were destroyed early in our solar system. Their parent bodies repeatedly-struck each other and repeatedly broke apart. These three categories do not resemble chondrites.

Achondrites differ from chondrites and are without chondrules. Chondrules were formed by the accretion of chondrules and matrix material into a parent body but without adequate heat to totally destroy the chonrule structure. Melting and recrystallization on a chondrite parent body forms achondrites, which are more coarsely crystalline and more closely resembling our terrestrial igneous rocks. Chondrites are stony meteorites that contain spherical millimeter-sized chondrules formed by re-melting of mineral grains.

Together with meteorites from the cores and crusts of planetary parent bodies, and broken mantle rocks, the stony-iron meteorites consist of silicate-minerals and nickel-iron. They are rare and only 2.8 percent of known meteorites are stony-irons. They melted and recrystalized, as also did the achondrites and irons.

The birth of our solar system was violent. There were endless collisions and re-collisions. Although major irons were never witnessed falls, and most are prehistoric, all of the great meteorites are irons.

Morris and Gould

Stars nightly shining are suns brightly burning, with their soft silver moons gleaming down upon worlds alien. Alien biochemical forms resemble ours—so, chemically and functionally, they will be recognizable to us—and it is no accident. Fed with a rain of complex organic compounds from endless comets and asteroids, Earth's ancient seas formed the early caldrons of life—as they do everywhere else under alien suns. And, because hydrogen is so ubiquitous, extraterrestrial water is plentiful.

Our earth's Cambrian explosion beginning 547 million years ago, saw the sudden and rapid emergence of most existing animals. Was it a bizarre sequence of accidents ("one damn thing after another"), as proposed by the late Stephen J. Gould? Or an inherent and inevitable "purposeful directionality" in evolution—an inherent manifest destiny of ascendancy in evolution that is championed by Cambrian paleontologist, Simon Conway Morris? Most likely, once evolution begins on alien worlds or ours it ascends the ladder of life, and the emergence of aliens resembling man is inevitable. There are few evolutionary accidents.

Even massive vulcanism (P-T boundary) and asteroids (K-T boundary) only reshuffled evolution's deck, and life went on. Evolutionary ascendancy and convergence continued, and wildly different creatures independently arrived endlessly at the same solutions to life's challenges. Biology has a limited number of solutions—to seeing, eating, reproduction; to motion, defense, attack, hiding, excretion, and so on—and so the same solutions reappear endlessly both here and on earths under alien suns. Inevitable adaptations wherever life arises throughout our galaxy and also on earth include bipedal hominid creatures with human intelligence, with color vision (including trichromacy) and advanced social systems that exploit agriculture and technology.

But, once life starts, evolution follows predictable trajectories. Inherent in evolution's biology, intelligent alien hominids are also endlessly inevitable—as were we. In fact, curiously, as is well known, several other erect hominids co-existed simultaneously on earth, independently of us and arriving at similar solutions—but we alone remain.

Is Gould or Morris correct? Is evolution divergent or convergent? It is a meaningless question, because bioorganic aliens replaced themselves with machine intelligence. All it took was just one civilization. These machines proliferated, occupying asteroid and comet belts of billions of stars. Where we now stand, aliens once stood long ago on their journey forward into robot machine intelligence. Our journey has begun.

Paradox of Physics

Whether ball lightning is a step in the path to a final theory, or in fact will lead directly to the final theory, we do not know. But I suspect that it contains a true—and not merely a mathematically synthetic edifice, but a natural—key to unlocking gravity's secrets that could lead to an honest and natural unification of gravity with the other forces.

Truly, I dread that this is not the end, just another platform—like a box enclosing a box that encloses another box—to a deeper theory. Or worse, perhaps there is no final theory, and like forever zooming deeper into colorful fractals that dance across our computer screens, could the search for a final theory go on endlessly? I do not believe this, but it is a disturbing thought.

Regarding quantum mechanics, we begin and end with Richard Feynman, who once remarked, "A paradox is not a conflict within reality. It is a conflict between reality and your feeling of what reality should be like." In a manner, quantum physics' journey of reality to solve this paradox that "defies" gravity has just begun.

As we have seen in this revolutionary book, by their mysterious propulsion, navigation, confinement and quasi (as if) intelligent behavior, both ball lightning and anomalous aircraft have one thing in common—both "violate" physics (as we know it) and "defy" gravity—and thus provide us with clues to the reality of the paradox of inertialess-negative inverse gravity—and to a final Theory of Everything in physics—the final paradox.

Appendices

Because the following important items could not be seamlessly integrated within the preceding chapters, and because they add important insights, they are included as appendices.

ENTROPY AND THERMODYNAMICS

Fireballs should obey the laws of thermodynamics. But nothing generates more heat with physicists than the second law, which is not a law in the true sense. Seemingly simple, the second law is not, and provokes debate among physicists. (Your author partially favors those views of Nobelist Illya Prigogine, and finds those of Stephen Wolfram curious—but others disagree.) There is something awry about the second law of thermodynamics. The first law requires engines to operate with an energy source but says nothing about how. There is no free lunch, no perpetual motion magic machine.

The second law limits our choice of heat sources to those that generate energy. Heat flows from higher temperature heat-reservoirs. No heat engine is perfectly efficient and wasteless. Regardless of heat content, heat flows from a higher temperature reservoir. The second law says we cannot even break even—it is all downhill. It is Murphy's Law of the cosmos.

The universe is a single energy system of subatomic particles powered by energy flowing from more energetic reservoirs (stars and so on) to expanding space-time. It does not apply to lonely particles, but idealized by dynamics, applies to the vast evolution of near-infinite assemblies of particles. It is not a basic principle in the usual sense of a law, but it results from combining mechanics and probability and cannot predict specific things. It does not correlate with the arrow or direction of time.

The history of vast cosmic systems of near-infinite particles is not revisitable. Legions of dead men will never rise to life, and broken eggs cannot reassemble themselves. Pool table billiard balls on a film run backwards are indistinguishable and for them the reversible arrow of time does make sense. Un-crushing eggs do not. Dumped off a wall, Humpty Dumpty's fall and sprawl was time-irreversible. Most events in the life sciences, inorganic and organic chemistry, and physics are also irreversible.

And the recent discoveries of dark energy and mass, and our flat cosmos re-accelerating outward spurns reversibility. Recent progress in Multiverse Theory—particularly Level III—is refreshing and potentially explains away many of these dilemmas and extracts physicists from quantum weirdness. Because of its unexplainable qualities, true ball lightning poses puzzling questions to this heated discussion.

MULTIVERSE FIREBALLS

As vast as we think our own universe is—some physicists now believe that it is but an infinitely tiny point in an infinitely larger multiplicity or multiple of universes—the Multiverse—or universes without end. In the infinite universes of the Multiverse model, everything is possible—somewhere, in the infinite universes are infinite worlds that are like ours—perhaps identical twin clones—endlessly repeating themselves, averaged every $10^{\wedge}10^{118}$ meters away from each other.

Your author is unhappy with anthropic and Multiverse-type ideas. Lately the theory has become better formulated. Some physicists and cosmologists now say there are an infinite number of parallel and multiple universes, or Multiverses, and that there are four possible candidates—Levels I through IV. If true, the key question is not whether there is a Multiverse, or if it can be tested (maybe), but which level is correct?

Multiverse Theory has a scientifically unsatisfying feeling, and even an unsavory odor. But it also has a legal weirdness. But, if true, Multiverse Theory will ask lawyers and moralists many questions. Is there free will or predestination? Should criminals be punished? Which of your clones will go to hell? In other Multiverses, did Jesus and Mohammed commit sins and not rise? In the infinite Multiverses, everything not impossible occurs. Could Multiverse Theory destroy human ethics? If so, will Multiverse Theory become legally dangerous?

But to ball lightning researchers, Multiverse Theory asks—Is there a connection between Multiverse Theory and fireball originations? First, in the infinite number of parallel universes, the identical laws of physics also work, so the answer is obviously—absolutely yes—a no-brainer answer. Infinite fireballs occur in infinite Multiverses—same as here.

Second, but more important and enigmatic—Are there interconnections between some Multiverses via fireball origination, behavior and termination? Perhaps through hyperspace wormholes, as envisioned by Kip Thorne and popularized by Carl Sagan? Or used to keep open a wormhole? There was one requirement—anti- or negative gravity, as proposed by Thorne—a quality exhibited by fireballs. By this, Thorne respectably opened the door to sidestepping the speed of light as a limit.

However, energy and resources required are massive. Physicists like Visser, Kar and Dadhich examine spacetime geometries that contain transferable wormholes supported by arbitrarily small quantities of "exotic matter." To travel through time requires opening wormholes with "exotic matter" which is repelled by gravity. It has negative energy—that is, less energy than empty space—which is the same as saying that it experiences gravity as a repulsive force. Exotic matter might arise from "averaging null energy condition" (ANEC).

Exotic matter lies in quantum fluctuations, which gives empty space a "fuzziness." Subatomic particles and corresponding anti-particles continually pop in and out of existence in the vacuum of empty space.

To avoid "fireball causality loop" problems, "time loops" threading though a wormhole along which backwards time travel occurs, but unable to alter the future. The problems are—how to open a wormhole and how to create a small amount of exotic matter—for a small amount is all that is needed. This hypothesized "exotic matter" that is repelled by gravity may indeed exist—perhaps as a quality exhibited by fireballs and extraterrestrial craft in their gravity defying behaviors.

For us skeptics, these are dangerous visions.

When the cosmic background radiation left over from the Big Bang was recently analyzed, from the COBE and WMAP satellite measurements, the conclusion was startling—our own universe's topology was neither positive nor negative, neither a sphere nor saddleback, or hyperbola, nor some convex curvature with unusual topology and interconnectedness, shaped like a pretzel or doughnut without edges and finite volume. Our universe is perfectly flat.

Level I Multiverses or parallel universes have the same laws but with different initial conditions. [See Wolfram on initial conditions.] Cosmologists calculate that our own Hubble volume has over a hundred trillion trillion inhabited planets, some resembling our Earth. But there are an infinite number of parallel universes forever beyond our sight. Since our cosmic expansion is accelerating, our distant descendants may never see these other Multivereses.

Pioneer Andre Linde proposed that each brief burst of inflationary expansion is infinitely being repeated, sprouting new universes that bubble out from black holes, which we cannot see through their event horizons. Lee Smolin avoided the limitations of the anthropic principle and suggested that in the new universes cosmic evolution, sort of a genetic mutation, occurred where the new universes will closely resemble, but not be identical to their parents. However, certain offspring might be more prolific in creating black holes and new universes even more fecund than their parents—until we have a runaway condition?

Predicted by the theory of chaotic eternal inflation, Level II Multiverses are also infinite, but some possess both different physical constants and spacetime dimensionality. First explained by Alan Guth, inflation caused rapid stretching in the early expansion of our universe. Although space continues to stretch forever, some regions stop and create bubbles—infinite embryonic Level I Multiverses, each infinite in size.

Symmetry breaking, in the creation of our own cosmos, created spacetime dimensionality and physical constants, and these are not sacrosanct or inherent in our universe's laws. Three of the nine initial dimensions drove expansion and so the other six remained in torus microtopologies or matter on the three-dimensional surface membrane (brane) in nine-dimensional space, sprouting new universes through brane physics. In other bubbles, different quantum fluctuations drove chaotic inflation to break symmetry differently.

Some suggest that the birth and death of universes generate a Level II Multiverse, with a second 3-D brane parallel to ours but offset in a higher dimension that interacts with our universe—sometimes via ball lightning? Rather than sprout new universes through brane physics, some physicists suggest that they mutate and sprout through black holes. [There are too many versions to examine, which will lead us astray.] Other Level II parallel universes can be inferred by miraculous coincidences in fortuitous fine-tuning of fundamental constants that make our universe hospitable to intelligent life.

In the many-worlds quantum interpretation of the Level III Multiverse, driven by random quantum processes, the universe massively branches into endless duplicates for each outcome—into infinite permutations for all the tiniest macroscopic or quantum variations.

Explaining quantum mechanic's strangeness, with wave functions that collapsed into classical outcomes only upon observation, did explain quantum theory observations, but with a nonunitary and inelegant interpretation that upset physicists like Einstein, Bohm and Bell.

In the many-worlds interpretations, which is the same as Level II, the other infinite universes are located in the abstract realm of all possible states elsewhere. By analogy, imagine a baccarat wheel at Las Vegas, or a "quantum lottery," or a rotating cage that spins and spits out numbered balls, with a googilian permutations each time. In our Newtonian world, this cage ejects only one random number. But quantum mechanics uses an inelegant paradox of Schrodinger's cat and Copenhagen Interpretation. The quantum cat is in neither state—dead nor alive—until the observer peeks inside that sealed box. But Multiverse Theory states that you are locked into one universe out of googilians of newly-minted *Cloneverses* and see only a tiny fraction of total quantum reality—and another "you" elsewhere wins the lottery cash prize.

Actually, the billions of chance outcomes all occur in googilians of different bifurcating pathlines, which clone new universes that in turn keep mutating and dividing into endless new parallel universes. But because you are locked into one universe out of billions of new bifurcating universes, branching into existence at each nano-fempotosecond, you see only a tiny fraction of outcomes. One of your clones won a fortune—you did not.

These parallel universes sensibly explain the infinite paths of wave interference and quantum computation. Although the Level III Multiverse adds nothing much beyond the two lower levels, it is the most controversial. Unlike Levels I, II and III, which posses different physical constants and initial conditions, in Level IV all the laws vary. [For further discussion, read the excellent articles by Max Tegmark.]

Physicists marvel how mathematics so closely mimics physical systems. Aristotelians (like Hawking) see mathematics as descriptions of reality. Platonists (like Penrose) see them as true reality that exists outside of spacetime. There may be a third view—the failure of equations to accurately model reality. Could Wolframian Science, as described in *A New Kind of Science,* explain issues that may modify Multiverse Theory—irreversibility, unduplicatablility, initial conditions, and time?

Multiverse theories explain time. Since all possible states exist at each instant, the passage of time is an illusion. [See Julian Barbour.] If there is no change there is no time, for time is change of position. Motion defines change defines time defines motion defines…This unsatisfying circular search goes forever around and around. Facing time, in their souls, relativity, classical and quantum mechanics all fail, and all explanations are unsatisfying. At between 13.5 to 13.9 billion years ago (according to WMAP), at the instant of our own Big Bang, time did not exist for us. Eternity is a timeless state without change, which neither computers nor man can truly comprehend, for without time (change) it is impossible to think.

Do wormholes naturally occur in our universe and other infinite parallel Multiverses; and between them? Does quantum entanglement ever apply to interconnections between Multiverses? As described elsewhere, some evidence suggests that certain fireballs exhibit evidence of possible quantum entanglement. Kip Thorne speculated that sporadic, transient and microscopic wormholes could briefly be enlarged and may go back in time, and receive and transmit information. Interconnecting singularities annihilate each other and create a wormhole that grows and then contracts and disconnects, destroying what is in it.

To hold wormholes open, Kip Thorne wrote that we need an exotic (yet unknown) means or material that anti-gravitationally repels a beam's light rays, so it pushes them away from itself and from each other, thus defocusing them. (This anti-gravitational and defocusing is the inverse of the gravitational focusing in astronomy that magnifies distant stars.) Does Thorne imply that such stealthy aircraft could "disappear" from view?

Because fireballs "defy" gravity—that is, manifest inertialess-negative-gravitational behavior—do they possess clues for holding open wormholes, as first described by Kip Thorne and Carl Sagan? Doubtful, but possible.

Anthropic ideas say dead-end things like there is no "me" over there because I am "me" here, where I can survive. Saying that things are this way because they could not have been otherwise is no fundamental explanation—and feels intuitively unsatisfying, as does the Multiverse. Time will tell.

What happens to causality when Kip Thorne's wormholes meet Multiverse Theory? If Kip Thorne is correct, and wormholes do exist, well, this creates temporal abnormalities and logical paradoxes. Then even sending back in time even one bit (one or zero) of information could conceivably, through the laws of chaos and complexity, irreversibly alter the nonlinear dynamics of local subsystems—or in the extreme, alter the entire system (our Multiverse)—via the Butterfly (Seagull) Effect.

This could introduce temporal feedback loops that are nonlinear. So, if Kip Thorne is correct, then could future physicists go back in the infinite cosmoses to alter the infinite different Big Bangs so that spacetime unfolded differently? And keep re-looping with infinite iterations of feedback loops? Like endlessly looping computer subroutines?

The Multiverse advocates will argue that this is fine, because there are infinite parallel branching copies of our cosmos. This may be beyond our abilities to grasp the concept of infinite parallel permutations and combinations, that is for sure, but that is a human shortcoming, and so that is the nature of infinity.

Mathematicians tell the tale of the six typing chimpanzees. Well, somewhere out there are infinitely branching Multiverses, where six chimpanzees each are keyboarding at word processors and some are typing error-free all the works of Shakespeare and the entire *Encyclopedia Britannica*—perfect, without error, all by random chance. Come to realize it—Is Multiverse Theory even "crazier" than quantum mechanics? It must be correct—it makes no sense.

MONDs Versus WIMPs

Perhaps the cosmos did not begin 13.7 billion years ago. Maybe it began farther back, or had no beginning—and our measurements are all in error because we assume Newton's Second Law is perfect. A slight nonlinearity in Newton's linear equation alters everything and the cosmos may not have had a Big Bang. Could this also alter the argument for inertialess-negative-gravity, and therefore that of the Sagan-Hill Hypothesis?

In the Modified Newtonian Dynamics (MOND) model, Moti Milgrom argues that a fudge factor like dark matter is unnecessary, since a slight change in Newton's law of force will suffice. Below an acceleration of one ten billionth of a meter per second each second, gravity is no longer directly proportional to its acceleration, but is slightly stronger, proportional to the square of acceleration.

With foresight that extends into modern physics, Newton defined force as $F=dp/dt=d(mv)/dt=mdv/dt+vdm/dt$, where p is momentum. Excluding rockets losing and relativistic particles gaining mass, in non-relativistic physics, the mass remains invariant (constant), so that its derivative is zero, and it reduces to the familiar $F=ma$—a linear relationship of slope m. Or is it really linear?

Although many (including your author) remain skeptical, some physicists modify Newtonian dynamics and claim that cold dark matter, allegedly 90 percent of the cosmos, does not exist and that Newton was wrong—that is, $F<ma$, or that $F<G(m1m2)/r^2$. It is slightly in error, that is, nonlinear. Or do vast amounts of a mysterious substance permeate space and hold galaxies together? Like Dracula, is a modern phlogiston arising to haunt cosmologists?

What is the universe made of? Our galaxy rotates at one revolution each 200 million years, so quickly that the centrifugal force of galactic gravity is too weak to hold the outer stars from flying away. This is true for other galaxies. For all galactic clusters with rapid galactic motions, something unknown holds them together.

Dark, burned-out stars cannot be dark matter because we already know how much total real matter exists. We know, because if there were more matter, after the Big Bang, matter would have fused into heavier elements and we would detect much more helium, boron and neon.

Whatever this dark matter is, it does not overcome fusion. A likely candidate is "cold dark matter" (CDM), a gas of heavy subatomic particles called "weakly interacting massive particles" (WIMPs) cannot feel electromagnetic forces or the strong nuclear forces that bind protons and neutrons together in nuclei, and thus are collisionless and pass through each other and are immune to fusion.

Computer simulations demonstrated realistic stellar clumping and galactic congregation into webs of walls, knots and voids. Unfortunately, the simulations plotted cold dark matter density across a galaxy as a pointed cusp (inverted "V"). But measurements show an inverted "U." Other astronomers question telescopes' spatial resolution.

The heated controversy spawns hypotheses that all generate correct large-scale structures, but fail. Each hypothesis conjectures an imaginary property of dark matter that is tweaked to fit. As you may guess, theoreticians look askance at these conjectures.

Could the long-discarded hypothesis that large black holes, each over millions of times the mass of our Sun, were born 13.7 billion years ago in the Big Bang and after the inflation, migrated into the center of galaxies. Involved in the violent and complex stellar dance, some are flung out of galaxies, limiting swarm density and solving the cusp dilemma. Recently, black holes are being discovered in our galaxy's nucleus, so this discarded model is being re-examined. Although black holes are not totally dark, as Hawking proved, they might be detected by lensing (bending) of starlight rays.

Theories proliferate. If these dark matter particles strongly interact, they may resist compression and explain the inverted "U" density in galactic centers. But their interactions and shock waves create spherical not ellipsoidal galaxies and their observed ellipsoidal dark matter haloes. Other cosmologists hypothesize that complex interactions that depend on particle velocities or that dark mater only interacts strongly with itself between weakly with ordinary matter. At this time, if future observations prove galactic dark matter cores are ellipsoidal, then strong interactions will be false.

A problem with slow-moving dark matter is that it should cluster densely at galactic centers, but does not. If the particles are hotter, that is, move faster—and assuming they obey the laws of energy as we know them—then they will escape the galactic cores, but make them undetectably weak. Worse for warm dark matter "gas," it will fail to help galaxies grow after the Big Bang by not dragging in enough gas clumps and will fail to grow, or do so too slowly. Warm dark matter is unlikely.

Spontaneous decaying dark matter would slowly turn into neutrinos, photons and other particles that shoot out of galaxies and dissipate central density spikes. In the early universe, the stronger, dark matter gravity would create greater structure and smaller galaxies. If true, after another several billion years, stars will be alone and the night sky will begin going dark.

If dark matter particles—if they are indeed particles—are ultra light (ten billion trillionths of an electron volt), they are positionally ill defined because of Heisenberg's uncertainty principle, and would be "smeared out" over a great volume several thousand light years across. Because there is no dense concentration of dark matter at galaxy centers, there can be no such dense concentrations.

ANCIENT ACCOUNTS WEAK

As for historical accounts of ancient fireballs—these old accounts offer no useful information. Etruscan art shows lighting, including fireballs. Ancients who described it include Aristotle, Seneca, Ovid, Posidonius, Lucretius, Pliny the Younger and Elder. [Singer, pp. 5-7.] In the sixth century A.D., Gregory of Tours described a blinding fireball that materialized above a procession of civil and religious dignitaries from Tours, attending the dedication of a chapel. The Anglo-Saxon Chronicle, in 793 A.D. described an electrical storm in which floated glowing "fire dragons." On the evening of March 3, 1557, a fireball came in through a window and into the bedroom of Diane of France and flew about the room, and then burned her hair and clothes. William Shakespeare described lights floating in a storm preceding Caesar's assassination. Ancient accounts are unreliable.

COLLOIDALS

Some witnesses report that after the departure of a fireball, that a smoke, dust or colloidal suspension remains that scatters light, and that may be the Tyndall Effect. When light shines through a fine suspension of particles in a solid, liquid or gas, light rays are scattered or deflected sideways from the surfaces of dust particles. If these particles are very small, the light scattered can appear blue. Light diffuses off these colloidal particles.

Colloidal suspensions are of colloids that range in size between ordinary molecules and microscopic particles. They range in diameter from 10 to 1,000 Angstroms. (One Angstrom is ten nano-centimeters.) Simple molecules are only several Angstroms in diameter, although protein macromolecules do reach 100 Angstroms. Extra-large protein molecules are in the colloidal range. On the small end, colloidals resemble solutions, but on the upper end, they resemble suspensions. Examples of colloidal suspensions are liquids in gas such as clouds, solids in gas such as smoke, solids in liquids such as India ink, and even a solid in solid such as Ruby glass.

The scattering power of dust is inversely proportional to the wavelength of the incident light, whereas the scattering power of gas varies inversely as the fourth power of the wavelength.

TESLA MEETS REALITY

After inventing AC motors and polyphase systems, Nikola Tesla generated Artificial Electric Balls in his Colorado Springs laboratory. Tesla was of the classical Newtonian school of physics. Fashionable, neat and fastidious in both behavior and attire, Tesla was the last Victorian scientist. Like Howard Hughes, Tesla also suffered from obsessive-compulsive disorder, albeit, not that extreme. Tesla liked fine foods and expensive restaurants. He was a perfectionist. He demanded a lot. He paid well. He dressed well. Tesla had style.

Tesla was flamboyant, egotistical and wrote poetically in a fluid literary style. He read prodigiously, including philosophy and all of Voltaire in the French. He spoke eight languages. He slept three hours every night. Tesla apparently understood James Clerk Maxwell's two-volume classic, *A Treatise on Electricity and Magnetism*. But unlike his contemporary, the electrical genius Charles Proteus Steinmetz at General Electric, Tesla rarely wrote down differential equations. He had an innate ability to transcend pages of advanced calculus. His ability to conceptualize and visualize astounded all. However, contrary to O'Neill's biography and Tesla mythology, Tesla was not a "lightning mental calculator" and exhibited only moderate synesthesia.

Before the Twentieth Century, Tesla unwittingly patented AND and OR logic gates—the building blocks (when combined with inverters and clocks) of combinational and sequential logic, and thus of all modern computers. Unknown to Tesla, his invention was secretly employed in the world's first vacuum tube computers, both named *Goliath* at Britain's Lesley Park in England during World war II and used by Alan Turing and Flowers to break the Nazi's Ultra Secret code—and save England. Wary of Nazi and Soviet spies, the British waited until the mid-1990s to tell the Americans the total story.

Tesla was gracious and charitable. But with age, he grew eccentric. A tortured Tesla in anguish languished for three decades, dying early in 1943. Several months later, the U.S. Supreme Court declared the Marconi patents invalid and handed down a landmark decision in favor of Tesla (also John Stone and Oliver Lodge) over Marconni as the primary inventor of the radio. It was too late. No one cared. Marconi and the robber barons had already massively robbed and gang-raped Tesla.

After a triumphant return from Colorado in 1900, Tesla had the successful seminal research upon which he intended to construct a large wooden transmitting tower to transmit information and wireless power with small attenuation to any location on earth by Zenneck "surface" waves. In Colorado, he created Artificial Electric.

Tesla adequately budgeted for the Shoreham project. But a nine-month delay in receiving the promised financial backing from magnate J.P. Morgan, rocketing labor fees, and ruinous prices—these and other unanticipated obstacles conspired to hamper his project. Finally, killing his project, further disaster struck—economic stagnation and inflation. Then, blamed by outraged investors, the fearful J. P. Morgan fled to his yacht, the *Corsair*, in New York harbor. With money worth less, Tesla sought more, which was not forthcoming. Angry, Tesla wrote to Morgan, comparing him to a venomous spider. Prior good relations went south fast.

Undercapitalized, Shoreham failed. Standing alone in mute testimony to what might have been, a forlorn relic, the abandoned ghost-like wooden skeleton tower on Long Island was finally dynamited. What was to become Tesla's crowning achievement became his demise. On the wreckage of Tesla's Wardenclyffe in New York's Long Island in Shoreham, ended everything. What Tesla accomplished there was "blowing in the wind"—many of his notes tossed into the wind by vandals. What could have been is speculative.

Screwballs love Tesla. Since 1965, eccentric zealots poisoned Tesla's reputation with their speculative myths so badly that it frightens off legitimate scientists. For example, back in the mid-1980s, a brilliant scientist (who will remain anonymous) heading a particle beam project for SDI secretly explored Tesla technology—until his project was suddenly defunded by a nosey Congressman who accidentally discovered that he was investigating Tesla's unpublished technology. Many weird myths resonate with zealots.

These following books dispel mystical Teslian "calculating myths," repeated by cultists—speculative myths that grow more Olympian with each retelling. Some flying saucer aficionados even allege Tesla was really a secret alien superman who came from another star and had supercomputer calculating skills. Not likely. Tesla was not even an amateur mental calculator, as the following books reveal.

Cutler, Ann. *The Trachtenberg Speed System of Basic Mathematics*. NY: Doubleday. 1960.

> Created by mental calculator engineer Jakow Trachtenberg while in Hitler's concentration camps, this Swiss system is taught to architects and students otherwise slow in arithmetic, and is superior to other lightning mental arithmetic in division. From examining Tesla's notes, your author concludes that Tesla did calculations the old fashioned way—by longhand, using the European version. He made occasional mistakes.

Cutler, Ann. "Juggle Your Numbers the Shorthand [Trachtenberg] Way." *Science Digest*. November 1960. pp. 36-43.

Lorayne, Harry. Most books by this legendary mnemonist are excellent. Although not possessing synesthesia, Lorayne excelled at mnemonic arithmetic and was better than Tesla.

Luria, Alexander. *The Mind of a Mnemonist*. NY: Avon Books (Hearst). 1969.

> An amazing story by a leading Russian psychologist who studied Alexsandr Venomenovich Shereshevenskii (Mr. S) who severely suffered from synesthesia and possessed an indelible, endless, unlimited memory for perfect recall of hours of conversation, and even nonsensical vast assemblages of equations,

pictures and information—a perfect memory which did not erode after several decades! This godlike gift the gods grant to rare humans, they curse. This "gift" gave Mr. S great trouble, as it did for Tesla in his youth. Tesla worked hard to overcome this gift-handicap.

Ramachandrian, Vilayanur and Hubbard, Edward. "Hearing Colors, Tasting Shapes." *Scientific American.* May 2003. pp. 53-59.

Gifted people such as Tesla who suffered from moderate synesthesia—their mingled senses blend together—they see sounds and hear colors, and can use this neurological ailment to enhance their creativity, metaphorical skills, visualization and memory.

Smith, Steven B. *The Great Mental Calculators: the Psychology, Methods, and Lives of Calculating Prodigies—Past and Present.* NY: Columbia University Press. 1983.

This is the best and most comprehensive, and most reliable book ever written on this amazing subject. Lists great human calculators, like Wim Klein of CERN, Hans Eberstark, Shakuntala Devi, John Wallis, Leonhard Euler, Andre Ampere, Karl Gauss, John von Neumann. As expected, Tesla is not listed.

By examining some of Tesla's previously unpublished notes, where he worked out calculations, your author discovered—and it is painfully obvious that contrary to biographer John J. O'Neil—that Tesla was not even a mediocre mental calculator, nor did he memorize a table of logarithms for use in mental multiplication and division, as O'Neil wrote. In fact, the late Isaac Asimov was better at mental arithmetic and wrote a book on it. And in his biography of famous scientists, Asimov criticized Tesla as so temperamental that others had difficulty getting along with him.

Wenger, Win. *How to Increase Your Intelligence.* NY: Laurel (Dell). 1975.

Describes innovative methods and touches upon Tesla and his synesthesia.

NEW VOCABULARY

Never underestimate how terms influence success. Every field creates its own terms. Ball lightning lacks a vocabulary. For it to grow, it needs them. For example, consider that in 1939, Oppenheimer and Snyder adapted Schwarzschild's and Birkhoff's geometry and pioneered black hole math, but that subject languished until later—and not simply because of Einstein's excoriations and reticence—until the term "black hole" was first adopted and popularized in 1967 by John A. Wheeler. (Read chapter six of Kip Thorne' *Black Holes And Time Warps*.) So this phrase, "black hole," suddenly captured the world's imagination and became the force pushing interest and research.

For the sake of aesthetics and appearances, I avoided padding the following list with commas, and instead indicated words that go together in phrases (that is, not standalone single words) with inter-word dashes. With marketing in mind, I offer the following terms that could jump-start research into ball lightning. Most are self-explanatory, but others list brief definitions in parentheses, as follows.

Floater Bouncer Slider Floor-roller Gutter-roller Line-roller Dazzler Railroad-balls Phone-line-roller (outdoors) Line-hugger Pole-hugger Mother-ball (one that breaks into smaller balls) Wire-slider Power-line-roller (outdoors) Clothesline-roller Surface-roller Pole (electric utility) Plunging-ball (drops out of sky) Descender Skyball Seaball Ballettes (tiny, sub-inch-sized) Shrinker Expander Twin-Balls (with inter-ball vertical striae or strand) Sizzler Hisser Exploder Monster (over 15 feet across) Big (five to ten feet across) "Merge-and-surge" (fireball enters wire and disappears, replaced by a huge current or transient) Chimney-seeker Digger (rare, digs hole or surface tunnel) Spook-lights Mystery-lights Cemetery-lights Ghost-lights Swamp-gas Drifter Riser Bolide Meteoride/ite Sizzler Hisser Tongues Horns Floater Exploder.

FIELD RESEARCHERS

Look for evidence of chemical changes. A fireball that hit or melted or merged and surged with a metal wire, such as copper, may alter the metal. For example, heated wire becomes copper-colored. On heating, copper and oxygen from the air combine chemically to make the wire black. Laboratory analysis shows copper (II) or cupric oxide will consist by weight of 20.1 percent oxygen.

A scanning tunneling microscope will show arrays of copper atoms and changes in the crystalline lattice structure. Physical changes often accompany chemical changes. Measure and weigh wood that burned. Take specimens. Place them inside sterile plastic bags or containers. Label everything in detail. It can provide data for calculations and energy, and times of reaction.

Hypnosis can improve recall. Your author studied hypnosis for years and took extensive seminars. As described in *Scientific American,* susceptibility to hypnosis varies widely and is innate within each person, and does not change throughout life. As an investigative tool, even when performed by reputable and skilled (rare) hypnotist-examiners, hypnosis still comes with dangers—as explained by skeptics at CSICOP.

Used to examine fireball memories, hypnosis is one thing; but to retrieve suppressed memories of extraterrestrial abductions is quite another. Extraterrestrial abductions, investigated with hypnosis by Harvard's prestigious John Mack, MD, probably have more to do with yet-unexplored internal psychological phenomena that have little to do with external reality.

Beware of systematic errors that skew data. Just because many witnesses report a certain quality, and your tests give contradictory results of a quality, this may not be evidence of those qualities. In fact, it may be only a systematic error that most witnesses make.

Field researchers and theorists resemble blind men examining an elephant, horse, hippopotamus, and rhinoceros—each blind man examines only one feature—leg, side, tusk, horn, hoofs, trunk or tail—and each offers different, but bizarre theories and descriptions, of the (they falsely think) single beast. Likewise, could witnesses be seeing several different phenomena, not just one? The isolated reports are anecdotal and differ in their characteristics. So, not surprisingly, many researchers wonder if all things reported as ball lightning may not really be several phenomena.

ACKNOWLEDGEMENTS

First, although my ancestors came from the same geographical region, except for being of kindred spirit, I am unrelated to Carl Sagan—whom I greatly admire—and am not associated with Akamai Technologies.

Second, I have not met the respected scientists from whom I have drawn my inspiration over the years—many of them referenced in the following bibliography—but have studied your writings and through them feel that I have in some deep way met you.

Third, I thank Bob Golka for his friendship over many years, for providing me with the original 630 ball lightning cases from the McNally study conducted at Oak Ridge National Laboratories, and for permission to publish them. Almost lost to history, 230 of these important cases—many from leading scientists—are revealed here for the first time.

Finally, I took great care and borrowed much from accepted physics. Over several years, I assembled this book—alone—and my theories are mine alone, and no one listed necessarily agrees or disagrees.

Bibliography 1—Fireballs

Ball lightning researchers must be familiar with many subjects, such as high voltage phenomena, General Relativity, quantum mechanics, Multiverse and string theory. If you want a refresher, or would like to examine related topics, then read the following books—most are in my personal library. These are not the only books, and there are certainly other quite good ones out there for you. My comments follow some listings.

Acheson, Dean. *Present At The Creation: My Years in the State Department.* New York: W.W. Norton & Co. 1969.

> The late James Tuck of Los Alamos in a paper on ball lightning delivered at MIT stated that in his voluminous autobiography, that while aboard Air Force One, Secretary of State Acheson saw ball lightning cross President Truman's table.

Aczel, Amir. *Entanglement: The Greatest Mystery in Physics.* NY: Four Walls Eight Windows. 2002.

> From Einstein's "spooky action at a distance" futile attacks on quantum mechanics, to Alain Aspect's entangling photons and to Nicolas Gisin's electrons communicating instantaneously over seven miles apart, examine this mystery that maybe only Multiverse theories can explain.

Adair, Robert. *The Physics of Baseball—3rd ed.*, revised and updated. New York: Harper Collins Publishers. 1994.

> Fireballs violate the aerodynamics of flying spheres, as described in this book.

Ames, Adelbert. *The Ames Demonstrations.* New York: Hafner Publishing Co. 1968.

> Rather than rehash traditional statistics, this landmark study considers individual visual perceptions, which are more important to fireball field investigators.

Anderson, Leland. *Ball Lightning and Tesla's Electric Fireballs.* Breckenridge, CO: Twenty-First Century Books. 1997.

This brief 32-page monograph provides a capsule summary, covers Golka, and includes John O'Neill's unpublished chapter from his 1943 book, *Prodigal Genius.*

Anderson, Leland. *Nikola Tesla: Lectures Before the New York Academy of Sciences—April 6, 1897.* Breckenridge, Co: Twenty-First Century Books.

From unpublished notes, Tesla's complete lecture of 1897 describes Tesla's independent discovery of x-rays with his vacuum high-field emissions by bremsstrahlung. Unlike Roentgen's use of gaseous discharge tubes utilizing electron avalanching, Tesla's cold single-electrode, cathode tubes worked best with high vacuum, and led the way into high-energy particle accelerators, with megavolt potentials, atmospheric accelerators, and intense beam techniques. Because Tesla did not interpret his discovery, Roentgen received the credit. After taking exposures at great distance and experiencing burns, Tesla warns about the dangers. At Colorado Springs and Shoreham, Tesla used single-electrode devices to charge the hemispherical dome.

"Ball Lightning Made In The Laboratory." *The Experimenter.* February 1925. pp. 237 and 255.

Barbour, Julian. *The End of Time: The Next Revolution in Physics.* New York: Oxford University Press. 1999.

Barbour hypothesizes that unifying gravity and quantum mechanics will explain motion as illusion and remove time from the final Theory of Everything, and alter our view of the spacetime continuum. Time has no role in the many worlds interpretation.

Barry, James D. *Ball Lightning and Bead Lightning.* New York: Plenum Publishing. 1980.

This excellent classic describes difficulties with existing theoretical models and provides a good bibliography.

Bass, Robert; Golka, Robert; and Bryner, John. *Investigation of the Laboratory Production of Ball Lightning: Proposal to the National Science Foundation.* Provo, Utah: Pyrosphere, Inc.

"Batteries and Battery Management." Diesel-Electric Submarines And Their Equipment. *International Defense Review Journal.* 1986. pp. 50-51.

Provides insight into the massive submarine batteries used by Golka.

Bergmann, Peter. *The Riddle of Gravitation—Revised and Updated.* New York: Dover Publications. 1992. First edition by Charles Scribner's Sons. 1968.

Bohm, David. *Quantum Theory.* Dover Publications. 1989 reprint.

Replete with advanced calculus, it is for readers with a physics background. A still-valid 1951 affordable classic praised by Einstein. Better explanation of the Copenhagen interpretation than by Bohr. Introduced a new version of the EPR experiment, now renamed EPRB. Is the EPRB effect involved in twin ball lightning?

Bohm, David. *Wholeness and the Implicate Order.* London, UK: TJ. Press (Padstow)/Routledge. 1980/1981.

Popular explanation of Bohmian Mechanics. (pp.65-213) Bohmian Mechanics, as later refined by Bohm's associate Basil Hiley, argues that particles are real—precise and with unfuzzy properties of position and momentum, and precise paths through spacetime, and are real waves—never probability waves.

Briggs, John and Peat, F. David. *Seven Life Lessons of Chaos.* New York: HarperCollins Publishers.1999.

One caveat—outside of mathematics and the physical sciences, chaos and complexity remain only analogies and metaphors—just the latest, imprecise management buzzwords.

Briggs, John. *Fractals: The Patterns of Chaos.* New York: Simon & Schuster. 1992.

This inspiring large, colorful, art-rich picture book blends beautiful fractal geometry, chaos and art to focus on turbulence—a quixotic quality challenging physicists investigating fireballs.

Bychkov, A.V.; Bychkov, V.L.; Abrahamson, John. "On the energy characteristics of ball lighting." *The Royal Society.* pp. 97-106.

Cade, C. Maxwell and Davis, Delphine. *The Taming of the Thunderbolts: The Science and Superstition of Ball Lightning.* London: Abelard-Schuman. 1969.

> Easily readable treatment that sometimes borders on the mystical.

Campbell, Stephen K. *Flaws and Fallacies in Statistical Thinking.* Englewood Cliffs, NJ: Prentice-Hall, Inc. 1974.

> This classic book illuminates the dangers of inherently faulty data, meaningless and "junk" statistics, improper comparisons, and faulty induction—errors inherent in all ball lightning statistical studies.

Cantor, Issie. Chief of Explosions Safety Function at Ogden AFLC Safety Office. AFLC document dated January 28, 1975.

> One page memo documents the dangerous, repeated formation of ball lightning inside the Hill AFB Liniac linear accelerator, used to x-ray solid-fuel Minuteman rocket motors.

Cawood, W. and Patterson, H.S. "A Curious Phenomenon Shown by Highly Charged Aerosols." *Nature.* No. 3221, Vol. 128. July 25, 1931. Pg. 150.

Cawood, W. Thesis for PhD Degree. *Leeds University.* 1932.

> This forgotten, pioneering 98-page thesis covers coagulation, experimentally counting and producing smoke particles, the structure and densities of particles, scattering light, electrification of smokes, and how materials affect rate of coagulation of smokes. A charged aerosol of p-xylene-azo-beta-naphthol in an electrified glass room formed a red ball that relates to Sim's fractal fireball model. Golka visited Leeds U. in England and textile chemists at U. of Massachusetts in Dartmouth, Massachusetts, and constructed a large room of thick glass walls to duplicate and exceed Cawood's experiments.

Chalmers, J. Alan. *Atmospheric Electricity. 3rd ed.* New York: Pergamon Press. 1967.

> This still-valid classic summarizes research in this arcane field and should be familiar to everyone studying ball lightning.

Chaston, Peter R. *Terror From The Skies.* Kearney, MO: Chaston Scientific Publishing. 1995.

> Unusual meteorological events, including ball lightning, explained in a popular style by a renowned meteorologist.

Cheney, Margaret. *Tesla: Man Out of Time.* Englewood Cliffs, NJ: Prentice-Hall. 1981.

> This balanced biography describes Golka's high-voltage coils at Wendover Air Force Base.

Cheng, D.K. *Analysis of Linear Systems.* New York: Addison-Wesley. 1959.

> Covers feedback systems and signal flow-graph transformations of systems with distributed parameters, useful to coilers.

Cobine, James Dillon. *Gaseous Conductors: Theory and Engineering Applications.* New York: Dover Publications. 1958, reprint of 1941 book.

> Remains essential technical reading and reference, even if somewhat dated, for fireball researchers and coilers.

Cole, K.C. *The Hole In The Universe: How Scientists Peered over the Edge of Nothing and Found Everything.* A Harvest Book, Harcourt Co. 2001.

> Examines recent theories on gravity, a topic important to any accurate theory of ball lightning.

Committee for the Scientific Investigation of Claims of the Paranormal. (CSICOP) Box 703, Amherst, NY. 14226.

> This excellent skeptics organization publishes the bimonthly journal, *The Skeptical Inquirer*, which debunks alien contactees and spontaneous human combustion. Their Website "Library and Research Center" states that ball lightning is a well-documented phenomenon that lacks a scientific explanation. CSICOP states that it passes through houses, vehicles, airplanes—even materializing through walls—or along telephone lines or wire fences. Typically baseball-to-basketball-sized, rotating, yellow or changing colors, it either fades or explodes. CSICOP states that ball lightning has not been reproduced in the laboratory and remains unexplainable.

Condon, Edward. "Atmospheric Electricity and Plasma Interpretations of UFOs," chapter 7, by Martin Altschuler. *Scientific Study of Unidentified Flying Objects.* New York Times and Univ. of Colorado. 1969.

Good coverage of ball lightning (chapter 7) and meteors (chapter 9). Sponsored by USAF.

Corum, Dr. James and Kenneth. Everything by the Corum brothers is recommended reading. The many technical monographs on Tesla and Artificial Electric Balls by these brilliant authors, although pricey, are worth the cost. Hopefully, one day they will collect them and publish them in a single volume. Please check Internet Web sites.

Most myths about Tesla are just that—only myths. The Corums brilliantly debunk Tesla mythology and correct errors of history. For example, they describe how Tesla probably detected singlet, doublet, and triplet pulses from Jupiter—not from Mars, as mistakenly thought.

Corum, J.F. & K.L. *Class Notes: Tesla Coils and the Failure of Lumped Element Circuit Theory.* 1999. [Internet paper]

Tesla coils have more in common with cavity resonators than with conventional inductors. It is a velocity-inhibited SWR with voltage magnification by standing waves. Lumped element transformers and lumped element LC resonating circuits fail.

Corum, J.F. & K.L. *Vacuum Tube Tesla Coils.* Windsor, OH: Corum and Associates.1987.

Corum, J.F. & K.L. and Edwards, D.J. *TCTUTOR: A Personal Computer Analysis of Spark Gap Tesla Coils.* Corum and Associates. 1988.

Includes 3.5-inch 1.44-MB diskette program written in advanced BASIC under MS-DOS that, for the first time, successfully simulates Tesla's three-coil Magnifier of 1899 and Golka's Wendover re-creation. Your author ran this program successfully under Windows Millennium and also printed out its coding. A must-own for every coiler.

Coveney, Peter and Highfield, Roger. *The Arrow of Time: A voyage through science to solve time's greatest mystery.* New York: Fawcett Columbine (Ballantine Books; Random House). 1990. Foreword by Ilya Prigogine.

> This readable classic includes novel insights into self-organization in chemistry, thermodynamics, Mandelbrot, Koch curves, cascades into chaos, and fractal geometry—subjects important to ball lightning.

Coveney, Peter and Highfield, Roger. *Frontiers of Compexity: The Search for Order in a Chaotic World.* Ballantine Books. 1995.

Cowgill, Werner. "Curious Phenomenon in Venezuela." *Scientific American.* 55:389. December 18, 1886.

> Fireball's loud humming and strange effects upon witnesses suggest radiation sickness—then unknown.

Davis, Sidney. *Feedback and Control Systems—Revised Ed.* New York: Simon and Schuster. 1974.

> Theory and detailed solutions to 355 problems that apply to the enigma of fireball flight, stability and navigation.

Davies, Paul and Gribbin, John. *The Matter Myth: Dramatic Discoveries That Challenge Our Understanding of Physical Reality.* New York: Simon & Schuster. 1992.

> Popular treatment of complexity, chaos, and soliton waves—topics relevant to fireballs.

Dickerson, R., Gray, H., Haight, G. *Chemical Principles, Second Ed.* Menlo Park, CA: W.A. Benjamin, Inc. 1974.

> Van der Waals Forces, pp. 601-607, are believed important in several recent ball lightning theories.

Dvali, Georgi. "Out of the Darkness." *Scientific American.* February 2004. pp. 68-75.

> Out-leakage of gravity and not dark energy may cause cosmic acceleration. Unlike traditional string theory and Randall-Sundrum infinite-sized extra dimensions with strong curvature to concentrate their volume around our

cosmos, the author proposes that just like ordinary three dimensions, that extra dimensions are infinite in size but uncurved.

Dwer, Joseph R. "A Fundamental Limit on Electric Fields in Air." *Geophysical Research Letters.* Vol. 30, No. 20, 2055. 25 October 2003.

By modifying the avalanche mode of runaway breakdown to include positive feedback from gamma rays and positrons, massive bursts of radiation are created in strong electric fields in air, with peak fluxes one billion times greater than with conventional models—bursts that generate so many runaway electrons that the electric field is quickly discharged. This results in a fundamental upper limit on the electric field strength achievable in air—a limit with important implications for electrification of thunderstorms and producing lightning.

Dyer, Alan. "Canadian Daylight [meteor] Fireball." *Sky & Telescope.* Cambridge, MA: Sky Publishing Corp. January 2002. p. 128.

Color photo of dramatic silvery meteor seen over western Canada on October 14, 2001, and like similar ones and bolides, cannot be mistaken for a lightning ball fireball.

Eddington, Sir Arthur. *The Mathematical Theory of relativity.* New York: Chelsea Publishing Co. 1923; 3rd ed., 1975.

Good coverage in this re-issued classic book by this famed author—that is, if you enjoy and are fluent with advanced calculus—of classical relativity, tensor calculus, relativity mechanics, space-time curvature, electromagnetics, world geometry, Weyl's theory, Gauge-invariance, Riemann math, etc.

Evans, B.C. Executive Secretary. *Central Intelligence Agency.* Letter dated 10 August 1976.

Responding to author's inquiry on ball lightning, Evans described how CIA is aware of ball lightning, its potential military applications and its strange behavior, and that no theory accounts for its stability, mobility and death.

Fermi, Enrico. *Thermodynamics.* New York: Dover Publications. Reprint of Prentice-Hall, 1937.

Covers Van der Waal's equation, pp. 69-75.

Ferris, Timothy. *The Whole Shebang*. New York: Simon & Schuster. 1997.

John Bell and David Bohm, pages 280-287.

Feynman, Richard. *The Feynman Lectures on Physics: Commemorative Issue*. Reading, MA: Addison-Wesley Publishing Co. 1964, 1989.

A joy to read, these provocative three volumes lure the reader with a fresh, breezy and enjoyable style. Although you should already be conversant in introductory physics and advanced calculus, if you only studied differential equations, you can still follow this updated Feynman classic. As it relates to ball lightning, there are several sections, including an excellent coverage on measuring atmospheric electricity in 9-1 to 9-11 of volume II.

If there is any flaw, it is in the first volume of this trio. Feynman mistakenly used simpler, more historical and primitive techniques, which should be in a sidebar; and, instead he should have used the more elegant mathematical approaches from the start, such as Hamiltonians and Lagrangians, which make classical mechanics so clean and concise, and can be more cleanly applied to ball lightning modeling. The red paperbacks are not inexpensive, but I would recommend the pricier blue-green boxed hardbacks. Hopefully, someone will add a fourth volume in the style of Feynman to cover recent developments.

Feynman, Richard. *QED: The Strange Theory of Light and Matter*. Princeton, NJ: Princeton University Press. 1985.

Chapter 4 covers loose ends in QED of interest to fireball researchers, particularly the discussions on getting rid of the infinities in quantum gravity.

Fischetti, Mark. "Flight Control: Golf Balls." *Scientific American*. June 2001. pp. 96-97.

Examining the surface of golf balls in flight is important not just to golf ball designers, but also to understanding the paradox of fireball flight.

"Flammable Ice." *Scientific American*, November 1999.

For a detailed description of vast stores of methane-hydrate ice crystals embedded in the seafloor, of interest to mystery lights.

Folger, Tim. "Nailing Down Gravity." *Discover.* October 2003. pp. 34-40.

In the Modified Newtonian Dynamics (MOND), Moti Milgrom argues that a fudge factor like dark matter is unnecessary, since a slight change in Newton's law of force will suffice. Below an acceleration of one ten billionth of a meter per second each second, gravity is no longer directly proportional to its acceleration, but is slightly stronger, proportional to the square of acceleration. Most astronomers remain skeptics.

Freau, John. "Comments on the Corum's Paper." *TCBA News.* Vol.20, No. 2. pp. 7-8.

The lumped analysis circuit coil design—which Golka used in the 1970s and 1980s in Wendover and Leadville—may have been more accurate than later believed. Freau argues that the Corum's use of the distributed parameters approach overlooked key aspects. Effort would be better directed at reducing gap losses instead of raising the secondary Q or quality factor.

Gaddis, Vincent. *Mysterious Fires and Lights.* Garberville, CA: Borderland Sciences. 1967, 1994.

A nonscientific but entertaining coverage of ball lightning, combustion, and mystery lights, some of it not easily found elsewhere.

Gardner, Martin. "The Guided Wave Theory of Louis de Broglie and David Bohm." *Skeptical Inquirer.* May/June 2000. PP. 9-14.

Bohm explained the EPR paradox by pilot-wave or guided wave theory. Does it explain paradoxical behavior, particularly "action at a distance" effects associated with fireballs?

Gell-Mann, Murray. *The Quark and the Jaguar: Adventures in the Simple and the Complex.* New York: W.H. Freeman and Co. 1994.

Examines complexity in a popular style and describes how Oppenheimer and ball lightning met at Princeton and generated many sparks.

Golka, Robert K. "Laboratory-produced ball lightning." *Journal of Geophysical Research.* Vol 99, No D5. May 20, 1994. pp. 10,679-10,681.

Golka, Robert K. Personal correspondence with author between 1973 and 1993.

Gleick, James. *Chaos: Making a New Science*. New York: Penguin Books. 1987.

This classic is still the best popular starting point for an introduction to chaos.

Gray, Harry. *Chemical Bonds: An Introduction to Atomic and Molecular Structure*. Menlo Park, CA: W.A. Benjamin, Inc. 1973.

Van der Waals forces, pp. 194-206.

Greene, Brian. *The Elegant Universe: Superstrings, Hidden Dimensions, and the Quest for the Ultimate Theory*. New York: W.W. Norton & Co. 1999.

Exceptional. Highly readable with a non-mathematical treatment by this leading physicist. Greene's pioneering loose grammatical style is beautiful and almost poetical. His examples are creative and his analogies, innovative. Ball lightning suggests new directions in gravity for string theory, which is incomplete and whose edifice keeps evolving.

Greene, Brian. "The Future of String Theory: A Conversation with Brian Greene." *Scientific American*. Vol. 289, No. 3. Nov. 2003. pp. 68-73.

This interesting article was abbreviated from a more detailed unpublished 37-page manuscript, available from *Scientific American*.

Gribbin, John. *In Search of Schrodinger's Cat: Quantum Physics and Reality*. New York: Bantam Books. 1994.

Gribbin, John. *Q is for Quantum: An Encyclopedia of Particle Physics*. New York: Fee Press/Simon & Schuster. 1998.

Grossman, Daniel. "Spring Forward." *Scientific American*. January 2004. pp. 85-91.

Warming climates decouple interdependent species.

Grotz, Toby. Internet-accessible articles.

A former member of Project Tesla in Leadville, Colorado, electrical engineer Toby Grotz studied ball lightning and advocates using a giant three- or four-coil Magnifier to transmit non-Hetzian electrical wireless power with Zenneck "surface" waves at eight hertz.

Guth, Alan. *The Inflationary Universe: The Quest for a New Theory of Cosmic Origins*. Reading, MA: Helix Books/Addison-Wesley Publishing Co. 1997.

Hall, Nina. *Exploring Chaos: A Guide to the New Science of Disorder*. New York: W.W. Norton & Co. 1993.

A collection of reports by recognized experts, edited by Hall.

Hawking, Stephen and Penrose, Roger. *The Nature of Space and Time*. Princeton, NJ: Princeton Univ. Press. 1996.

Collection of 1994 lecture debates with friendly sparks flying between these two giants. Requires reader to be intimate with advanced physics.

Hawking, Stephen and Penrose, Roger. *The Nature of Space and Time. Scientific American*. July 1996. pp. 60-65.

Excellent summary of 1994 lecture debates at U. of Cambridge on General Relativity.

Herbert, Nick. *Quantum Reality: Beyond the New Physics*. New York: Anchor Book (Bantam Doubleday Dell Publishing Group). 1985.

Covers EPR, Bohmian Mechanics, and Bell's Theorem.

Horgan, John. *The End of Science: Facing the Limits of Knowledge in the Twilight of the Scientific Age*. New York: Broadway Books. 1996.

Will we reach the end of physics, cosmology and chaoplexity?

Huff, Darrel. *How to Lie With Statistics*. New York: W. W. Norton & Co, Inc. 1954.

All ball lightning statistics "statisticulate," sampling with many known and unknown built-in biases that skew and corrupt the results.

Hull, Richard. *The Tesla Coil Builder's Guide to The Colorado Springs Notes of Nikola Tesla. Second edition*. Breckenridge, CO: Twenty First Century Books. 1999.

This daily annotation of Tesla's Colorado Springs Diary by a member of the Tesla Coil Builders of Richmond in Virginia is excellent and is a must-own for every coiler. Very highly recommended is all material published by the TCB, which has extensive experience with constructing state-of-the-art large coils. They continue to construct many different Magnfier three-coil systems, refine them, and then construct better ones. The TCB holds monthly meetings and yearly Teslathons, and openly shares its discoveries through its numerous videotapes, still photographs, and printed documentation. Some members recreate and discuss Tesla Artificial Electric Balls. Recently, emphasis has shifted to pursuing the Fusor, an experimental fusion device pioneered by Farnsworth.

Hunt, Inez and Draper, Wanetta. *Lightning in his Hand: The Life Story of Nikola Tesla.* Omni Publications 1977.

Emphasizes the Colorado years, since the authors lived nearby. Adds details not located elsewhere.

Ivins, Molly. "Rent and Ire Rise at a Utah Laboratory." *The New York Times.* February 23, 1980. p. 6.

Conflicts and new financial expenses haunt Golka's Project Tesla and his attempt to re-create electric fireballs and hopefully a clue to controlled fusion at the historical aircraft hanger, where the Enola Gay was once retrofitted with the first atomic bomb for Hiroshima. Will the "expense explosion" end Golka's project? A large article and photo document his dilemma.

Jorgensen, Neil. "Swamps and Bogs." *A Guide to New England's Landscape.* Chester, CT: The Globe Pequot Press. 1977.

Explains how our cold bogs differ from those in the south, and thus generate less methane (swamp gas), which explains why we New Englanders never see wil o' the wisp.

Journal of Scientific Exploration. A publication of the Society for Scientific Exploration. Stanford Univ., CA. Pergamon Press.

Publishes novel and authoritative articles on ball lightning.

Kaku, Michio. *Hyperspace: A Scientific Odyssey Through Parallel Universes, Time Warps and the 10th Dimension.* New York: Anchor Books/Doubleday. 1994.

Kaku, Michio. *Beyond Einstein: The Cosmic Quest for the Theory of the Universe.* New York: Doubleday Publishing. 1995.

Kane, Gordon. *Supersymmetry: Squarks, Photinos, and the Unveiling of the Ultimate Laws of Nature.* Cambridge, MA: Helix Books (Perseus Publishing). 2000.

> All fundamental particles should have superpartners—an electron has a selectron; a quark, a squark; and so on. If true, this may validate string theory and justify the Higgs boson, and further confirm dark matter.

Kane, Gordon. "The Dawn of Physics Beyond the Standard Model." *Scientific American.* June 2000. Pp. 68-75.

Kauffman, Stuart. *At Home In the Universe: The Search for the Laws of Self-Organization and Complexity.* New York: Oxford U. Press. 1995.

Kerner, B.S. and Osipov, V.V. *Moscow, Russia: Autosolitons: A New Approach to Problems of Self-Organization and Turbulence.* Kluwer Academic Publishers. 1994.

> Covering solitary intrinsic states (autosolitons), this text on nonlinear physics compares autosolitons, which occur in highly non-equilibrium regions in slightly non-equilibrium systems, to atmospheric ball lightning. The authors treat self-organization and turbulence as a result of spontaneous formation and evolution of autosolitons. Self-organization comes from autosolitons interacting, but autosolitons that randomly appear and disappear create turbulent patterns. The authors cover autosolitons in active systems, and ball lightning in gaseous hot spots, regions within plasmas of low or high electron temperatures, and so on. For experts only.

Klein, Cornelus and Hurlbut, Cornelius. *Manual of Mineralogy, 20th ed.* New York: John Wiley & Sons, Inc. 1985. pp. 214-218.

> By using piezoelectricity, pyroelectricty, luminescence, thermoluminescence, and triluminescence, do earthquakes and other phenomena generate electric lights and fireballs?

Lauwerier, Hans. *Fractals: Endlessly Repeated Geometrical Figures.* NJ: Princeton U. Press. 1991.

Lemonick, Michael. "Before the Big Bang." *Discover.* February 2004. pp. 35-41.

> Challenging inflation, maverick cosmologists contend that our cosmic creation was only part of an infinite cyclic of titanic collisions between our cosmos and a parallel cosmos.

Levin, Simon A. *Fragile Dominion: Complexity and the Commons.* Reading, MA: Perseus Books.

Lewis, Harold W. "Ball Lightning." *Scientific American.* Vol. 208, no. 3. March 1963. pp. 107-116.

> Suggests BL is a stable configuration of highly ionized gases that may enable us to confine plasma in a fusion reactor. Erroneously states that a 10-inch singularly ionized plasma ball possess one megajoule, but it is 0.858 megajoule. Proposed Hill and Kapitza models, since rejected.

Libbrecht, Kenneth and Tanusheva, Victoria. "Electrically Induced Morphological Instabilities in Free Dendrite Growth." *Physical Review Letters.* Vol. 81, No. 1. pp. 176-179.

Lightman, Alan. *Great Ideas In Physics.* New York: McGraw-Hill. 2000.

> Do fireballs violate conservation of energy? No, but they present puzzles of reversible and irreversible flow of energy and may tweak the second law of thermodynamics.

Lindley, David. *The End of Physics: The Myth of a Unified Theory.* Basic Books (Harper Collins Publishers). 1993.

Loeb, Leonard. *Fundamentals of Electricity and Magnetism.* New York: Dover Publications. 1961. Reprint of original by John Wiley & Sons in 1947.

> This classic is valid today and is well-worth studying by fireball researchers and coilers.

Lorenz, Edward. *The Essence of Chaos.* Seattle: U. of Washington Press. 1993.

> The Father of Chaos explains chaos.

Lowell, S and Shields, Joan. *Powder Surface Area And Porosity, Third Edition.* Chapman & Hall: New York. 1994.

> Of relevance to Tesla electric ball fractal cluster theory.

McNally, James R. *Preliminary Report on Ball Lightning.* ORNL-3938, May 1966. Oak Ridge National Laboratories.

McNally, James R. *Ball Lightning—A Survey.* Bull. Amer. Phys. Soc. 6.1962. p. 202

Marshall, Ian and Zohar, Danah. *Who's Afraid of Schrodinger's Cat? An A-to-Z Guide to All the New Science Ideas You Need to Keep Up with the New Thinking.* Quill, William Morrow. New York. 1997.

> Bell's Theorem, pp. 64-67. Concise encyclopedic explanations of quantum mechanics, chaos and relativity.

Martin, Thomas Commerford. *The Inventions, Researches, and Writings of Nikola Tesla.* New York: The Electrical Engineer. 1894.

> From this reprinted classic we can see how Tesla already had the basics in place for 1899.

Miele, Frank. "A Quick & Dirty Guide To Chaos and Complexity Theory." *Skeptic* journal. Skeptics Society.

> Although chaoplexity is important to physics, ball lightning, meteorology, and mathematics, scientists must look skeptically at its clever misuse in "fuzzy" nonscientific fields used as metaphors to cloak their unpredictability in respectability.

Moin, Parviz and Kim, John. "Tackling Turbulence with Supercomputers." *Scientific American.* January 1997. pp. 62-68.

> Discusses aerodynamic drag on golf balls, which should also be important to fireballs in flight, but perhaps not—and no one knows why.

Moore, Arthur. *Electrostatics: Exploring, Controlling, and Using Static Electricity—including the Dirod Manual.* 2nd ed. Morgan Hill, CA: Laplacian Press. 1997.

> This readable classic's intuitive insights into static electricity include coronas and ball lightning for students and experimenters. Explains construction and operation of high voltage disk-rods.

Murray, Peter and Dawson, P. *Structural and Comparative Inorganic Chemistry.* London: Heinemann Educational Books. 1976.

> Applications to methane-fueled, so-called mystery lights.

Nadis, Steve. "When Branes Collide." *Astronomy.* March 2003. pp. 34-39.

Nadis, Steve. "Will dark energy steal all the stars?" *Astronomy.* March 2003. pp. 42-47.

Naeye, Robert. "New Binary Neutron Star Will Further Test Einstein." *Sky & Telescope.* March 2004. pp. 22-23.

> A curio in reference to start of first chapter. A newly-discovered binary neutron star advances 150,000 times faster than the advance of Mercury's perihelion from General Relativity. In the binary neutron star of PSR J0737-3039 (off the tail of Canis Major), with its regular blipping, this new (there are only six known) pulsar exhibits a precession (rotation) of its periaston (marks long axis of elliptical orbit) changing at an awesome 16.88 degrees per year!

Nature. This eminent journal once frequently published many articles—too numerous to list here—on ball lightning, starting in the mid-1880s. Your author photocopied all of these articles at the MIT Library in Cambridge, Massachusetts.

Nori, F.; Sholtz, P.; Bretz, M. "Booming Sand." *Scientific American.* September 1997. pp. 84-89.

> Does booming sand cause electrical effects that sometimes create strange lights.

Norton, O. Richard. *Rocks From Space (second ed.).* Missoula, Montana: Mountain Press Publishing. 1998.

> Sometimes mistaken for ball lightning, meteors require familiarity among investigators.

O'Neill, John J. "Tesla Tries to Prevent World War II." *Prodigal Genius: The Life of Nikola Tesla.* New York: David McKay Company. 1944.

> Unpublished chapter 34 of Tesla's biography printed in Tesla Coil Builder's Association News, vol. 7, 1988, no. 3, pp. 13-15 by Leland I. Anderson. A copy of O'Neill's original chapter was in my possession in 1975 and your author submitted a copy to McKay asking for permission to publish it.

O'Neill, John. *Prodigal Genius: The Life of Nikola Tesla.* Ives Washburn, subsidiary of David McKay Co. 1944.

Most Olympian myths about Tesla are just that—myths. This classic biography of Tesla by this former science editor of the *New York Times*, who knew Tesla in his later years, contains erroneous mystical personality information that spawned cultism. Although he greatly exaggerated Tesla's eidetic and synesthesic mnemonic memory, and erroneously described the Magnifier, it was the first biography and became a landmark. An alleged spiritualist, O'Neill launched unfortunate mystical associations of a mysterious nature that resonated with enthusiastic fans.

Pagels, Heinz. *The Dreams of Reason: The Computer and the Rise of the Sciences of Complexity.* New York: Bantam Books. 1989.

Good writing, great author. Unfortunately, Pagels died mountain climbing and there will be no sequels.

Pagels, Heinz. *Perfect Symmetry: The Search for the Beginning of Time.* Bantam Books. 1985.

Pauling, Linus and Wilson, E. Bright. *Introduction to Quantum Mechanics: With Applications to Chemistry.* New York: McGraw-Hill Book Co. 1935.

Good coverage, and still valid. Read this only if you are very intimate with advanced calculus and already know some quantum mechanics. Rich in math that avoids concise-but-difficult elegant derivations. Schrodinger wave equations, perturbation theory, variation and other approximation methods, Pauli exclusion principle, rotation and vibration of molecules, statistical quantum mechanics, etc. Not as oriented to chemistry as the title suggests. Amazing how sophisticated they were back then.

Pauling, Linus. *General Chemistry.* New York, Dover Publications. 1988. Reprint of original from W.H. Freeman and Co., Sand Francisco in 1970.

Innovative and unusual coverage. See Van der Waals radii.

Peat, F. David. *Synchronicity: The Bridge Between Matter and Mind.* Bantam Books. 1987.

David Bohm, pp. 168-176.

Peat, F. David. *Superstrings and the Search for The Theory of Everything.* Contemporary Books. 1988.

Explores superstrings and Penrose's Twistor gravity.

Peat, F. David. *Infinite Potential: The Life and Times of David Bohm.* Addison-Wesley Longman. 1997.

For explanations of Bohmian Mechanics, see pages 107-19, 137-42, 150-1, 159-60, 168-9, 181-5, 201-8, 219-22, 242-51, 256-81.

Penrose, Roger. *The Emperor's New Mind: Concerning Computers, Minds, and The Laws of Physics.* New York: Oxford University Press. 1989.

Excellent. Laws deeper than quantum mechanics explain the human mind—and fireballs? Penrose is not easy reading, and takes re-reading and deep thought to really understand this super genius' thoughts.

Penrose, Roger. *Shadows of the Mind: A Search for the Missing Science of Consciousness.* New York: Oxford Press. 1994.

Brilliant. Penrose asserts that "something" in the human mind's conscious activity—and also for fireballs?—transcends computation, which can never completely simulate human thinking. He attacks artificial intelligence and concludes that profound changes are necessary for physics to accommodate this "something" that is foreign to our current physical picture. Penrose argues that not even future computers (as projected) could perform the operations required for human conscious understanding. Not too mathematical, but heavily technical, and not for amateurs.

Peterson, Gary L. "Rediscovering the Zenneck Surface Wave." Newsletter. Twenty First Century Books. Breckenridge, Co.

Pickover, Clifford. *Black Holes: A Traveler's Guide.* New York: John Wiley & Sons. 1996.

Can black holes be used as interstellar gates for fireballs and others? Maybe, but with difficulty. (pp. 120-121) If advanced civilizations could prevent the sensitivity of Kerr tunnels to outside disturbances, they might prevent their decay. In this action-packed (for physicists) fictional journey, you follow the journeys and Socratic exchange between two galactic explorers. Explains black hole physics. Entertaining.

Preiter, H.; Saupe, Dietmar; Jurgens, H.; Yunker, L. *Chaos and Fractals: New Frontiers of Science.* New York: Springer-Verlag. 1992. 984 pp.

> This outstanding volume describes how feedback and interators [sic] wipe out all computers, self-similarity, complexity, scaling, transformations, and everything else you could possibly ever want to know about chaos—and, which also applies to ball lightning. With its 14 chapters, this hefty (984-page) rigorous volume contains many spectacular illustrations.

Prigogine, Ilya and Nicolis, Gegoire. *Exploring Complexity: An Introduction.* New York: W.H. Freeman and Co. 1989.

Prigogine, Ilya. *The End of Certainty: Time, Chaos, and the New Laws of Nature.* New York: Simon & Schuster (The Free Press). 1997.

> The arrow of time is asymmetric, and cannot flow backwards.

Proceedings of the 19XX International Tesla Symposium. Colorado Springs: International Tesla Society. (For years 1984, 1986, 1988, 1990, 1992.)

> These large volumes are out of print, and the ITS (an IEEE SIG) was disbanded, so these large volumes are only available on out-of-print web sites. There is talk that these sought-after volumes will be re-issued as trade paperbacks. If so, they would be best sellers. Some of these classic papers were first rate; others, historical; but some others, like some Tesla fans—well, rather strange. Overall, thumbs up.

Ramo, Whinnery and Van Duzer. *Fields and Waves in Communication Electronics.* John Wiley & Sons. 1965.

> Classic text, with excellent treatment of electromagnetism and lots of advanced calculus.

Rauscher, Elizabeth and Van Bise, William. *Earth-Ionosphere Resonant Cavity Soliton.* Reno, NV: Technic Research Laboratories.

Rees, Martin. *Before The Beginning: Our Universe and Others.* Reading, MA: Addison-Wesley. 1997.

Resnick, Robert and Halliday, David. *Fundamentals of Physics, revised printing.* New York: John Wiley & Sons. 1970, 1974.

> Excellent undergraduate text, still available at used book stores—I own three copies.

Ricketts, L.W., Bridges, J.E., Miletta, J. *EMP Radiation & Protective Techniques.* New York: John Wiley & Sons.

> Covers mathematical technical topics relevant to fireball formation—and prevention.

Ritchie, Donald. *Ball Lightning: A Collection of Soviet Research in English Translation.* New York: Consultant's Bureau. 1961.

> Historical papers plus annotated bibliography.

Roberson, John and Crowe, Clayton. *Engineering Fluid Mechanics, second ed.* Boston: Houghton Mifflin Co. 1980.

> This text knowledge is needed to appreciate the rapid flight of fireballs and calculate the dynamic air resistance, which can be enormous, and how this greatly increases estimates of energy content.

Royal Society. *Philosophical Transactions of the Royal Society: Ball Lightning.* Compiled and edited by J. Abrahamson. 15 January 2002, Volume 360, Number 1790. 152 pp.

> Still no explanation, but nice attempts and descriptions.

Savant, C.J. *Control System Design, second ed.* New York: McGraw-Hill Book Co. 1958, 1964.

> Controlled fireball flight requires linear and non-linear control system servo-mechanisms—no easy feat.

Schaefer, Vincent and Day, John. *Atmosphere: Clouds, Rain, Snow, Storms.* Boston: Houghton Mifflin Peterson Field Guides. 1981.

Schonland, Basil. *The Flight of Thunderbolts.* Oxford: Clarendon Press. 1964.

> Covers ball lightning by a renowned lightning researcher. Readable.

Schroeder, Manfred. *Fractals, Chaos, Power Laws*. New York: W.H. Freeman & Co. 1991.

Science News: The Weekly Newsmagazine of Science; 1719 North St NW, Washinton, DC.

> Summaries of recent journal articles cover physics, cosmology, distant exoplanets, and astronomy—and occasionally ball lightning.

Seifer, Mark J. *Wizard: The Life and Times of Nikola Tesla, Biography of a Genius*. Birch Lane Press. 1996.

> Balanced historical and psychological coverage in this most thorough book to date that provides new material on negotiations, consulting work, and friendships. Graphologist and psychologist Dr. Seifer correctly noted Tesla's character faults, such as greed, vanity, paranoid tendencies, and megalomania, but continued the tradition of exaggerating Tesla's achievements. Although not a technologist, Dr. Seifer's book is the best. The only weakness is that some of Seifer's engineering claims are incorrect, as later reviewed in *Scientific American*.

Seifer, Mark J. "Tesla's Magnifying Transmitter." (pp. 324-363). *Nikola Tesla: Psychohistory of a Forgotten Inventor*. Saybrook Institute of San Francisco. 7.10.1986.

Senapathy, Periannan. *Independent Birth of Organisms*. Genome Press. 1994.

Shunaman, Fred. "12-Million Volts." *Radio-Electronics*. June 1976.

> In this feature article, Shunaman wrote that small discharges resembling ball lightning lasted for sub-second life spans at Golka's Project Tesla in Wendover, Utah.

Smirnov, B.M. "The Properties of Fractal Clusters." *Physics Reports*. 1990. Vol. 188, pp. 1-78.

Smolin, Lee. *Three Roads to Quantum Gravity*. NY: basic Books. June 2002.

> Promising roads include string theory, loop quantum gravity, twistor theory, non-commutative geometry, and holographic principles.

Smolin, Lee. "Atoms of Space and Time." *Scientific American.* January 2004. pp. 66-75.

> If loop quantum gravity is true, then space and time are ultimately grainy and discrete, not smooth. First proposed by Penrose, spin networks describe spacetime and play a role in quantum gravity. If true, might LQG lead to a final unification? No one knows.

Snigier, Paul. "Fusion Energy: Will Experiments in Ball Lightning Provide the Key?" *EDN* (a journal for electrical engineers). Newton, MA: Reed Electronics Group (Cahners) Publishing. 4.20.76.

> Description of Golka's Project Tesla that again re-created the 1899 Magnifier at Wendover AFB. Can true ball lighting provide a workable fusion containment scheme? Perhaps a remote candidate for fusion—but not directly, and from relevance in a completely unexpected manner.

Singer, Stanley. *The Nature of Ball Lightning.* New York: Plenum Press. 1971.

> Although out of print, this excellent classic landmark summarizes cases and lists an extensive bibliography.

Singer, Stanley. "Great Balls of Fire." *Nature.* 1991. vol. 350, March 14. pp. 108-109.

Skeptics Society. Box 338, Altadene, CA 91001. Publishes *Skeptic* journal and Michael Shermer's excellent *Skeptics Corner* in *Scientific American.*

Stenhoff, Mark. *Ball Lightning: An Unsolved Problem in Atmospheric Physics.* New York: Kluwer Academic/Plenum Publishers. 1999.

> This excellent scholarly text lists an extensive bibliography. It is wisely cautious and finds no evidence ("very weak") that BL damages buildings (chapter 6), aircraft (chapter 7), or trees (chapter 8). That some BL may not exist is supported (p. 178) by other phenomena that resemble it, because eyewitnesses are unreliable, and because scientists cannot explain it. There is a paucity of physical evidence and shortage of reliable photographic evidence. Overestimates of energy occur because associated lightning fools observers and does the real damage. Damage is mistakenly reported to come from massive currents within BL, which is not likely because a lightning ball cannot contain sufficient charge to deliver such massive current surges. Therefore these observers are mistaken. BL has two possible energy sources—from

inside or outside—and is a low-energy phenomenon that is not dangerous, but precedes lightning, which really does the damage.

This book concludes that BL cannot be explained, and many sophisticated theories with impressive mathematics rest on weak foundations, and says, "At present there is no fully satisfactory or generally accepted scientific theory of ball lightning." It does not mention Tesla and the Corums' recreating Artificial Electrical Balls, or Tesla's description of their explosive power. Only once (p. 186) does it mention Golka in a small experiment—oddly not the Magnifier in Wendover and Leadville.

Stewart, Ian and Golubitsky, Martin. *Fearful Symmetry: Is God a Geometer?* Cambridge, MA: Blackweall Publishers. 1992.

Fascinating book covers many aspects that relate to ball lightning. Anything by Ian Stewart and Martin Gardner is enjoyable and highly recommended.

Stewart, Ian. *Another Fine Math You've Got Me Into...* Cambridge, MA: Blackwell Publishers. 1992.

Stewart, Ian. *What Shape Is a Snowflake?* Freeman. 2001.

Too rigid to explain everything, symmetry must be combined with chaos and complexity to explain all of nature's regularities—perhaps including fireball structures.

Stewart, Ian. "Crystallography of a Golf Ball." Mathematical Recreations. *Scientific American.* February 1997. pp. 96-98.

Stratton, Julius Adams. *Electromagnetic Theory.* New York: McGraw-Hill Book Co. 1941.

A former professor of physics at MIT, the author uses rigorous advanced calculus in an excellent 616-page treatment that includes a respectable coverage of Zenneck "surface waves."

Suess, E.; Bohrmann, G.; Greinert, J.; Lausch, E. "Flammable Ice." *Scientific American.* November 1999. pp. 76-83.

Vast seafloor deposits of methane-hydrate-laced ice crystal release methane gas, which could raise global temperatures. Methane fuels some mystery lights.

Talbot, Michael. *The Holographic Universe*. New York: Harper Collins Publishers. 1991.

> Covers EPR, Bell's Theorem, and Bohmian Mechanics. David Bohm and Karl Pribram propose that the universe may resemble a giant super-"hologram." Suggests a construct created in part by human consciousness.

Tatham, Dr. Elaine L. *Same Size for Random Sampling Without Replacement: Formulas and Examples*. Customer Satisfaction Research Institute. Shawnee Mission, Kansas: CSRI. December 1979.

> Properly gathered and interpreted, a small random sample sometimes may supply information to a large population and provide meaningful associations and exclude trivial ones. Important to ball lightning statisticians.

Tegmark, Max. "Parallel Universes." *Scientific American*. May 2003. pp. 40-51.

> Excellent summary of Multiverse Theory. A valid Theory of Everything in physics must explain gravity and fireballs—and dovetail with Multiverse Theory. Parallel universes form a four-level hierarchy of Multiverse candidates, each of progressively greater diversity. Multiverse Theory poses questions for any theory that unifies gravity with the other three forces. Fireballs "violate" certain laws of physics, including gravitation. Until true fireballs are explained, do not ignore possible connections.

Tegmark, Max. "Is the 'Theory of Everything' Merely the Ultimate Ensemble Theory?" *Annals of Physics*, Vol. 270, No. 1, pages 1-51 November 20, 1998.

Tesla, Nikola. Nikola Tesla: Lectures, Patents, Articles. Nolit. Nikola Tesla Museum, Beograd, Serbia. 1956.

> The patents from Shoreham experiments describe the explosive power of Tesla's Artificial Electric Balls.

Tesla, Nikola. *Nikola Tesla: Colorado Springs Notes 1899-1900*. Nikola Tesla Museum. Nolit, Beograd, Serbia.

> This important diary contains lab notes that Tesla wrote. Unfortunately for posterity, Tesla omitted crucial lab entries—such as the unpublished chapter 34 in O'Neill's biography, the details of how he ionized a path up to the lower ionosphere, and how he converted massive damped sinusoids of currents and voltages into a pulsed 8-hertz direct current. He described light-

ning balls but failed to explain how he created them. These omissions thwarted Golka and Corum. The museum possesses unpublished materials and photographs.

Tesla, Nikola and Childress, David. *The Fantastic Inventions of Nikola Tesla.* Stelle, IL: Adventures Unlimited. 1993.

Tuck, James. *Ball Lightning Summary to Novmeber 1971.* LA-4847-MS. Los Alamos Scientific Laboratory of the U. of Calif. [Delivered at MIT.]

Thorne, Kip. *Black Holes and Time Warps: Einstein's Outrageous Legacy.* New York: W.W. Norton & Co. 1994.

In Chapter 14 (pp. 483-521), Thorne describes his epochal paper on wormholes and time machines, which has implications for Multiverse Theory—and perhaps fireball theories?

Uman, Martin. *All About Lightning.* New York: Dover. 1986. Reprint of Bek Technical Publications in 1971.

Uman is a leading authority on lightning and has repeatedly met with Golka to talk about ball lightning.

Uman, Martin. *Lightning.* New York: Dover. 1980. Reprint of McGraw-Hill of 1969.

Van Valkenburg, M.E. *Network Analysis, 2nd ed.* Prentice-Hall. 1955, 1964.

Excellent introduction to Laplace functions and circuit analysis, but weak on Fourier. One caveat—on page 143, he omits derivation for a parallel RLC network, for which the parallel QL=BL/GL. A classic, still important to coilers.

Venexiano, Gabriele. "The Myth of the Beginning of Time." *Scientific American.* Vol. 290, No. 5. May 2004.

The Big Bang was only an outcome of a preexisting state, and string theory suggests we live in a universe without beginning or end—timeless.

Visser, Matt; Kar, Sayan; Dadhich, Naresh. "Transversable Wormholes with Arbitrarily Small Energy Condition Violations." *Physical Review Letters.* 23 May 2003. 90, 201102 (2003).

> Addressing the massive energy problems that Kip Thorne recognized, authors examine spacetime geometries that contain transferable wormholes supported by arbitrarily small quantities of "exotic matter" which violate the averaged null condition. This hypothesized exotic matter is repelled by gravity—and may indeed exist—perhaps as a quality exhibited by fireballs' "gravity-defying" behaviors?

Wait, James R. Director of Theoretical Studies Group. Memo to Golka from U.S. National Oceanic and Atmospheric Administration, Boulder, CO. Room 242, RB 1. 18 Nov. 1975.

Waldrop, M. Mitchell. *Complexity: The Emerging Science at the Edge of Order and Chaos.* New York: Touchstone/Simon & Schuster. 1992.

Weinberg, Steven. "A Unified Physics by 2050?" *Scientific American.* December 1999. pp. 68-75.

Weinberg, Steven. *Dreams of a Final Theory.* New York: Pantheon Books. 1992.

Weiss, Peter. "Anatomy of a Lightning Ball: An aerial wonder, pondered for ages, no longer seems so ghostly." *Science News.* Vol. 161, No. 6. pp. 87-89. Feb. 9, 2002.

Wheeler, Michael. *Lies, Damn Lies, and Statistics.* New York: Dell Publishing (Laurel Edition). 1976 (1977).

> Although focusing on errors in polling people that leads to false pictures, which manipulate public opinion, many of the same biases skew data gathered and mis-interpreted by ball lighting researchers.

Wick, David. *The Infamous Boundary: Seven Decades of Heresy in Quantum Physics.* Copernicus, imprint of Springer-Veriag: New York. 1995.

> Covers Bohm's version of EPR experiment, Bohmian Mechanics and Bell's Theorem.

Willem, Jan. "Spontaneous Human Confabulation: Requiem for Phyllis." *Skeptical Inquirer.* March/April 2001. pp. 28-34.

Wilson, Daman. *Spontaneous Combustion: Amazing True Stories of Mysterious Fires.* New York: Sterling Publishing Co. 1997.

Wolfram, Stephen. *A New Kind of Science.* Champaign, IL: Wolfram Media, Inc. 2002.

> Written to be readable, this excellent and enjoyable self-published monster-sized book on ANKOS avoided pre-publication peer review. Not without surprise, Website reviews are unkind. But if Wolfram's predictions are half-true, physics will change greatly and affect both ball lightning research and Theories of Everything in physics. Until the fallout clears, your author is cautious of the hubris, but hopeful.

Wu, Corrina. "Tiny icicles grow in electric fields." *Science News.* Vol. 154. July 11, 1998. pg. 23.

Yost, Charles. *The Tesla Experiment: Lightning & Earth Electrical Resonance—A Commentary on 1983 Research & Test Results.* Leicester, NC: Dynamic Systems. 1983.

Bibliography 2—Aerial Phenomena

Most authors rewrite only what they read elsewhere. Few go out and investigate their cases, as your author did, and thus most just perpetuate errors—as with the many errors of fact repeated about Flatwoods. It is easy to write, hard to investigate. Thus I am skeptical of most cases. Here is some literature.

Acampora, Anthony. "Last Mile by Laser." *Scientific American.* July 2002. pp. 49-53.

> Earth is gradually going electromagnetically dark, as did earlier star civilizations. Short-range infrared lasers will beam advanced broadband multimedia services directly into homes and offices. Accidental emissions cease, and hence alien civilizations become electromagnetically "invisible."

Barclay, David and Therese. *UFOs: The Final Answer? Ufology for the 21st Century.* London, England: Blandford Imprint of Villiers House. 1993.

> Anthology covers UFOs, ball lightning, Vallee Hypothesis.

Barlowe, Wayne; Summers, Ian; and Meacham, Beth. *Barlowe's Guide to Extraterrestrials: Great Aliens from Science Fiction Literature.* New York: Workman Publishing. 1987.

> Can science fiction writers and artists contemplate what intelligent life might look like and their civilizations? This fanciful book is rich in diverse artwork that makes us wonder—as did Carl Sagan—why Earth's alleged UFO extraterrestrials breath our air and germs, and are reported so boringly alike?

> The late Stephen Jay Gould (championed divergent evolution) would say that extraterrestrials are all different. If Gould is right, then extraterrestrials are diverse, but their robot interstellar ships—using matter mined from our system's comets and asteroids—bioengineered intelligent "so boringly alike" hominids (from earth's DNA) to be compatible with Earth and familiar to us.

Alternatively, as Simon Conway Morris (champions convergent evolution) proposes, emergence is inevitable here and elsewhere. Thus re-running Earth's evolutionary "tape" of the Cambrian (and also that of extraterrestrial evolution) will create the same designs. If so, Barlowe's fanciful creatures will always be just that. But it may matter little, because extraterrestrial offspring were long ago succeeded by intelligent machines. These robots live unobserved among comet and asteroid belts and slowly spread across our galaxy.

Becker, Robert, MD and Selden, Gary. *The Body Electric: Electromagnetism and the Foundation of Life.* NY: William Morrow (Quill). 1985.

Classic work on Becker's pioneering medical research in bioelectricity, and how electromagnetism affects life. So, why do close encounters with extraterrestrial craft sometimes lead to health problems for some nearby witnesses but not to others? Since extraterrestrial craft stop the flow of current in wires, why do they not cause neurological and biological damage? For reverse engineering, this is another clue.

Bramley, William. "Can the UFO Extraterrestrial Hypothesis and Vallee Hypotheriss Be Reconciled?" [or are they both accurate but mutually exclusive?] Stanford, CA: *J. of Scientific Exploration.* Vol. 6, No 1 (1992)

Castellano, Tim. Detecting Transiting Exoplanets." *Sky & Telescope.* March 2004. pp. 77-81.

For equipment under $5,000, amateur astronomers can now detect planets outside our solar system.

Clarke, Jerome. *The UFO Book: Encyclopedia of the Extraterrestrial.* Detroit, MI: Visible Ink Press (Gale Research). 1998. ("Washington National Radar/Visual Case." pp. 653-662.)

Committee for the Scientific Investigation of Claims of the Paranormal. CSICOP. Box 703, Amherst, NY. 14226.

This excellent skeptics group publishes a bimonthly journal, *The Skeptical Inquirer,* which debunks extraterrestrial contactees and flying saucers. One successful author of ancient gods in saucers—who spent time in prison—attempted to drag down Carl Sagan, who then compared such saucerian authors to imaginary leprechauns. Both CSICOP and Michael Shermer's Skeptics Society have an unstated purpose—to protect legitimate scientists and keep them from being harassed by mystics and "saucer kooks" who will drag good reputations down.

Condon, Edward. *Scientific Study of Unidentified Flying Objects.* New York Times and Univ. of Colorado. 1969.

Crawford, Ian. "Searching For Extraterrestrials—Where Are They? Maybe we are alone in the galaxy after all." *Scientific American.* July 2000. pp. 38-47.

And thus logically *Scientific American* recommends defunding SETI?

Dershowitz, Alan M. *The Genesis of Justice.* NY: Time Warner. 2000.

Ten stories of Biblical injustice that led to the Ten Commandments and modern law. The origin of human morality predates man and originates with earliest hominids. However, modern morality of our civilization, forming its genesis of justice, begins with the Bible. But would extraterrestrial robot machine intelligences (especially those resident within our asteroid and comet belts) share this Biblical sense of injustice?

Dewdney, A.K. *Yes, We Have No Neutrons.* New York: John Wiley & Sons. 1997.

See chapter 4, "The Search for Extraterrestrial Intelligence," for serious errors in SETI and Drake's equation. Addresses mistaken and bumbling science. Unlike induction, deductive science and math are clean and there is no controversy in the rare cases when they go bad. Ideally, we should pose a question and frame a hypothesis that we can test with experiments, or sufficient observations, that lead to conclusions and critiqued publication. SETI has not done this.

Dobbins, Thomas and Sheehan, William. "The Canals of Mars Revisited." *Sky & Telescope.* March 2004. pp. 114-117.

Recent discoveries prove that Percival Lowell glimpsed real surface features when he charted Martian canals—artifacts of edge enhancement of the boundaries of adjoining regions of different albedo corresponding physically to adjoining surfaces strewn with bright or dusky surface materials.

Easton, Stewart C. *The Heritage of the Past: From the Earliest Times to 1500 (revised ed.).* NY: Holt, Rinehart and Winston. 1966. 795 pp.

> Expert insights by a brilliant historian who contrasts static and dynamic civilizations, which provides deep insights into what extraterrestrial civilizations may be like before they evolved into machine civilizations. Although his description of Egyptian history omits a few details of recent discoveries, Stewart's descriptions and observations remain essentially correct and valid today.

Ely, J.J.; Leland, M.; Martino, M.; Swett, W.; Moore, C.M. "Technical Report: Chromosomal and mtDNA analysis of Oliver." *Am. J. of Physical Anthropology. [JAPA]* 105(3):395-403. 1998.

Falk, Dan. "The Anthropic Principle's Surprising Resurgence." *Scientific American.* March 2004. pp. 43-47.

> Is life here because it's not there? (in another Multiverse) Is this more than amusing philosophy? Experts and skeptics argue pro and con.

Feschino, Frank. *The Cover-Up of the Flatwoods Monster Revealed.* Forward by Stanton T. Friedman. 2004.

> First book solely devoted to this encounter, by this dedicated researcher who spent several years investigating.

Fisher, David and Marshall. *Strangers in the Night: A Brief History of Life on Other Worlds.* 1998: Cornelia & Michael Bessie Counterpoint (Perseus Books Group)

> Describes obstacles that SETI faced and overcame. Too bad SETI is a massive failure—at least, so far. SETI will not receive signals unless extraterrestrials wish to reveal their existence.

Hill, Paul R. *Unconventional Flying Objects: A Scientific Analysis.* Charlottesville, VA: Hampton Roads Publishing Company. 1995.

> From his experience at NASA, this high-ranking rocket engineer evaluates unconventional extraterrestrial aircraft propulsion, flight control and thrust vectors—applicable to explaining propulsion of flying saucers and ball lightning. Hill adheres to the laws of physics. Some calculus. Hill had a major UFO sighting (Ruppelt, chapter 12, July 16, 1952) and then analyzed sightings at NASA for clues to reverse-engineer "inertialess-negative-gravity" pro-

pulsion. Hill was NASA's unofficial UFO in-house expert. Highly recommended.

Journal of Scientific Exploration, Box 838, Stanford, CA. Hardcopy and also electronic Web subscriptions available.

Refereed journal explores unusual subjects, such as ball lightning and Vallee Hypothesis.

Jung, Carl. *Man and His Symbols.* NY: Anchor Books. 1964.

A popular overview of Jungian psychology of symbolism, with artwork, and excellent for understanding Vallee Hypotheses—that extraterrestrial craft are not alien, but "something else."

Jung, Carl. *Flying Saucers" A Modern Myth of Things Seen in the Skies.* Princeton U. Press. 1978.

Explores symbolism of saucers. Jung never seriously interviewed witnesses, but it did not stop him from having opinions.

Keel, John. *The Mothman Prophecies.* NY: Signet. 1975. Reissued in 2002 to coincide with movie.

A good read that makes no sense. Ditto for the movie. But, truly spooky, if true. So, check it with CSICOP first.

Keyhoe, Maj. Donald (USMC). *Flying Saucers From Outer Space.* New York: Henry Holt & Co. 1953.

Decent (for the time) description of Flatwoods Monster on pages 116-120. In Chapter 8, "The Canadian Project." (pp. 128-149) describes early attempts to reverse-engineer extraterrestrial propulsion by Canada's Wilbur Smith. The most interesting and gifted writer in UFO history, Keyhoe graduated from West Point with Eisenhower and Macarthur, and flew with Admiral Byrd.

Keyhoe, Maj. Donald. *UFOs: A New Look.* Washington, DC: National Investigations Committee on Aerial Phenomena (NICAP). 1969.

Keyhoe, Maj. Donald. *The UFO Evidence.* Washington, DC: National Investigations Committee on Aerial Phenomena (NICAP). 1964.

Excellent. After Keyhoe's death, Paul R. Hill wrote a sequel.

Keyhoe, Maj. Donald. *Aliens From Space: The Real Story of Unidentified Flying Objects.* New York: Doubleday & Co. 1973.

His final book, it is more detailed than his other books. I gave a copy to Kathy May.

Klass, Phillip. *UFOs: Identified.* Random House. 1968.

Noted CSICOP skeptic correctly believes fireballs account for many alleged UFO sightings and does excellent research to discredit sloppy claims. Klass correctly claims that the media is more interested in good ratings and fails to sufficiently examine the side of the critic.

Unfortunately, skeptics' books do not sell. Those fans that get their thrills or earn their living from sensational UFOs have good motivation, and they certainly are more interesting to viewers—but do great harm to serious cases and frighten off legitimate scientists.

Kurland, Michael. *The Complete Idiot's Guide to Extraterrestrial Intelligence.* Alpha Books/Simon & Schuster Macmillan.1999.

Overall balanced treatment with humor and some skepticism.

Kurzweil, Ray. *The Age of Spiritual Machines.* New York: Viking. 1999.

What happens when future earth's computers far exceed human intelligence? It is what extraterrestrial civilizations encountered millions of years ago and what we might encounter—when (and if) extraterrestrials ever decide to meet or talk to us, although they will be way ahead of that, know it, and have bioengeered a "buffer interface" (hominids) for future contact, after millions of years residing hidden inside our comet and asteroid belts. For more, examine Roger Penrose's excellent books on machine intelligence.

Levy, David. H. *The Quest For Comets: An Explosive Trail of Beauty and Danger.* Avon Books (Hearst). 1994.

Lorenz, Konrad. *On Aggression.* New York: Harcourt Edition. 1966.

> In man there exists a "killer instinct" less controlled than in most savage animals that affects technology. Is this applicable to extraterrestrial bioorganic and robot machine intelligence psychologies? Or to our non-human successors? Hopefully not.

Mack, John (MD). *Abduction: Human Encounters With Aliens.* (revised ed.) NY: Ballantine Books. 1995 (Simon & Schuster/Scribner 1994)

> Mack suspects that perhaps "something else" and not true extraterrestrials are involved in these encounters—that correspond to real internal but not external reality—a legitimate psychological phenomena little recognized until now, which must be studied.

Masters, Roy. *The Adam and Eve Sindrome* [sic]. Foundation of Human Understanding. Grants Pass, Oregon.

> Describes the conflicted man-woman relationship, which extraterrestrial robot machine civilizations without sexes, organic bodies or eating food will not face—and the machine extraterrestrials will have vastly different natures from ours, and may have bioengineered hominids to later meet with us and have something in common—always a good way to start a meeting with strangers.
>
> Several of Roy's books went out-of-print—which could be satisfied by print-on-demand. Harvard, Yale and Cambridge are using POD to bring back many of their own out-of-print books.
>
> Soon, E-paper and OLED sheets will replace paper books (and many flat panel terminals) and revolutionize publishing—breakthroughs equaling Gutenberg's development—and make POD with single-sheet E-paper and OLEDs far cheaper than traditional paper books, which will become obsolete. [*Scientific American,* Feb. 2004] This dramatic upheaval is a financial warning threatening traditional publishers and their agents.

McCampbell, James. *UFOLOGY.* Millbrae, CA: Celestial Arts. 1973, 1976.

> Examines types of craft, radiation emissions, interference with electrical circuits, and paralysis and shock experienced by witnesses that resembles some fireball behavior.

Morris, Simon Conway. *Life's Solution: Inevitable Humans in a Lonely Universe.* Cambridge University Press. 2003.

> Although skeptical of intelligent extraterrestrial life (who needs that bad press?), famed Cambrian paleontologist Morris thinks it will look and think very much like us. Driven by "purposeful directionality," evolution once begun continues unabated, and the emergence of man was inevitable. (Deists find some comfort in this.) Extraterrestrial worlds are also driven by evolution, and create carbon-based creatures similar to earth.

Morris, Simon Conway. *The Crucible of Creation: The Burgess Shale and the Rise of Animals.* Oxford University Press. 1998.

Nadia, Steve. "Using Lasers to Detect ET—Seeking intelligent life in the universe, SETI begins a new optical search." *Astronomy.* Sept. 2002. pp. 45-49.

> Unfortunately, because Earth is quarantined, SETI may never detect laser signals—unless they are purposely beamed our way—something extraterrestrials seem reluctant to do—thereby, exposing their existence. SETI's dreams forever remain schemes—unless extraterrestrials get sloppy and SETI gets lucky.

Nash, Michael. "The Truth and Hype About Hypnosis." *Scientific American.* July 2001. Pp. 47-55.

> Not all persons and witnesses to extraterrestrial abductions are equally hypnotizable—even by the best hypnotists, such as Dr. Benjamin Simon of the Betty and Barney Hill abduction—and this hypnotizability is an inherent trait, that we are born with, and remains constant throughout life.

Perino, Kojo. "Cracking the Harvard X-Files." *Psychology Today.* March/April 2003.

Randle, Kevin and Estes, Russ. *Faces of the Visitors.* NY: Warner Books Co. 1980.

Randle, Capt. Kevin, USAF, Ret. *The UFO Casebook.* NY: Warner Books Co. 1980.

Reinmould, Robert & Queen, William. *Ecology of Halophytes.* NY: Academic Press, Inc. 1974.

Opening a window into extraterrestrial-like plants that thrive in harsh salt desert and salt marsh estuarine ecosystems where other plants die, halophytes possess hairy or fleshy leaves and stems.

Reynolds, Mike D. *Falling Stars: A guide to Meteors & Meteorites.* Mechanicsburg, PA: Stackpole Books. 2001.

Ruppelt, Capt. Edward. *The Report On Unidentified Flying Objects.* New York: Doubleday & Co. 1956.

If you read only one general UFO book, read this one. It is the classic and authoritative book by the founder of Project Blue Book. One caveat—the odd, last chapter was added later under duress.

Sagan, Carl. *The Demon-Haunted World: Science as a Candle in the Dark.* New York: Random House. 1995.

In his final book, Sagan provides gold standard tests by which all enthusiastic investigators should cautiously examine all contactees and UFOs—with skepticism. Re-read and deeply ponder Chapter Ten: *The Dragon in My Garage.* Despite his heartfelt lonely memory for his departed parents—and how he wishes he could meet them for ten minutes each year—we are left wondering if Sagan was really a true agnostic (as are most skeptics), or more likely a spiritual agnostic. May his soul forever in seas of uncharted thought, boldly voyage the stars that he so loved. He has left for us his keen insights into UFOs. Strongly recommended for all investigators.

Saunders, David R. and Hawkins, R. Roger. *UFOs? Yes! Where the Condon Committee Went Wrong.*

Scott, Lee. *Magician's Arsenal: Professional Tricks of the Trade.* Boulder, CO: Paladin Press. 1993.

It takes a magician to catch a magician. Sincere UFO field researchers must be magicians. As famous magician, "The Amazing Randi" of CSICOP warns, some hoaxers perpetuate clever frauds. Small cells within the French and other governments, as investigated by Jacque Vallee, used brilliant magical illusions to create despicable UFO hoaxes to destabilize witnesses, create disreputable sightings, and create bad publicity to discredit honest witnesses and to discourage scientists. Scott explains unusual tricks that combine illusions with explosives, incendiaries, firearms, electronics, flamethrowers, sleeve canons, psychokinetic scramblers, radio jammers, and other magic that were modified and adapted by hoaxers.

Shermer, Michael. *Why People Believe in Weird Things: Pseudoscience, Superstition, and other Confusions of our Time.* NY: MJF Books. 1997.

In chapter 6, "Abducted! Encounters with Aliens," Shermer correctly exposes hoaxed alien autopsy films.

Shermer, Michael. "Skeptic: Why ET Hasn't Called." *Scientific American.* August 2002. p. 33.

Although Shermer claims to be an "unalloyed enthusiast," and correctly points out the vast unknown toleraces in Drakes equation, he concludes that maybe only three civilizations exist in our entire galaxy—or near-zero odds for SETI.

Sherwood, John C. "Gray Barker's Book of Bunk: Mothman, Saucers, and MIB." *The Skeptical Inquirer.* May/June 2002. Vol. 26, No. 3. pp. 39-44.

Shostak, Seth. "The Future of SETI." *Sky & Telescope.* April 2001. pp. 42-53.

"Shrinking Toward The Ultimate Transistor: Electronic devices go atomic." *Science News.* 8.10.02 Vol 162. pp. 88-89.

Even without massively-parallel quantum computers, single-atom transistors will be awesome. Extraterrestrials achieved them hundreds of millions of years ago. Extraterrestrial intelligences may not be biological.

Sigurdsson, Steinn, et al. "A Young White Dwarf Companion to Pulsar B1620-26: Evidence for Early Planet Formation." *Science.* July 11, 2003. [For description, see Villard.]

Steele, Dianna. "Surviving In Space." *Astronomy.* October 1999. pp. 58-63.

Before humans can reach Mars, engineers must determine if humans can safely withstand space that long (500 days). Serious medical problems exist. And at 5 percent of light-speed, for years of interstellar flight, even with centrifugal wheels, it may be catastrophic. Intelligent machines are superior.

Steel, Duncan. *Target Earth: The Search For Rogue Asteroids and Doomsday Comets That Threaten Our Planet.* Pleasantville, NY: Readers Digest Association. 2000.

Understanding asteroids is interesting to ball lightning—to rule out certain mistaken sightings—and to understand how a swarm of extraterrestrial robot probes arriving here long ago mined asteroids and comets, and grew bioorganic humanoids compatible with Earth's atmosphere and life.

Sturrock, Dr. Peter A. *The UFO Enigma: A New Review of the Physical Evidence.* New York: Warner Books. 1999.

Under leadership of this Stanford astrophysicist, nine impartial scientists examine evidence and come to inconclusive findings.

Sturrock, Dr. Peter A. "Composition Analysis of the Brazil Magnesium." Stanford, CA: *J. of Scientific Exploration.* Vol. 6, No. 1 (1992) pp. 69-95.

"When Failure Is Not An Option." *MIT Technology Review.* July 1997. pp. 38-45.

Extraterrestrial craft must be ultra-reliable, so they never provide our engineers with clues to their inertialess-negative-gravity propulsion, or even evidence of extraterrestrial existence. Success is harder to analyze than failure. Aircraft carriers are ultra reliable. For catastrophic failure, a chain of events must all line up.

Terrry, Sara. "Alien Territory." *The Boston Globe Magazine.* October 11, 1992. pp. 21-27.

About Harvard's eminent Dr. John Mack, MD, a ufologist.

Teets, Bob. *West Virginia UFOs: Close Encounters in the Mountain State.* Terra Alta, WV: Headline Books. 1995.

Interview with Kathy May Horner about her encounter with Flatwoods Monster in "Chapter Three: Great Balls of Fire in the Sky."

Vallee, Jacque. *The Invisible College: What a Group of Scientists Has Discovered About UFO Influences on the Human Race.* NY: E.P. Dutton & Co., Inc. 1975.

Borrowing from the dark ages of science, when scientists were harrassed, John Allen Hynek revived "the invisible [secret] college" for legitimate scientists—by which they now secretly study flying saucers, but escape detection and ridicule. Vallee dismisses extraterrestrial origins—no extraterrestrials come from space—and instead proposes a psychic reality similar to historic religious revelation. Explores archaeological aspects.

Vallee incorrectly dismisses religious miracles as part of this psychic reality. But he was conned by fake magician Uri Geller, as later exposed and discredited by Johnny Carson and CSICOP's magician the Amazing Randi. It takes a magician to catch a magician! Perhaps ufology needs more skeptical magicians. Ponder this: there are clever magicians around faking flying saucer contacts—so be skeptical.

Vallee, Jacque. *Dimensions: A Casebook of Alien Contact.* NY: Ballantine Books. 1988.

Explains cases in Jungian and historical contexts. Describing the impossibly vast number of sighting, contactees, weird medical experiments, and contradictory information, this astrophysicist argues that not extraterrestrials, but "something else" far greater is behind this—which he calls the Vallee Hypothesis.

Vallee, Jacque. *Confrontations: A Scientist's Search for Alien Contact.* NY: Ballantine Books. 1990.

Vallee Hypothesis denies extraterrestrials, but suggests a yet-unexplained phenomenon.

Vallee, Jacque. *Revelations: Alien Contact and Human Deception.* NY: Ballantine Books. 1991.

Encountering fraudulent kidnapping of duped witnesses into a fake saucer by a rogue cell of psychological researchers inside the French government, Vallee exposes deceit. Also exposes a cleverly simulated case of UFO fraud on a USAF base.

Vallee, Jacque. "Anatomy of a Hoax: The Philadelphia Experiment Fifty Years Later." *J. of Scientific Exploration.* Vol. 4: No. 1: Article 9.

Using this as a classical model for UFO hoaxes, Vallee extracts thirteen parameters to expose questionable UFO cases. Vallee interviewed an eyewitness who contradicted the fraudulent claims of this hoax.

Vallee, Jacque. "Five Arguments Against the Extraterrestrial Origin of Unidentified Flying Objects." *J. of Scientific Exploration.* Vol. 8: No. 1: Article 2.

Villard, Ray; Schaller, Adolf. "Genesis Planet." *Astronomy.* June 2004. pp. 45-49.

> Recent discovery of a gas-giant planet (only 2.5 Jupiter masses), the "Methuselah Planet," orbits a tight binary star 5,600 LY away in the globular star cluster M4 (pulsar PSR B1620-26) in Scorpius, a region devoid of heavy elements needed to build planets. It was born 13 BY ago, under a billion years after the Big Bang, and thrice the age of our earth.

> This suggests that planets and extraterrestrial life were formed far earlier than we thought; and early in its history, our universe was more efficient at making planets than realized. Other recent discoveries demonstrate that our galaxy is teeming with planets—and probably alien life.

> The M4 discovery proves there are several recipes to making gas giants, including rapid contracting of cooling clumps of gas (disk instability)—a speedy process of planet creation. Earth-like moons could circle such gas giants. Research indicates life is aggressive and opportunistic, using basic elements to create ascending complexity.

Walter, Malcolm. *The Search for Life on Mars.* Cambridge, MA: Perseus Books. 1999.

> A leading astro- and palaeobiolobist for NASA on exobiology, Walter mirrors Michael Shermer's and Paul Davies' assertions that technological intelligence in our galaxy may be very rare, perhaps only two other planets, if that, and that the variables in Drake's equation are total unknowns. In fact, we may be the only extraterrestrials, and SETI's schemes may forever remain dreams! This dismal view of SETI is also shared by skeptic Michael Shermer. (And thus logically SETI should be defunded?)

> Or, maybe Walter, Shermer and Davies are all wrong, and maybe extraterrestrial civilizations use highly-directional lasers in their quite-active extraterrestrial communications network—only Earth is not allowed on their party line! And they exist as a machine civilization amongst our comet and asteroid belts. Logically, if only one civilization existed, its robots long ago would have colonized all stars comet and asteroid belts. They probably did.

Webb, Stephen. *Where Is Everybody?* Copernicus Books. NY. 2002.

> In 1950 Enrico Fermi proposed the Fermi Paradox—if extraterrestrials and saucers exist, then why do they fail to contact us? (Would you?) Webb offers 49 solutions—one that they are already here, or prefer not to communicate with us—yet.

Wood, Robert M. "The Extraterrestrial Hypothesis Is Not That Bad." *J. of Scientific Exploration.* Vol. 5: No. 1: Article 5.

Zimmer, Carl. "What came Before DNA?" *Discover.* June 2004. pp. 34-41.

> Harvard biochemist Jack Szostak attempts to recreate first pre-DNA life with novel RNA experiments and evolve RNA replicase to discover how life formed prior to DNA before 4 BY ago.

> Szostak's "natural selection" evolved RNAs aptamers that performed many tasks, then evolved specialized ribozymes. Without enzymes, vesicles grew, divided, and grew again under early-earth conditions in a montmorillonite clay.

> Such vesicle-laden water oozed though filtering pores of rocks near hydro-thermal vents, around which the floors were coated with such clay, as duplicated by Szostak's team. Some montmorillonite grains end up inside the vesicles. Did RNA mix with vent clay, mix with fatty acids and yield RNA on clay particles within the vesicles? Can a ribozyme carry on biochemistry inside a vesicle? If so, Szostak will prove that some alien life may not be DNA based.

0-595-31394-9

www.ingramcontent.com/pod-product-compliance
Lightning Source LLC
Chambersburg PA
CBHW020719180526
45163CB00001B/33